IET CONTROL, ROBOTICS AND SENSORS SERIES 122

Sensors in the Age of the Internet of Things

The IET International Book Series on Sensors–Call for Authors

The use of sensors has increased dramatically in all industries. They are fundamental in a wide range of applications from communication to monitoring, remote operation, process control, precision and safety, and robotics and automation. These developments have brought new challenges such as demands for robustness and reliability in networks, security in the communications interface, and close management of energy consumption. This Book Series covers the research and applications of sensor technologies in the fields of ICTs, security, tracking, detection, monitoring, control and automation, robotics, machine learning, smart technologies, production and manufacturing, photonics, environment, energy, and transport.
Book Series Editorial Board

Proposals for coherently integrated international multi-authored edited or co-authored handbooks and research monographs will be considered for this Book Series. Each proposal will be reviewed by the IET Book Series Editorial Board members with additional external reviews from independent reviewers. Please email your book proposal to: vmoliere@theiet.org or author_support@theiet.org.

Other volumes in this series:

Volume 8	**A History of Control Engineering, 1800–1930** S. Bennett
Volume 9	**Embedded Mechatronics System Design for Uncertain Environments: Linux-based, MATLAB® xPC Target, PIC, Ardunio and Raspberry Pi approaches** C.S. Chin
Volume 18	**Applied Control Theory, 2nd Edition** J.R. Leigh
Volume 20	**Design of Modern Control Systems** D.J. Bell, P.A. Cook and N. Munro (Editors)
Volume 28	**Robots and Automated Manufacture** J. Billingsley (Editor)
Volume 33	**Temperature Measurement and Control** J.R. Leigh
Volume 34	**Singular Perturbation Methodology in Control Systems** D.S. Naidu
Volume 35	**Implementation of Self-tuning Controllers** K. Warwick (Editor)
Volume 37	**Industrial Digital Control Systems, 2nd Edition** K. Warwick and D. Rees (Editors)
Volume 39	**Continuous Time Controller Design** R. Balasubramanian
Volume 40	**Deterministic Control of Uncertain Systems** A.S.I. Zinober (Editor)
Volume 41	**Computer Control of Real-time Processes** S. Bennett and G.S. Virk (Editors)
Volume 42	**Digital Signal Processing: Principles, devices and applications** N.B. Jones and J.D.McK. Watson (Editors)
Volume 44	**Knowledge-based Systems for Industrial Control** J. McGhee, M.J. Grimble and A. Mowforth (Editors)
Volume 47	**A History of Control Engineering, 1930–1956** S. Bennett
Volume 49	**Polynomial Methods in Optimal Control and Filtering** K.J. Hunt (Editor)
Volume 50	**Programming Industrial Control Systems Using IEC 1131-3** R.W. Lewis
Volume 51	**Advanced Robotics and Intelligent Machines** J.O. Gray and D.G. Caldwell (Editors)
Volume 52	**Adaptive Prediction and Predictive Control** P.P. Kanjilal
Volume 53	**Neural Network Applications in Control** G.W. Irwin, K. Warwick and K.J. Hunt (Editors)
Volume 54	**Control Engineering Solutions: A practical approach** P. Albertos, R. Strietzel and N. Mort (Editors)
Volume 55	**Genetic Algorithms in Engineering Systems** A.M.S. Zalzala and P.J. Fleming (Editors)
Volume 56	**Symbolic Methods in Control System Analysis and Design** N. Munro (Editor)
Volume 57	**Flight Control Systems** R.W. Pratt (Editor)
Volume 58	**Power-plant Control and Instrumentation: The control of boilers and HRSG systems** D. Lindsley
Volume 59	**Modelling Control Systems Using IEC 61499** R. Lewis
Volume 60	**People in Control: Human factors in control room design** J. Noyes and M. Bransby (Editors)
Volume 61	**Nonlinear Predictive Control: Theory and practice** B. Kouvaritakis and M. Cannon (Editors)
Volume 62	**Active Sound and Vibration Control** M.O. Tokhi and S.M. Veres
Volume 63	**Stepping Motors, 4th Edition** P.P. Acarnley
Volume 64	**Control Theory, 2nd Edition** J.R. Leigh
Volume 65	**Modelling and Parameter Estimation of Dynamic Systems** J.R. Raol, G. Girija and J. Singh
Volume 66	**Variable Structure Systems: From principles to implementation** A. Sabanovic, L. Fridman and S. Spurgeon (Editors)
Volume 67	**Motion Vision: Design of compact motion sensing solution for autonomous systems** J. Kolodko and L. Vlacic
Volume 68	**Flexible Robot Manipulators: Modelling, simulation and control** M.O. Tokhi and A.K.M. Azad (Editors)
Volume 69	**Advances in Unmanned Marine Vehicles** G. Roberts and R. Sutton (Editors)
Volume 70	**Intelligent Control Systems Using Computational Intelligence Techniques** A. Ruano (Editor)
Volume 71	**Advances in Cognitive Systems** S. Nefti and J. Gray (Editors)
Volume 72	**Control Theory: A guided tour, 3rd Edition** J.R. Leigh
Volume 73	**Adaptive Sampling with Mobile WSN** K. Sreenath, M.F. Mysorewala, D.O. Popa and F.L. Lewis

Volume 74 **Eigenstructure Control Algorithms: Applications to aircraft/rotorcraft handling qualities design** S. Srinathkumar
Volume 75 **Advanced Control for Constrained Processes and Systems** F. Garelli, R.J. Mantz and H. De Battista
Volume 76 **Developments in Control Theory towards Glocal Control** L. Qiu, J. Chen, T. Iwasaki and H. Fujioka (Editors)
Volume 77 **Further Advances in Unmanned Marine Vehicles** G.N. Roberts and R. Sutton (Editors)
Volume 78 **Frequency-domain Control Design for High-performance Systems** J. O'Brien
Volume 80 **Control-oriented Modelling and Identification: Theory and practice** M. Lovera (Editor)
Volume 81 **Optimal Adaptive Control and Differential Games by Reinforcement Learning Principles** D. Vrabie, K. Vamvoudakis and F. Lewis
Volume 83 **Robust and Adaptive Model Predictive Control of Nonlinear Systems** M. Guay, V. Adetola and D. DeHaan
Volume 84 **Nonlinear and Adaptive Control Systems** Z. Ding
Volume 86 **Modeling and Control of Flexible Robot Manipulators, 2nd edition** M.O. Tokhi and A.K.M. Azad
Volume 88 **Distributed Control and Filtering for Industrial Systems** M. Mahmoud
Volume 89 **Control-based Operating System Design** A. Leva *et al.*
Volume 90 **Application of Dimensional Analysis in Systems Modelling and Control Design** P. Balaguer
Volume 91 **An Introduction to Fractional Control** D. Valério and J. Costa
Volume 92 **Handbook of Vehicle Suspension Control Systems** H. Liu, H. Gao and P. Li
Volume 93 **Design and Development of Multi-lane Smart Electromechanical Actuators** F.Y. Annaz
Volume 94 **Analysis and Design of Reset Control Systems** Y.Guo, L. Xie and Y. Wang
Volume 95 **Modelling Control Systems Using IEC 61499, 2nd Edition** R. Lewis & A. Zoitl
Volume 96 **Cyber-physical System Design with Sensor Networking Technologies** S. Zeadally and N. Jabeur (Editors)
Volume 99 **Practical Robotics and Mechatronics: Marine, space and medical applications** I. Yamamoto
Volume 100 **Organic Sensors: Materials and applications** E Garcia-Breijo and P Cosseddu (Editors)
Volume 102 **Recent Trends in Sliding Mode Control** L. Fridman J.P. Barbot and F. Plestan (Editors)
Volume 104 **Control of Mechatronic Systems** L. Guvenc, B.A. Guvenc, B. Demirel, M.T Emirler
Volume 105 **Mechatronic Hands: Prosthetic and robotic design** P.H. Chappell
Volume 107 **Solved Problems in Dynamical Systems and Control** D. Valério, J. T Machado, A.M. Lopes and A.M. Galhano
Volume 108 **Wearable Exoskeleton Systems: Design, control and applications** S. Bai, G.S. Virk and T.G.Sugar
Volume 111 **The Inverted Pendulum in Control Theory and Robotics: From theory to new innovations** O. Boubaker and R. Iriarte (Editors)
Volume 112 **RFID Protocol Design, Optimization, and Security for the Internet of Things** A.X. Liu, M. Shahzad, X Liu and K. Li
Volume 113 **Design of Embedded Robust Control Systems Using MATLAB®/Simulink®** P.H. Petkov, T.N. Slavov and J.K. Kralev
Volume 114 **Signal Processing and Machine Learning for Brain-machine Interfaces** T. Tanaka and M. Arvaneh (Editor)
Volume 117 **Data Fusion in Wireless Sensor Networks** D. Ciuonzo and P.S. Rossi (Editors)
Volume 118 **Modeling, Simulation and Control of Electrical Drives** M.F. Rahman and S.K. Dwivedi (Editors)
Volume 119 **Swarm Intelligence Volumes** 1–3 Y. Tan (Editor)
Volume 121 **Integrated Fault Diagnosis and Control Design of Linear Complex Systems** M. Davoodi, N. Meskin and K.Khorasani
Volume 123 **Data-Driven Modeling, Filtering and Control: Methods and applications** C. Novara and S. Formentin (Editors)

Sensors in the Age of the Internet of Things

Technologies and applications

Edited by
Octavian Adrian Postolache, Edward Sazonov
and Subhas Chandra Mukhopadhyay

The Institution of Engineering and Technology

Published by The Institution of Engineering and Technology, London, United Kingdom

The Institution of Engineering and Technology is registered as a Charity in England & Wales (no. 211014) and Scotland (no. SC038698).

First published 2019

The Institution of Engineering and Technology
Michael Faraday House
Six Hills Way, Stevenage
Herts, SG1 2AY, United Kingdom

www.theiet.org

British Library Cataloguing in Publication Data
A catalogue record for this product is available from the British Library

ISBN 978-1-78561-634-1 (hardback)
ISBN 978-1-78561-635-8 (PDF)

Typeset in India by MPS Limited

Contents

Chapter 1

The Internet of things: a survey and outlook

Giovanni Perrone[1], Massimo Vecchio[2], Javier Del Ser[3], Fabio Antonelli[2], and Vivart Kapoor[4]

The recent history has witnessed disruptive advances in disciplines related to information and communication technologies that have laid a rich technological ecosystem for the growth and maturity of latent paradigms in this domain. Among them, sensor networks have evolved from the originally conceived set-up where hundreds of nodes with sensing and actuating functionalities were deployed to capture information from their environment and act accordingly (coining the so-called wireless sensor network concept) to the provision of such functionalities embedded in quotidian objects that communicate and work together to collaboratively accomplish complex tasks based on the information they acquire by sensing the environment. This is nowadays a reality, embracing the original idea of an Internet of things (IoT) forged in the late twentieth century, yet featuring unprecedented scales, capabilities and applications ignited by new radio interfaces, communication protocols and intelligent data-based models. This chapter examines the latest findings reported in the literature around these topics, with a clear focus on IoT communications, protocols and platforms, towards ultimately identifying opportunities and trends that will be at the forefront of IoT-related research in the near future.

1.1 Introduction

The term 'Internet of things' (IoT), coined in 1999 by Ashton [1], has attracted and attracts a multitude of research and industrial interests. Whatever domain we take into consideration, from housing to precision agriculture, from retail to transportation, from infrastructure monitoring to personal healthcare, from urban mobility to autonomous vehicles, just to mention a few, is going to be supported by each-day-smaller and smarter devices (i.e., things) able to collect data and to push them

[1]SMARTEST Research Centre at eCampus University, Novedrate (Como), Italy
[2]OpenIoT Research Unit at FBK CREATE-NET, Trento, Italy
[3]ICT Division, TECNALIA, Derio, Bizkaia, Spain; Department of Communications Engineering, University of the Basque Country (UPV/EHU), Bilbao, Spain; Data Science group, Basque Center for Applied Mathematics (BCAM), Bilbao, Spain
[4]Endress+Hauser Process Solution GmbH, Dept. IoT DE, Freiburg, Germany

to the Internet. Gartner Inc. expects that by 2020, 26 billion objects will be connected to the Internet and, by 2022, a typical family home will contain more than 500 connected smart objects [2].

K. Ashton says:
If we had computers that knew everything there was to know about things using data they gathered without any help from us we would be able to track and count everything, and greatly reduce waste, loss and cost. We would know when things needed replacing, repairing or recalling, and whether they were fresh or past their best. We need to empower computers with their own means of gathering information, so they can see, hear and smell the world for themselves, in all its random glory. RFID and sensor technology enable computers to observe, identify and understand the world without the limitations of human-entered data. [1]

However, around 20 years after its coinage [3], it would be reductive to translate the IoT vision into the routine process of physically connecting everyday objects to the Internet. Indeed, the original promise of the IoT was about connecting such everyday objects to the Internet with the goal of making them autonomous, hence smart. Loosely speaking, this smartening process relies on the implied assumption of the need of 'closing the loop' that consists of sensing, communicating, reasoning, taking decisions and actuating back into the physical side [4]. As already noticed by the authors of [5], this can be seen as the specification of the more general Monitor-Analyse-Plan-Execute plus Knowledge (MAPE-K) reference model for Autonomic Computing, as initially proposed by Horn as far back as 2001, and later formalized in [6] (the interested reader is also invited to refer to [7]).

P. Horn says:
It's time to design and build computing systems capable of running themselves, adjusting to varying circumstances, and preparing their resources to handle most efficiently the workloads we put upon them. These autonomic systems must anticipate needs and allow users to concentrate on what they want to accomplish rather than figuring how to rig the computing systems to get them there [. . .] it is the self-governing operation of the entire system, and not just parts of it, that delivers the ultimate benefit. [8]

Overall, the number of devices connected to the Internet is exploding, as well as the volume of data produced and pushed to the Internet by these devices. To support this uncontrolled proliferation of physical devices and associated, hence tremendous, digital data production, IoT embraces a number of architecture, protocols, standards, services and applications for ubiquitous data acquisition (at the physical level), as well as large-scale data storage and analysis tools (at the digital level). On the one hand, this guarantees a genuine evolution of the IoT towards a truly open, highly pervasive and multivendor ecosystem. On the other hand, too often we observe how this lack of well-established, horizontal, truly scalable and consistent IoT architecture translates into the sharp tendency of each new application (even worse in some cases: each new device) that enters the IoT arena to bring its own (sometimes proprietary!) technical implementations and manuals. It is clear that this tendency is one of the causes of the today IoT market fragmentation. The obvious – though incumbent – statement is that the IoT market is fragmented because of its very nature. Indeed, very specific end-users' requirements cannot be always homogenized. This is one of the reasons why, sometimes, even apparently simple IoT use cases are built on top of complex system architecture that cannot be promoted as the reference one [9]. However, the whole truth is that major counterforces are slowing down the exploitation of the full potential of the IoT. The most critical challenge that the IoT has tackled since the beginning – and not completely defeated, yet! – is this sort of mistrust shown by well-established large players: the IoT implementation may involve radical structural changes and drastic shift in value creation. It follows that for such players, it is difficult to quickly adapt to new business models and engage in new types of alliances. This inertia explains also why Small and Medium Enterprises (SMEs – esp. entrepreneurs and start-ups) are today considered to have the full potential to seize new opportunities brought up by the IoT. Nevertheless, to have such many actors acting unilaterally promotes the proliferation of proprietary system architecture and technology, as well as business models. To complicate things in this Babylon, we still certify the lack of common standards and interoperable solutions throughout the products and services life cycles, while interoperability is still considered essential to ensure seamless flow of data across sectors and value chains. In the end, this reality does not encourage cross-cutting approaches, risks reinforcing silos, and prevents innovation across areas. However, the opposite situation would be also undesirable: in the end, monopolizing the IoT would be then an insurmountable obstacle to the development of these markets, and to the development of truly open digital platforms.

Thus, it is not a coincidence the current proliferation of such many platforms, standards and technologies claiming to be able to work across multiple vendors, industries and sectors, as well as such a multitude of SMEs focusing efforts on very specific vertical segment use cases and applications [10]. The objective of this chapter is to take stock of the present of the IoT, starting from the enabling radio technologies (see Section 1.2), through the most common communication protocols (see Section 1.3), until the state-of-the-art backend and frontend services and features provided by the modern IoT platforms available on the market and within the open-source communities. Finally, Section 1.5 concludes this chapter by outlining research challenges related to IoT communications, protocols and services that are throwing up manifold questions in need of further investigation. We will argue that these challenges, which lie at the core of foreseen evolutions of IoT architecture and systems, serve as a stimulating incentive and a propelling driver for research efforts conducted in this vibrant field in the future.

1.2 Communication and transport technologies

An essential requirement for implementing a large and cooperating network of devices is the ability to have a flexible, inexpensive and adaptive connectivity layer. The restrictions and limitations of the physical cabling, both in terms of implementation costs and lack of flexibility in adapting to the ever-changing dynamics of the environment being monitored, make fixed cabling a solution that, in practical terms, is not viable for several IoT scenarios. Not to mention applications where the devices to be connected are moving, or are simply too far away from each other to be physically connected by wires. For these reasons, during the years, several wireless connectivity standards have emerged, ranging from very short-range, low-power connections to long-range connectivity solutions based on cellular technologies. This section will provide an overview of some of the most common and widely adopted radio technologies used for IoT implementations, emphasizing advantages and disadvantages of each protocol for typical IoT scenarios.

1.2.1 Short range

Short-range solutions are aimed at creating and managing networks of devices that are physically located in a confined area. Such networks are typically called Wireless Personal Area Network, also to emphasize the fact that they operate within a short radius and are separated from other networks. We will see that it is possible to partially overcome the limitations related to the distance between the nodes using meshed and cluster-tree networks (i.e., a sort of cascade of networks connected together) but in any event two elements belonging to the same network must be within a few hundred metres of distance (maximum) in order for connectivity to be established. While short range may seem a disadvantage at first sight, for several applications (e.g., smart home), a maximum connectivity range of 200 metres can be more than enough. To counterbalance the limitations on range, very low power consumptions (directly corresponding to long battery lifetime) and relatively easy protocol stacks allow implementation of these standards on small and inexpensive devices.

1.2.1.1 802.15.4

The 802.15.4 standard has been created and is maintained by the IEEE WPAN Task Group 4, which was chartered to investigate a low data rate solution with multi-month to multiyear battery life and very low complexity. It is operating in an unlicensed, international frequency band. Potential applications are sensors, interactive toys, smart badges, remote controls and home automation. The previous definition embeds a large part of the most important characteristics of the 802.15.4, which can be summarized as follows:

- low data rate (maximum 250 kbps);
- low power consumption;
- uses unlicensed bands;
- allows implementation in very simple devices (e.g., microcontrollers);

- suitable for timing critical applications (access to the network in maximum 30 ms) and
- supports star, meshed and cluster-tree networks.

The standard defines only the first two levels of the OSI protocol stack (i.e., the Physical and the MAC layers), leaving to others to define how to manage network addressing, transport, flow control (if any) and so on. For these reasons, on top of 802.15.4, a number of different standards have been devised in order to implement specific functionalities and features.

A network based on 802.15.4 can be composed of two types of devices: 'end devices' and 'full function devices' (FFD) that allow the creation of meshed and cluster-tree types of networks. The possibility of connecting multiple FFD together allows for a great degree of freedom in shaping dynamic networks that can grow (and shrink) based on the specific needs of the application.

Physical layer uses channels in the unlicensed (but regulated) frequencies for Industrial, Scientific and Medic (ISM) bands (2.4 GHz worldwide, 868 MHz in Europe and 915 MHz in the Americas), allowing for a maximum data rate of 250 kbps if the 2.4 GHz band is used. BPSK or O-QPSK modulations are used. The media access control (MAC) layer takes care of addressing, collision avoidance and access to transmission channels as well as security. Security is implemented through 128-bit AES symmetric cryptography (keys can be specified by the upper layers of the protocol stack).

1.2.1.2 ZigBee

ZigBee offers a full protocol stack that can be easily implemented in small, battery-operated devices to create WPANs – the protocol has been designed and is maintained by the ZigBee Alliance, a nonprofit organization that has more than 150 companies, including big names like GE, Huawei, Amazon, etc. ZigBee is based on 802.15.4 and implements the Physical and MAC layers, as specified in the IEEE standard. As such, ZigBee is able to reach a maximum of 250 kbps data rate; considering however that the typical use cases for a ZigBee network require the exchange of payloads of a few hundred bytes, it is easy to realize that the limited data rate is not actually an issue.

One of the key concepts of ZigBee is the Key-Value Pair (KVP): similarly to what happens in publish-subscribe protocols, information is distributed within the network based on the contents to be communicated (and not on a possible list of recipients). To give an example, this means that, when an end node wants to notify a change of state of one of its sensors, it will notify the network coordinator of a change for that specific KVP associated to that sensor. The coordinator will know how to distribute this information based on binding tables, that is, lists of end nodes that have subscribed to receive updates for that KVP. ZigBee protocol stacks add, on top of the Physical and MAC layers, two more layers as follows:

- Network layer: it takes care of addition/removal of nodes into a network and routing of packets between the nodes. The routing algorithm is ad hoc on-demand distance vector (AODV) for meshed networks, while for the other types of networks only static routing is allowed. It is interesting to note that meshed networks, complemented by AODV routing, improve the reliability and resilience of the ZigBee networks, allowing dynamic routing of packets even when one or more nodes are not available and

- Application Framework layer: it is responsible for maintaining the KVPs and for the discovery functions of networks. The application framework (AF) layer implements also the ZigBee define object (ZDO), that is, a software object that provides access to a number of functions used to manage the ZigBee network by the application. One of the biggest advantages of ZigBee is the wide availability of plug-in modules that can be easily connected to microcontrollers and similar small devices, effortlessly adding wireless capabilities with limited expenses.

1.2.1.3 Bluetooth low energy

Bluetooth low energy (BLE), sometimes also called Bluetooth Smart, is a WPAN standard based on Bluetooth (BT), and specifically designed to meet the reduced power requirements of those IoT devices equipped with reduced energy availabilities. BLE is defined in release 4.0 of the Bluetooth specifications and uses the same 2.4 GHz radio frequencies of the previous Bluetooth releases, even if the two technologies are not compatible with each other (i.e., a BLE device is not BT-compatible).

BLE offers higher data rate than 802.15.4-based protocols, giving to the end applications the possibility of accessing to bursts of up to 1 Mbps, with however a reduced coverage radius when compared to other solutions. BLE theoretical coverage is, in fact, typically under 100 m, but in real-life typical coverage drops to a few tenths of metres. BLE implements a mesh type of network, that can be used to implement also other network topologies. One of the key advantages of BLE against other protocols is the fact that it is already supported by almost all modern mobile devices; this means that by using BLE it is relatively easy to add support for mobile-based interactions.

Similarly to what we have seen for the 802.15.4-based protocols, devices included in a BLE network may have different roles; such roles are defined in the generic access profile (GAP) and define a distinction between peripheral devices (typically the constrained things to connect) and central devices (typically mobile devices or, however, devices with more processing power than typical peripheral devices).

On top of the already mentioned network topologies supported by other protocols such as point-to-point, star and mesh, BLE supports also a Beacon type of network, where a device sends periodic advertisement messages to let other devices (typically mobile phones) know that they are in proximity of that beacon. This is the technology used, for instance, by Apple iBeacon service and that can allow implementing presence detection, targeted marketing and other applications where it is important to pinpoint with accuracy the indoor location of the user.

1.2.1.4 Z-Wave

Z-Wave is a protocol originally developed by a company called Zensys, in 2001, whose specifications have been later released to the public domain (Zensys donated the specifications to ITU who released them in recommendation G.9959 in 2015). It has been designed from the beginning to provide a low data rate wireless communications protocol for home automation applications, creating therefore what is called by the specifications a 'home area network'. Z-Wave is now maintained by Sigma Designs (who acquired Zensys) and the Z-Wave Alliance, the latter is the entity that administers the Z-Wave Standard while the former is the subject that releases interoperability certificates – in fact, interoperability for commercial products is a mandatory requirement for all products that want to support the Z-Wave standard.

The PHY and MAC layers defined in Z-Wave are equivalent, from a logical standpoint, to IEEE 802.15.4. The radio frequencies used are the unlicensed ISM 868–915 MHz spectrum; actual frequencies vary based on the region of operation and, based on the region, different RF profiles (defining data rate and data accuracy values) may be available.

Z-Wave supports data rate up to 100 kbps for the highest performance R3 profile, while the R2 and R1 profiles define data rate of 40 kbps and 9.6 kbps, respectively. Reported average indoor coverage of approx 10 m – this apparent limitation can be overcome using one of the key functionalities sported by Z-Wave, that is, the capability of automatically creating meshed networks. Using this capability, a node can use other nodes (up to four) to transfer a message to another node that cannot be reached directly (either because of distance larger than the maximum allowed coverage, or because of multipath interface or radio holes phenomena). The maximum number of devices per each network is limited to 232 units, but it is possible to connect nodes belonging to different networks (called domains' in the ITU recommendation) using interdomain bridges, therefore actually extending the coverage of a single domain (Table 1.1).

Table 1.1 A summary of the short-range radio technologies analysed until now. Caution should be taken when analysing data regarding coverage and data rates since the values reported are the best possible ones, which obviously are significantly influenced by several factors (e.g., interference from other radio sources, especially for systems operating at 2.4 GHz)

Protocol	PHY	Band	Data rates	Coverage	Max num. devices	Topology	Notes
ZigBee	IEEE	2.4 GHz ISM	250 Kbps	20 m	65,000	Star, mesh and cluster-tree	Radio modules widely available
	802.15.4	868–915 MHz ISM					
BLE	BSIG proprietary	2.4 GHz ISM	1 Mbps	10 m	unlimited	Mesh	Not compatible with normal BT[1]
Z-Wave	ITU 9959	868–915 MHz ISM	100 Kbps	100 m	232	Mesh	Proprietary

[1]But typically implemented in mobile phones.

1.2.2 Long range

While short-range protocols can be successfully used for a number of applications (e.g., smart homes, wearables, etc.), for many other applications larger coverage radii become mandatory. For this reason, several protocols to implement wide area networks (WAN) using different radio technologies have spread in the past few years, implementing different specifications and technologies, but with the basic key requirements of low power consumptions, relatively low data rates and large coverage. Within this framework, an important role could be played in the near future (i.e., starting from 2018) by the introduction of IoT-specific cellular standards that could potentially take advantage of the existing cellular networks infrastructure and equipment.

1.2.2.1 Low-power wide area networks (LPWANs)

Long range (LoRa) and long-range wide area network (LoRaWAN)

Long range (LoRa) and long-range wide area network (LoRaWAN) are two protocols commonly used in IoT applications, whose luck is due to a mixture of aspects, one of them certainly being the fact that everyone can create a LoRa-based network.

The name LoRa refers to a proprietary PHY layer developed in 2008 by a French company called Cycleo, later acquired by Semtech, and that is used by a number of other higher-level protocols to implement LPWANs. The complement of LoRa, developed and maintained by the LoRa Alliance, is LoRaWAN, a standard that specifies the MAC level and that implements the functionalities required to set up a network and manage accesses to the physical level.

While it is possible to use the LoRa PHY layer with other protocols, the most common combination is LoRa with LoRaWAN and, for this reason, the two are often – erroneously! – used as synonyms.

LoRa uses unlicensed frequencies in the ISM spectrum (867–869 MHz in Europe and 902–328 MHz in the USA) together with adaptive spread spectrum technology based on chirped-FM modulation. The frequency used for the spread spectrum modulation is adapted based on the retransmission rate in order to automatically adjust radio parameters based on the actual error rates. The technology allows enabling longer coverage ranges and at the expenses of data rates. In fact, data rates for LoRa may range between 250 bps and 50 kbps and coverage may reach several kilometres.

The LoRaWAN MAC layer specifies the network topology, that in this case is a star of star, with each node communicating with multiple gateways that are communicating with a network server; interestingly enough, mesh networks are not supported, and therefore it is not possible for an end node to relay messages to other end nodes. In a LoRaWAN network three different types of elements can be found:

- end nodes: typically sensors and endpoints implemented on constrained devices. End nodes can be of three different classes (A, B and C) with different performances, in terms of data rates and latencies made available by the network and, correspondingly, different battery durations;

- concentrators/gateways: acting as transparent bridges between the end nodes and the network server. Connection with the network server is implemented typically through secure TCP/IP links; application data coming from the end nodes is encrypted as well and since concentrators do not implement higher-level protocols, they only act as a pass-through device and
- network server: it is where the actual receiving point of the application is implemented – before reaching the application point packets from concentrators are analysed and possible duplicates discarded.

One last interesting feature worth to mention is the recent introduction of device location capabilities without the utilization of GPS (that may have a consistent impact on the battery duration), notwithstanding the unquestionable advantages that such a functionality would introduce, there are doubts about its real effectiveness, since location of a device on radio signals (e.g., triangulation of the position using distance and signal strength) has proven several times to be extremely complex to obtain with good (or even usable) precision (e.g., below 500–1,000 metres).

The combination of LoRa and LoRaWAN is used in several IoT applications and one of the advantages is that, even if the technology used for the PHY layer is patented, LoRaWAN is an open-source standard that is freely available to whoever subscribes to the LoRa alliance. Therefore, it is possible for everybody to set up and maintain an independent LoRa-based network, without the need for specific licences from Semtech or from the LoRa alliance.

Sigfox

SigFox is the name of a proprietary technology owned by SigFox, a French company that in 2009 started to work in the LPWA area with a peculiar business model. While, in fact, other solutions like LoRa enable users to deploy and build private LPWA network, SigFox business model is closer to what a typical communication service provider (CSP) or mobile network operator (MNO) would do, leaving SigFox as the only owner of the networks being deployed all over the world. So, while it is possible to buy SigFox-compatible modules at a very low price (SigFox-compatible modules are actually often cheaper than LoRa-compatible ones) and deploy a set of end nodes, the only subject that can create and operate a SigFox network is SigFox itself, who sells access to the network and relevant additional functionalities on 'as-a-service' basis.

Deciding which model is better (between 'open' networks like LoRa and SigFox) is very much dependent on the specific solution that is being implemented: from one side with SigFox all the burden of deploying, maintaining and supporting the network is moved to a third party. This sort of outsourcing can become a competitive advantage when use cases involving very large areas to be covered are considered and, since it not possible to simply extend autonomously the SigFox network, monitoring and controlling on the actual coverage on the targeted installation area is mandatory. On the other side, other solutions like LoRa give the possibility of creating private, personally owned LPWAN networks, without having to share resources with anybody else (and without the need to pay recurring fees to a third party). Also, business continuity should be taken into consideration since, in case of severe problems to the mother company, availability of existing SigFox networks could be at stake.

From a technical standpoint, SigFox uses binary phase shift keying modulation (BPSK) in an ultra narrowband located within the ISM frequencies ranges. The utilization of an ultra narrowband allows for a simplified design for the radio components, and in particular the antenna, reducing the noise levels and the power consumptions. The downside is the available data rates: uplink data rate is limited to 100 bps with an additional limitation (due to the licensing model of the ISM band) of no more than 140 uplink and 4 downlink messages per day; the difference between uplink and downlink means that acknowledgement of the messages, as well as other functions like remote firmware update of the end nodes, is not possible. Coverage is, on the other side, a strong benefit of the radio technology used by SigFox: several studies have shown coverages ranging from 10 km in urban environments to 50 km for rural areas.

SigFox networks are only based on star topology: an end node can be connected to more than one base unit, but no mesh or combined topologies are supported. SigFox has recently enhanced its offering with a set of additional services, including data processing, big data analytics and devices geo-location. Just like what LoRa claims to do, also in this case, the position of the devices is calculated using a probabilistic calculation of the device position done using radio parameters. SigFox claims accuracy of this solution to range between 1 and 10 km depending on the base station density where the device is located. Whether this is enough or not will, of course, depend on the specific application being deployed.

1.2.2.2 Cellular technologies

Cellular communications have been growing at very high rates from the beginning of the 2000s; voice and short messaging services have progressively left to faster and faster data connections, the flag of being the driving force for the on-going expansions of the cellular networks.

Nowadays, three main families of cellular technologies are available:

- Global system for mobile communications (GSM): the first real widespread technology for cellular communications, mainly focussed on voice messaging, also known as 2G;
- Universal mobile telecommunications system (UMTS): still very focussed on voice and traditional messaging (through SMS), it introduced significant improvements to cellular data utilization in terms of available bandwidth and connection reliability, also known as 3G and
- Long-term evolution (LTE): an IP-only based cellular platform allowing even higher data rates and faster connection times, also knows as 4G.

Because of their nature, the cellular networks technologies that we are used (e.g., GSM, UMTS and LTE) are not a particularly well fit for IoT applications. Thus, power requirements and complexity of a GSM modem, for instance, are not compatible with scenarios where constrained devices would be used. On the other hand, most of the strengths of cellular data connections characteristics (e.g., support for high-speed mobility and high data rates) are losing their value in IoT scenarios that are considered. Even if cellular networks have been used in several M2M applications (for instance using SMS for data transfers, instead of proper data connections), many of the most important IoT requirements were left uncovered, leaving space, as we have seen, to other LPWAN technologies to emerge and take the field. For these reasons, in 2016 the third generation partnership project (3GPP, i.e., the organization that generates and maintains standards for cellular networks) released a series of specifications for IoT-specific standards, with the clear goal of making possible for the MNOs to be competitive in the IoT arena. LTE-mobile (LTE-M) and narrowband IoT (Nb-IoT) are the two IoT-specific standards, based on LTE technology, released by 3GPP in 2016. They have been both specifically designed for IoT applications, but with different specifications in terms of available data rates, support for mobility and link coverage.

The biggest competitive advantage for both LTE-M and Nb-IoT relies on the fact that both standards have been designed to be implemented on existing LTE networks with very limited or no changes to the existing network hardware. This feature will allow MNOs to upgrade their networks introducing LTE-M and Nb-IoT, without having the need of physically changing any of their equipment, thus reducing the implementation costs, making at the same time advantage of the existing installations and network coverage.

At the moment of writing, several MNOs have deployed pilot or commercial LTE-M and Nb-IoT networks, but for Nb-IoT there are very few modules commercially available that can use them, and the ones that are available are quite expensive, especially when compared to other LPWAN, noncellular technologies (100 US$ for an Nb-IoT module against 5–15 US$ for a SigFox or LoRa-compatible module).

LTE-M

LTE-M has been the first IoT-oriented standard released by 3GPP and it is closer to LTE in terms of functionalities, hence representing a sort of the first step into the IoT domain for LTE-based cellular networks. It supports extended battery life through the implementation of specific functionalities like the power saving mode (PSM) and the extended discontinuous reception (eDRX), introduced in order to preserve battery life. Considering all the available power saving functionalities, the system autonomy could reach almost 10 years. LTE-M offers higher data rates (up to 1 Mbps) that could support also voice-over-IP applications, making it different from the other LPWAN technologies that we have seen until now. On top of that, support for mobility (i.e., seamless handover of communication between nodes when the device is moving) is supported. While LTE-M was originally intended to be a sort of temporary solution to enable IoT on cellular networks, its capabilities allow for more flexibility than Nb-IoT and some mobile operators are actually considering LTE-M as an interesting solution for a wide range of applications in a future-proof way.

Nb-IoT

Nb-IoT is the narrowband standard specifically released by 3GPP to support IoT applications. It reuses a good number of LTE characteristics, removing at the same time other functionalities like Inter-RAT handover (i.e., the possibility for an LTE device to jump on GSM or UMTS networks) in order to allow for simplified, and therefore cheaper, device design. Nb-IoT uses LTE licensed frequencies and can be deployed as stand-alone, in-band or in guard-band, making possible to the MNOs to adjust frequencies utilization based on the specific conditions of the area where the network is deployed. The design goals for the radio interface of Nb-IoT are indoor coverage, high number of connected devices and large coverage radius per eNodeB (i.e., the transmitting base station for an LTE network). Being based on LTE, radio access is more complex than what is required for other technologies like SigFox and LoRa; complexity in this case also means more expensive hardware, but at the same time better performances in terms of data rates. Nb-IoT can, in fact, reach 250 kbps which is significantly higher than what is offered by many other competitive technologies. Nb-IoT embeds other functionalities that are designed to minimize access latencies to network and to speed up connectivity set-ups, simplifying some of the procedures typically required by cellular networks to set up a data link (specifically attach and pdp context activations).

Being based on an LTE design, Nb-IoT requires a subscriber identity module card (SIM) to be included in the device. While standards and functionalities to implement software-based SIMs are available, at the moment of writing SIMs are mainly physical and this is a factor that could introduce additional complications in an IoT scenario (Table 1.2).

5G

With massive tenders going on around the world to assign the different frequencies bands to the MNOs, it is worth spending a few words on the next evolution of

Table 1.2 A summary of some of the characteristics typically considered when comparing long-range communication protocols. For coverage, typical utilization scenarios have been used to make a qualitative evaluation. Notice, however, that for some technologies (e.g., Nb-IoT) limited information on real-life applications are available at the time of writing

Protocol	Packet size	Data rates	Coverage	Topology	Notes
SigFox	12 bytes defined by user	100 bps[1]	3–10 km (urban); 30–50 km (rural)	Star	Networks managed by SigFox
LoRa/ LoRaWAN	–	0.25–50 Kbps	2–5 km (urban); 15 km (rural)	Star of star	Network can be created and maintained privately
LTE-M	Defined by user	1 Mbps	2.5–5 km	Cellular	Network access regulated by contracts with MNOs
Nb-IoT	Defined by user	250 Kbps	15–30 km	Cellular	Network access regulated by contracts with MNOs

[1]With limitations on the number of messages/day.

cellular technology, named fifth generation (5G) in order to distinguish it from 4G. The main focus of 5G will be on even higher data rates and lower latencies than what is currently offered by 4G networks. 5G main applications will be therefore in areas where either very high throughputs are required or very reliable and fast connections are needed for security reasons, like for autonomous driving and vehicular traffic management. 5G is therefore born having IoT in mind, but it will take a significant amount of time before large commercial applications may be available using this technology.

1.3 Data protocols

Besides the radio and networking protocols that allow the interconnection and logical organization of the devices in the networks, the protocols that stay on top of the protocol stack are extremely important as well, since, at the end of the day, it will how they access the transmission layer to determine the actual dynamic characteristics of the IoT application. Similarly to what we have seen until, there is no simple answer, nor one catch-all solution that fits for all applications, but rather different options that may fit better one use case instead of another.

1.3.1 Hypertext transfer protocol

Hypertext transfer protocol (HTTP) is one of the most famous protocols in the world, well known also to the ordinary persons, being the most visible protocol used to browse resources on the web. The first version of HTTP was released 0.9, dated 1991 and even at this early stage the foundation principles of that made of HTTP such a huge success were already present: simplicity and the response-request structure.

HTTP is, in fact, based on a client–server paradigm, where a client (or more properly, a user agent) wants to access to resources located on a remote machine (the server). In order to access those resources, the client must formulate a request to the server, specifying which resource it wants to access using a resource identifier (called uniform resource identifier, or URI in short). The server must respond to this request specifying if the request has been successful or not and, in case of success, providing the resource itself.

HTTP specifies few methods that can be used by the client to interact with the server: these methods are used to instruct the server on the operation that the client wants to accomplish with the specified resource (e.g., retrieve an HTML page). Resources can be static, that is, present in the server in a permanent (or semi-permanent) way, or dynamic, that is generated dynamically following an explicit request from the client. In this way, it is possible for the server to generate user or session-specific contents that can, for instance, embed data coming as a result of a query from a database. Another important aspect of HTTP is the fact that it is a stateless protocol: the server does not store information on the current session with a specific user agent and it is for the latter to keep track of the history of its interactions with the server (cookies have been introduced for this purpose in more recent releases of the protocol).

HTTP evolved from version 0.9 to 1.1, released in its first version in 1997, which has been (and still is) the standard used for the world wide web (WWW); notwithstanding the various improvements that have been introduced over the years, the key principles outlined above have remained more or less unchanged. Using HTTP for IoT applications is possible (indeed, HTTP is already used by several IoT applications) but the utilization of this protocol in this domain comes with a number of concerns:

- the client–server architecture implemented in HTTP does not support server push – there is no support for a spontaneous data transfer from the server to the client;
- headers (i.e., the commands and the responses used by the two parties to communicate with each other) are transferred in plain text and are quite verbose. While this can be an advantage for a human being looking at an HTTP interaction, for constrained devices utilization of this type of headers can easily cost valuable resources both in terms of memory and bandwidth utilization;
- HTTP only allows one request at the time to be processed, generating unnecessary delays when processing multiple requests at the same time and
- HTTP is designed to run over TCP, which can be a resource-intensive protocol for constrained devices (and may introduce unnecessary features).

In order to overcome these limitations, HTTP 2.0 (in short HTTP/2) has been designed. The new version, introduced in 2015 and already supported by the most important web browsers, overcome almost all limitations listed above, with the only exception of the forced utilization of TCP, which remains the network protocol of choice also for HTTP/2. In particular, release 2 of HTTP introduces server push (essential for publish-subscribe architecture, see MQTT) and headers compressions in order to save bandwidth in constrained devices. Encryption is not considered as mandatory in the standard but several implementations (including all the most popular web browsers) will not support HTTP/2 without an encrypted layer like TLS.

Using HTTP/2 for IoT has interesting benefits in the form of using a widely adopted protocol that is supported by a large number of platforms and software systems, at the same time for use cases where bandwidth and processing power are scarce other options (like CoAP or MQTT) may be more efficient.

1.3.2 Constrained application protocol

As mentioned in the previous paragraph, HTTP/2 alleviate most of the major drawbacks that prevented HTTP/1.x from being successfully used in IoT applications. Nevertheless, it still requires TCP as its transport layer, despite admitting that the latter was not specifically optimized neither for IoT applications nor for constrained devices. In order to overcome these limitations, the constrained application protocol (CoAP) has been designed. Officially defined in RFC 7252, CoAP has been designed from the very beginning to be

- compatible (i.e., operational on) with constrained devices, typically 8-bit microcontrollers with 10 KB of RAM and 100 KB of code;
- based on UDP (much lighter in terms of resources utilization than TCP). CoAP is based on the REST paradigm (REpresentational State Transfer), an architectural

style based on the typical client–server paradigm of HTTP extended to web applications. In simpler terms, CoAP and HTTP share the same request-response messaging scheme, making interconnection between HTTP-based systems (such as web servers or web applications) and CoAP domains a relatively easy task. The basic scheme of operation of CoAP can be divided into two logical layers (even if the protocol layer remains only one);

- Messaging: this layer provides the tools for message transfer between the different nodes. Messages can be transferred in a reliable (i.e., with an acknowledge) or unreliable way – it is up to the application to decide which option to take (differently from TCP which 'forces' acknowledgement of every message). The availability of reliable message dispatch overcomes the lack of the error and flow control in UDP (present instead in TCP) and
- Request-response: this layer provides a way for the parties to publish (i.e., post) a value to the server, to obtain from the server information (identified by a URI) and to subscribe to a specific information, implementing a server-push mechanism.

An interesting functionality included in CoAP is the inclusion of discovery functions that can be used by a node to discover other nodes currently belonging to the same CoAP domain. This functionality can be useful when building dynamic networks of devices over, for instance, unstable or not continuously available connections.

It is interesting to note that even if CoAP is designed to run on UDP, it can actually be implemented on any network protocol since the CoAP messaging layer implements the required functions of flow and error control; this way, it is possible to use SMS (for instance) to relay CoAP messages over a cellular network saving the client from the additional burden of handling UDP.

Security is not defined in the CoAP specifications and, when implemented over UDP, security functions are implemented through the utilization of the datagram transport layer security (DTLS) – this may, however, represent a drawback since implementing DTLS in constrained devices may hinder some of the benefits portrayed by CoAP.

1.3.3 Message queuing telemetry transport and MQTT-SN

Message queuing telemetry transport (MQTT) is a lightweight message-based communication protocol based on a publish-subscribe paradigm and especially developed having in mind constrained devices. MQTT, in fact, has been designed to use as little bandwidth as possible: specifically, the primary requirement for this protocol was to use less bandwidth than that required to carry out the same actions using HTTP or similar protocols. Tests in this direction have been carried out, highlighting that utilization of MQTT results in more messages being transferred (in a given period of time) with lower power consumption when compared to HTTP/1.x. When it comes to HTTP/2, the header compression mechanism introduced in that release of the protocol actually limits the bandwidth used by the nodes, but increases the computational requirements, leaving MQTT to remain the most sensible choice for constrained devices.

As mentioned above, MQTT is based on a publish-subscribe paradigm; this means that the available information is classified in topics. For ease of information access, each topic may be positioned inside a hierarchy (similarly to what happens in a normal file system with folders). Distribution of the information is not therefore based on point-to-point (or even point-to-multiple points) model, but rather on the contents of the information itself.

In a basic MQTT network, there are two different types of elements: nodes and brokers. Using a typical client–server metaphor, the broker acts as a central server for all nodes (acting as clients) and all nodes communicate (both in read and write modes) only with the broker. It is the broker's responsibility to distribute the information accordingly to the 'interests' of the nodes.

A node can, in fact, be both an information producer and a consumer, even at the same time: nodes can subscribe to specific topics of interest, in order to make sure that the broker will send them all the messages pertaining to those topics. Similarly, nodes can create contents by publishing information to specific topics. The broker does not carry out any processing on the transmitted message, as it simply relays the payload to all nodes that previously subscribed to the topic.

An interesting advantage in respect of more generic protocols, like HTTP/2, is the inclusion of specific functionalities for message and telemetry transfer over troubled connections, like the 'last will and testament' (LWT) functionality, which allows to specify a message that the broker has to broadcast to all interested parties when a client is disgracefully disconnected from the network. The LWT allows therefore to implement fall-back strategies in specific circumstances.

MQTT relies on TCP as transport protocol, and as such requires a TCP layer in the lower protocol stack to operate – for applications where the implementation of a TCP stack can be problematic, MQTT-SN (MQTT for sensor networks) has been designed, introducing solutions to allow the application protocol to be agnostic from the lower levels.

Even if MQTT is widely used in a number of applications, one of its main limitations is about the lack of security measures specified into the standard, leaving the definition and the implementation of security measures to the specific implementation. While this leaves freedom and helps to implement the protocol on particularly constrained devices, it is important to note at the same time that security is becoming more and more an issue for IoT applications and as such inclusion of security mechanisms such as TLS is as a matter of fact almost mandatory.

1.3.4 Extensible messaging and presence protocol

Extensible messaging and presence protocol (XMPP) was born as an open-source effort to standardize interoperability between different systems using the eXtensible Markup Language (XML), providing a way to implement messaging and presence system using XML files. XMPP has been later standardized by IETF which has issued several RFCs that define various aspects of the protocol, including specific implementation notes for IoT applications.

Each node belonging to an XMPP network is identified by a unique identifier, composed by a user id and the identification of the XMPP domain. Each user can also have multiple devices connected to the same account with different priorities, making possible selective dispatch of messages to the different devices in a specific order (e.g., the 'main' device can receive all messages while others can receive only specific ones).

The key-characteristic of XMPP is, as already mentioned, the adoption of XML files to execute all operations, from nodes binding to information request and subscription. Requesting information from a node is carried out by sending an XML file containing the requester node identifier, the recipient identifier and an instruction to read a certain information element. The recipient node will then answer with another XML file with the response, being that the information itself or a message describing the status of the node and why the information could not be provided.

The key advantage of using XML is the possibility of implementing a much more complex hierarchy of data and queries with respect to what is allowed by other protocols using 'flat' data representations (e.g., MQTT). This makes possible, for instance, to specify different data types (e.g., integer, Boolean, string, etc.), to simultaneously read (or write) multiple information with a single operation, and to provide additional information about the data being read or written (e.g., if the data are an actual read from a sensor, an averaged value, an estimate, and so on).

This added level of flexibility brings an inherent increased level of complexity when comparing XMPP to other message-based protocols like MQTT: each node must be able to properly read, parse and manage XML files, and possibly manage the implemented data hierarchies. Even if specific measures have been defined for implementing XMPP in IoT scenarios, the added requirements in terms of bandwidth and computational power need to be taken in account when designing the application, especially if constrained devices are being used.

XMPP per se is defined as a very generic and extensible protocol and can be therefore adapted to a number of applications. Recommendation XEP-0232 explicitly targets utilization of XMPP for IoT and covers a number of areas such as definition of XML tags that preserve readability but reduce the overhead required to transfer XML files (typically larger in size than binary files). Even though XMPP foundation claims that XML parsers are typically as efficient as JSON parsers (or even more efficient), the very descriptive nature of XML requires a certain overhead for transmitting data.

Considering all of the above, many authors concur in considering XMPP an interesting alternative when scalability, data complexity and interoperability are key-requirements and where computational capabilities of the nodes being used are not a major design concern; for other applications, more simple and straight-to-the-point protocols (like MQTT or CoAP) may represent a better alternative (Table 1.3).

Table 1.3 A summary of the main features of each data protocol analysed in Sections 1.3.1–1.3.3, from an IoT application standpoint

Protocol	Type	Pros	Cons
HTTP/1.1	Generic	Widely available Easy interoperability with web applications	Uses TCP Bandwidth overhead because of headers
HTTP/2	Generic	More efficient than HTTP/1.x Maintains interoperability with web applications	Uses TCP Headers compression requires computational power
CoAP	Messaging	Compatible with constrained devices Can use UDP or other lower transmission protocols Specific for IoT	Security not included in the standard
MQTT	Messaging	Compatible with constrained devices Specific for IoT	Security not included in the standard Several different implementations may cause interoperability issues Uses TCP
MQTT-SN	Messaging	Compatible with constrained devices Specific for IoT Can use UDP or other transport protocols	Security not included in the standard Several different implementations may cause interoperability issues
XMPP	Generic	Extensible XML allows for more complex data management	Overhead introduced by XML More complex w.r.t. other protocols

1.4 IoT platforms

As lots of IoT platforms are currently available in the market, it is difficult to decide which one to go for. Almost all of them are capable of saving sensor data into the cloud and performing basic functions like predictive analytics, machine learning, visualization, etc., on top of it. However, having a deep dive into these platform solutions one will understand the core difference between the well-established and the basic products available in the market. To be precise, there are certain technical as well as commercial aspects which define the maturity of an IoT platform and must be taken into consideration while selecting an appropriate solution for a specific IoT use case.

1.4.1 Technical aspects

1.4.1.1 Services offered

Almost all platforms offer additional services on top of the basic data acquisition functionalities. These services typically include analytics functionalities as well as standard plug-ins or applications such as location tracking, error notification, real-time monitoring and so on. Considering therefore the set of the potential vertical functionalities, all platforms try to offer the same menu of main options. What differentiates a mature platform solution to others is the actual type of analytics (real-time, batch, predictive, etc.) and the number and diversity of the plug-ins already available and embedded into the platform. Some well-established platforms, for instance, offer industry-specific (aerospace, shipping, manufacturing) plug-ins in their portfolio that simplify the whole process of setting up the system.

Another important feature of an IoT platform is its openness (i.e., being open source or not) to allow the users to launch their own personalized plug-ins or APIs. Not every platform is based on an open-source environment and having to deal with closed, protected platforms may represent a significant issue when trying to make the platform evolving with the real-life application behind it. The ones which allow the users to launch their own APIs offer different Software Development Kits (SDK) to program these APIs. Certain mature platform solutions have SDKs which includes open-source libraries in Java, C, Python, etc., giving developers options to program their own solution using a platform of their choice [11].

1.4.1.2 Supported data protocols

IoT applications works on various data protocols depending upon the type (continuous, sensor data, audio/video, etc.) of data to be monitored and the required degree of security (TLS/DTLS) that needs to be achieved. As we have seen in Sections 1.3, among the most common data protocol for IoT cases are MQTT, HTTP and CoAP. In contrast to HTTP, MQTT and CoAP are protocols best-suited for a constrained environment where the bandwidth and sensor battery are at a premium. MQTT is used extensively for event-based monitoring and CoAP is a preferred protocol for continuous data transmission [11].

Not every IoT platform supports all of these protocols, which also means that the one covering all of them may have a significant competitive advantage in situations where flexibility and systems' interoperability is a key requirement.

1.4.1.3 Security

Security is a factor which cannot be ignored or underestimated. Since IoT is all about data and data management, the platform, as well as the underlying infrastructure, must ensure data integrity and safety. At the same time, no loose points should exist in the IoT infrastructure which can later be used as a backdoor entry for hackers which may then not only try to steal the data but can also manipulate the system. The balance between data integrity and data protection is defined by the specific application, but the best approach is to choose a solution that enables both aspects at the same time.

IoT security should therefore be implemented at every level in the IoT stack starting from the sensor-gateway and up to the cloud-data centre. A good platform solution uses unique identity-based authentication methods such as X.509 certificates at sensor/device level and offers encrypted (TLS/SSL) data transfer within the IoT network. Constrained devices may use other dedicated solutions that preserve the required data security functionalities while at the same time avoiding the overhead caused by some of the already mentioned security solutions (e.g., TLS/SSL). Some IoT platform solutions offer special authentication methods on the top of what is typically considered to be a standard set of features and can offer compliance with security management systems such as ISO 27001.

1.4.1.4 Device management

Device management is an important feature of an IoT platform that, interestingly, is not offered by some of the solution providers. It is a feature which maintains a list of devices connected to the IoT network and keeps track of their operational status. Moreover, this feature often includes remote devices provisioning and configuration management, allowing remote configuration of a device, firmware updates and faults management (detection and first-level troubleshooting) in the device.

Another important aspect of the IoT platform which is directly associated with Device Management is its degree of scalability. Adding more devices to the network increases its complexity and demands scalable storage along with load balancers (to distribute data load over several servers). Maintaining these servers, load balancers and subsequent data nodes is not easy. While Device Management is certainly a functionality that needs to be taken into account for large and complex systems, for smaller ones it may not be as essential. For this reason and its inherent implementation complexity some IoT platform solution providers do not offer device management capability [11].

1.4.1.5 Data storage

Data storage is an aspect obviously related to data security but that extends also to other specific (and very important) aspects that need to be considered, such as data availability, backup and easy recovery capabilities. All these capabilities derive directly from the IT infrastructure being used, that may include concentrated or distributed storage systems with different levels of resilience implementation. Backup solutions can also vary from simple backups (complete or incremental) to full disaster recovery solutions, with copies of data in multiple physical sites and complete solutions in place for fast and error-free system-wide recovery. Cloud platforms further extend this number of available solutions, including also the possibility of integrating cloud platforms with on-premises storage (hybrid cloud). The hybrid cloud option could allow the user to process and store business sensitive information locally and keep the remaining mass data inside the IoT platform cloud, obtaining at the same time interesting results both in the data security and data reliability domains.

In order to minimize the data load, certain platform supports decentralized, offline data computation, better known as edge computing (see Sec. 1.5). This feature will play a very important role in the future when IoT projects will get bigger and more complex: in this scenario it will be important for a redistribution of the computational and storage demands in order to ensure continuous scalability of the platforms. Edge computing not only allows for lower operation costs by reducing the data load being pushed to the cloud but also enhances the application performance by achieving lower latency levels.

1.4.2 Commercial and business aspect

1.4.2.1 Size of the company

Thumb rule says, the bigger the brand, the firmer they have to stand for their reputation. Which means that they strive to provide a quality solution. The same applies to the support and maintenance services, that are now more and more considered as an integral part of the features offered by a solution. Service level agreements (SLAs) offering global, 24/7, multilingual tech support are something which really differentiates big, well-established companies from newcomers or start-ups and while additional services come with an extra price tag, for complex and large systems, a well-performing support and maintenance system can make the difference between a usable and an unusable system.

1.4.2.2 Data centre

Certain big IoT platform providers manage their own set of data centres worldwide. This is particularly relevant if you are considering *'one-stop'* solution (e.g., platform + cloud) rather than dealing with two or more parties in order to host your application. Dealing with one party is of course commercially beneficial and easier than negotiating deals with two or more parties.

The idea behind one-stop platform plus data centre concept is also important if you are looking for a long-term, stress-free and reliable solution. The platform providers who host their solutions on a rented cloud/data centre may change their opinion with regards to the choice of data centre provider later thus resulting in troublesome data migration.

A third important aspect of the data centre is the availability of data worldwide with as low latency as possible. Some of the bigger IoT platform solution providers have data centres installed in every continent, with automated data replication functionalities included in the solution, thus addressing this issue [12].

1.4.2.3 IoT ecosystem

IoT ecosystem is currently one of the most important selection criteria for a platform solution. An ecosystem in IoT terms is actually the capability of an IoT platform to bring the partners of the value chain (e.g., stakeholders, market players, etc.) together or to integrate their solution in the platform in order to broaden the overall digital product portfolio. This is especially important if a company would like to integrate its solution and the solutions of its suppliers and service providers with customers at one place. Here again, the bigger the brand of an IoT provider, the larger its business network thus offering bigger ecosystems.

Another important aspect of IoT ecosystem is the interoperability. A good IoT platform solution should support integration with open-source ecosystems and should provide an interface for third-party services or applications. This aspect becomes particularly important in cases when the platform solution is needed to integrate with the existing ERP or other already existing database management systems [12].

1.4.2.4 Pricing model

The most common pricing model offered by platform providers is subscription based on a *pay-as-you-go* scheme (these platforms are also often labelled as *PaaS*, or Platforms as a Service). This model can be effective and attractive for small systems or for portions of larger solutions (imagine for instance a section of a monitoring solution that needs to use satellite links for a part of the data transfer), but of course does not scale particularly well when the size of the system increases, since the costs will easily and quickly become non-negligible. Another option is represented by a subscription model (i.e., with a fixed monthly fee) that offers an easy way for managing the cost and provides effective revenue projection. To summarize, the pay-as-you-go model is good for users who are well aware of the amount of data load they want to handle for a long period of time, whereas the subscription model is a better alternative for highly unpredictable, complex and fast-growing IoT use cases.

Some platform solution providers offer both pricing models in order to give the customer a choice of matching cost structure based on his individual use case. Flexible pricing models are more typical of newcomers or start-ups, that are using flexibility as a competitive advantage to gain access to the market.

1.4.3 Commercial or open-source

As already mentioned, many of the commercial IoT platforms are based on a *pay-as-you-go* scheme, based on parameters that scale with the amount of traffic (e.g., number of messages, number of connected devices, etc.). This of course means that the price of the solution will scale (linearly or not depending by the specific business model in place) with the size of the system hosted by the platform. As an alternative, there are several (literally hundreds!) of IoT open-source platforms that could represent a viable alternative to commercial solutions.

1.4.3.1 Evaluation criteria

However, there are at least three factors that need to be taken into account when making a decision about the platform to be used:

- technical aspects;
- projected growth of the system and
- commercial considerations.

The technical aspects include the basic requirements that the platform needs to fulfil, including, but not limited to

- reliability: how reliable is the platform? how large is the installed base?
- scalability: how easy is it to scale up (or down) the platform?
- protocols: which are the supported protocols? How easy is it to add a new protocol to the platform?
- support: what are the available SLAs? Is the support localized (i.e., available in your local language)?

- supported cloud platforms: which are the cloud platforms/interfaces compatible with the platform? In other words, does the choice of the platform 'lock' you to a specific cloud solution?
- security: what are the available security protocols/mechanisms? and
- ease of use: how easy is it to manage the platform and the devices? Does the platform include a provisioning or O&M section to allow for simplified devices management?

It is clear that the list could be further expanded, but the above points are most probably the most important ones to take in consideration. In particular, reliability is a point that is typically in favour of commercial platforms since large IoT providers invest a significant amount of money in their data centres, offering therefore availability and reliability figures that can be difficult to be matched by open-source alternatives. On the other hand, choosing a commercial platform may complicate future migrations to other solutions, since data export and migration functionalities may be limited (or completely absent).

The projected growth of the system has an impact, obviously, on the already mentioned scalability factor, but also on the other supported protocols. To make an example, for a small system, the lack of a specific communication protocol or the lack of a device management function may not be a problem, but the difficulties and the overhead caused by these missing functionalities may explode at a later stage when the size and the complexity of the system will increase. It is therefore important to have (even a minimal) vision of the potential growth directions of the system, both in terms of size (e.g., number of devices) and complexity (e.g., number of protocols to be supported).

Last but certainly not least, the commercial aspects may play an important role in the choice. We have already analysed the pricing model for the commercial platform; for open-source, there are essentially two factors to be taken into account:

- self-hosted or server-less and
- pricing model.

The first point refers to how the platform will be hosted: in a self-hosted the owner will have to sustain the initial costs for the hardware and storage, connectivity fees (i.e., Internet access) and all the related operational costs (e.g., maintenance, power, air conditioning, etc.). Considering that the upfront costs associated with the acquisition and set-up of a hardware platform that supports redundant data storage can be significant unless there are specific requirements (e.g., security or data confidentiality) that prevent the utilization of the cloud, a server-less implementation becomes almost a mandatory solution. In this case, the servers are hosted by a cloud provider; also in this case we typically have the choice between two business models:

- *pay-as-you-go*: in this case the price is calculated using a fixed based on the size of the storage, the number of CPUs, RAM, etc. plus a variable amount calculated on the number of data being transferred and
- fixed subscription: in this case, the fee covers the virtual hardware configuration and a *flat rate* of data per a certain period of time (e.g., a month).

There may be additional costs to be added to the basic subscription, like enhanced availability, extended support SLAs, etc. In any event, it is clear that, once again, having a view of the initial size and projected growth of the platform becomes a fundamental aspect, since it will not initially allow to choose between the two models but will help to predict about the future costs and possible migration directions. The *pay-as-you-go* model will in fact scale with the system's size and there will be a cut-off point when the fixed subscription model will become more financially attractive. Predicting when this transition may happen can become an important factor to ensure growth sustainability of the platform.

1.4.3.2 Open-source platforms

As already mentioned above, there are literally hundreds of open-source IoT platforms available, with a business model that is typically based on support and professional services (e.g., training and consultancy) fees. Choosing a specific platform may therefore be a complicated task, made even more complex by the fact that not all the platforms have comparable attributes in terms of the requirements listed in the previous paragraph. We will nevertheless mention here a limited set of the most common open-source solutions, with the goal of presenting the ones that bring particular functionalities in one or more areas.

Kaa: Kaa is an open-source platform that supports a wide range of endpoint systems (including Raspberry Pi, Intel Edison and ESP8266, but not Arduino family) [13]. It is possible to extend the support to the platform by using SDK and porting it to the target platform in case this is not included in the list of systems supported natively. Kaa uses Apache ZooKeeper to create and coordinate Kaa clusters, and SQL and NoSQL databases for configuration and data management. The implementation of a cluster structure allows the creation of high-availability (HA) solutions, that have to include also the database instances. The entire code is available for download on a git repository and there are active communities supporting and using Kaa, including a dedicated forum on Stack Overflow. In terms of supported protocols, Since Kaa is intended to be a middleware component, analytics components are not included in the distribution but can be easily connected to the database using, for instance, Apache Zeppelin. In terms of supported protocols, communication between endpoints and the Kaa cluster is managed through the Kaa SDK, but the platform supports MQTT, CoAP, XMPP and HTTP (all over TCP).

SiteWhere: SiteWhere uses a number of well-established technologies to provide a platform that facilitates devices management and different aspects of data generation, storage and distribution [14]. An interesting aspect of SiteWhere is the utilization of an In Memory Data Grid (IMDG) solution to provide subscription-based access in real-time to the data feed being generated by the devices. Permanent storage is achieved thanks to the utilization of nonrelational (e.g., MongoDB, Apache HBase) and/or time-series database components, that can be installed, together with the key components of SiteWhere, in a cluster configuration in order to implement HA. Communication between the end device is carried out through two pipelines (one for inbound messages and data and one for outbound commands to the devices), and the list of supported protocols include MQTT, AMQP, Stomp and WebSockets. SiteWhere does not include analytics components but rather relies on external ones that can be connected using MuleSoft AnyPoint connectors.

Eclipse IoT projects: Eclipse is a very well-known name among developers, particularly for being the house of one of the most used IDEs for Java. Within the Eclipse foundation several other projects are hosted and developed, including a set of IoT-related projects supported by a vibrant community. The home page of the Eclipse IoT project includes in fact several projects, all open-source, that cover almost all aspects of the lower layers of a typical IoT project, from devices and protocols to gateways [15]. Worth mentioning, among all the others, are as follows:

- Paho: an implementation of MQTT and MQTT-SN in several languages, including Java, C, C++ and C# [16];
- Californium: an implementation of the CoAP protocol in Java for nonconstrained devices [17];
- Hawkbit: a tool for roll-out of software updates to remote devices [18];
- Kura: a container for M2M applications running in service gateways – includes a native MQTT-based messaging solution to allow communication with devices and remote management [19] and
- Agail/Agile: a modular hardware and software gateway for the IoT, with support for protocol interoperability, device and data management, IoT apps execution and external Cloud communication, featuring diverse pilot activities, Open Calls & Community building [20,21].

While the Eclipse IoT initiative does not have a self-contained platform project, it is possible to use one of the existing projects and build on top of that. A key advantage of such a solution would clearly be the very large community of Eclipse developers around the world that provide through different means support and examples.

ThingSpeak: ThingSpeak is the 'open IoT platform with MATLAB analytics' [22]. The platform includes support for several endpoint devices, including Arduino, ESP8266 and Raspberry, offering REST API for communication with external additional components. On top of that, it offers the possibility of analysing the data coming from the devices using the MATLAB analysis application, one of the components of the MATLAB suite particularly suited for analysis on mathematical and statistical data. Interestingly enough, you do not need to own a MATLAB licence to use its analytics module in ThingSpeak. ThingSpeak is actually not a completely open-source system since the code for the server and analytics is not available; there are instead git repositories for the communication libraries to be used in the remote devices. Another interesting functionality offered by ThingSpeak is the possibility of triggering a reaction (e.g., sending a Tweet) when specific conditions arise. This functionality is implemented through a rule engine that processes the data coming from the device and, when the conditions specified in the rules are met, triggers the programmed reaction.

DeviceHive: Distributed under the Apache 2.0 licence, DeviceHive comes in a ready-to-be-deployed docker with support for virtualization and provides a 'digital playground' for experimentation [23]. All code is available from a git repository,

Table 1.4 A summary of the most common open-source solutions, together with some of their key features

Platform name	Device management	Supported devices	Protocols	Analytics	Database
Kaa	Yes	Raspberry, Intel Edison, 8266, ...	MQTT, CoAP, XMPP and HTTP	Through external tools	SQL and NoSQL
SiteWhere	Yes	Unclear	MQTT, AMQP, Stomp and WebSocket	Through external tools	IMDG and NoSQL
Eclipse	Yes	Several	MQTT, CoAP	Through external tools	External
ThingSpeak	No	Arduino, Raspberry, ESP8266, ...	REST and MQTT	MATLAB analytics	MySQL
DeviceHive	No	ESP8266	REST and MQTT	Through external tools	PostegreSQL and SAP

including device libraries that allow connection with the server using REST API, WebSockets or MQTT. Tutorials and documentation are available on the website, explaining also how to connect the platform with external modules to implement batch analytics and machine learning on top of the data coming from the devices. Though DeviceHive is not natively provided with an analytics solution, dataset interface for utilization with Grafana (an open-source software for time-series analytics) is provided (Table 1.4).

1.5 Future directions and challenges

Despite the huge research efforts invested so far within the IoT research field, there still persist several open issues that are under hot debate and actively investigated by the community, gathering diverse skills and competences from Computer Science, Electrical Engineering, Electronics and Physics, among many others. We heretofore list such challenges, along with a short summary of the current developmental status of each line in regards to the latest findings.

1.5.1 Fog/Edge/Cloud Computing

Given the *in crescendo* variety and complexity of tasks to be undertaken by IoT devices, there has prevailed a strong interest in the computational implications of the deployment of such tasks on different levels of the IoT architecture. This has been of utmost relevance when dealing with highly constrained use cases demanding particularly highly efficient communication protocols and/or processing functionalities. This can be exemplified by delay-sensitive communications in which, should the processing capabilities of IoT nodes be lower than required, the design criterion of the IoT architecture should prioritize the deployment of all resource-consuming processing tasks upstream in the cloud infrastructure, thus leaving IoT nodes mostly dedicated to monitoring, communication and/or actuating procedures. On the contrary, if IoT nodes are empowered with more processing capability, a fraction of the tasks could be performed as close to the IoT node as possible, giving rise to the so-called edge/fog computing paradigm [24]. This option should also be selected when the network traffic is a limited asset, due to either stringent radio propagation conditions of the communication channel (as in e.g., industrial M2M communications) or the aggregate amount of data captured by the IoT mesh. Furthermore, these computing alternatives find their rationale not only in exemplificative general use cases as the ones used above, but also in other casuistry such as applications dealing with time-sensitive data, private information that should be stored and kept as locally as possible, or federated computing and distributed inference/learning aiming at extracting knowledge from the collected information in a distributed manner [25].

All in all, several questions remain unanswered in this active field: to begin with, the capillarizations of processing functionalities to the edge of the IoT architecture comes along with a complex, decentralized management of possibly heterogeneous devices. This poses fundamental challenges regarding interoperability, programmability, service migration under mobile roaming, network virtualization and scalability, far beyond and more involved than the usual management aspects yielded by distributed device management (namely, updating, patch versioning, etc). Other challenges related to the implementation of fog computing in IoT environments include those related to data governance, data privacy, data marshalling and data delivery among nodes for distributed computation, and the consideration of economic criteria in the choice of one processing implementation or another under different cloud pricing schemes, among others [24].

1.5.2 Ultra-reliable low latency communications (URLLC) and tactile Internet

By these terms, closely related to each other, we refer to all communication scenarios demanding extremely low latencies in combination with high availability, reliability and security levels [26,27]. Tactile Internet extends further this envisaged scenario with haptic interaction and visual feedback so that objects can be manipulated remotely by the user, who feels the touch and strength of his/her action. These aspects have evolved from simple design optimization objectives to restrict themselves, especially in fields where the interactivity between the human user and the deployed devices is crucial for the success of the application itself, such as industrial robotics, transport systems, healthcare/surgery, education, augmented reality (AR), virtual reality (VR) and gaming, among others. In the case of IoT systems, the need for meeting ambitious goals in these processing domains impacts directly on the design on the whole architecture so as to improve the access delay and reliability in both directions of the communication link (uplink/downlink) and the provision of intelligence to predict contents, anticipate communication failures and in essence, improve the efficiency under which information is delivered to the network.

Unfortunately, even though the activity in this field is notable – particularly in what regards to provisioning such demands over future generations of cellular communications – URLLC and Tactile Internet are still far from maturity [28], with scarce practical prototypes using ad hoc underlying communication technologies. There is a long road ahead to be driven by the community towards exploring how IoT environments can support massive scenarios demanding URLLC/Tactile Internet, with a major focus placed on how to meet well-known operational thresholds imposed by the physics of the haptic interaction and visual feedback (e.g., the Motion-to-Photon 20 ms latency constraint in VR applications). Key elements to be explored will include the inclusion of new network elements for decentralized computation offloading (cloudlets), which in turn will unleash new issues such as online resource management, provisioning and task allocation [29].

1.5.3 Embeddable artificial intelligence

To provide machines with learning capabilities lies at the very core of the future of IoT environments, which has become a reality by virtue of the practical proliferation of IoT devices and the provision of a digital data substrate from which to learn, construct models and exploit the information acquired by IoT nodes. Although the history of Artificial Intelligence (AI) dates back to the early twentieth century, the evolution of learning algorithms has not grown in the last years towards more sophisticated models but has instead steered towards traditional models leveraging the availability of more powerful computing resources. This is, in fact, the case of the celebrated Deep Learning, which departs from traditional Artificial Neural Network models to implement highly dimensional neural nets capable, not only of capturing more complex data patterns from the collected data, but also to override any need for preprocessing and extracting predictors from the data themselves [30]. This has been accomplished partly by modified neuron models (allowing for fine-grained, recurrent information feedback). However, most of the success of Deep Learning models corresponds to the implementation of their learning algorithms efficiently on devices with unprecedented computation capabilities (e.g., Graphical Processing Units, GPUs) and unbounded consumption of other resources and assets such as energy or storage.

This is not the case for IoT devices, particularly when the processing architecture opts for deploying the learning algorithm as close to data as possible. In this case, not only the learning algorithm must run efficiently on the IoT node not to drain the battery of the device, but it must also allow for an incremental operation that dismisses any need for locally storing large amounts of data. From an application-agnostic point of view, the study of incremental online learning algorithms for stream processing has been at the forefront in the last decade, capitalizing on how models should be retrained when dealing with nonstationary data sources [31]. However, most studies do not consider how the learning algorithm should also take into account the remaining battery level of the device in which it is executed, so as to properly balance the between model performance and complexity. Some attempts have been reported in this regard, such as the Embedded Learning Library developed at Microsoft Research [32]. This recent big leap in embeddable AI suggests that we face an unexplored niche of research gravitating on the design of machine-learning algorithms tailored for resource-constrained systems as those in IoT environments.

In this same line of reasoning, a major momentum has been lately gained by the so-called Federated Learning concept coined by Google [33]: this technology enables to learn locally a prediction model and share it collaboratively among other devices in the network, while retaining all training data on the device in which the model was built. The motivation behind the creation of federated learning models concentrates again on eventual bandwidth and latency constraints in the network collecting the data, which pushes the processing paradigm to an edge computing architecture. When the computing power of a single node is limited and the network undergoes such bandwidth/latency constraints, there is no other way to proceed than to deploy intermediate processing platforms (e.g., *cloudlets*) or, instead, resort to collaborative model learning technologies such as the ones currently pursued in the literature. Albeit certainly promising, research lines around the so-called transfer learning and domain adaptation [34] – which could make the knowledge learned in one IoT node or group of nodes transferable to other IoT nodes – are still in their infancy and have not grown mature enough for their widespread adoption in severely constrained application domains.

In summary, research on AI model is opening a wide spectrum of possibilities and applications in IoT environments, which are closer to practicality than ever. Data availability is not a problem any longer for machine-learning models, hence processing constraints should grasp the attention of the research community in order to realize fully operational machine-learning functionalities in IoT applications.

1.5.4 Secure communications

It is clear that the proliferation of more and more connected IoT devices has ignited the number and capabilities of services and applications, but unfortunately also unchains an exponential increase of security threats and backdoors for malicious users to capture sensitive data and hack systems and processes monitored/actuated by IoT nodes. Such security holes, when discovered by cyber-criminals, can cause damage at a grand scale, with consequences that include economic losses and eventual human casualties. Examples abound: among them, denial of service attacks to critical infrastructures have been specially noted worldwide (power grids and nuclear plants), but IoT environments at lower scales are also at risk, including wearables (health information leaks), smart meters (presence at home), connected vehicles, unmanned aerial vehicles and drones and other systems alike [35–37]. Design factors should include a proper risk quantification of the asset to be protected, considering as critical the acquired data for the operated process or infrastructure, and the degree of resiliency of the asset taking into account the attack occurrence. Overall, a thorough understanding of the motivations behind the attacks should be attained to detect recurrent security holes, infer threat patterns and design IoT protocols according to such diagnosed vulnerabilities and thereby prevent future incidences. The fullest potential of IoT systems is strongly subject to its properly protected design against all kind of threats and vulnerabilities, by minimizing the number of security flaws from the very beginning of the design process. If the attack cannot be predicted nor avoided anyhow, to its resilience and capability to recover. To this end, further analyses and efforts must be made to evolve current access control, authentication and identity management mechanisms towards other alternatives encompassing more robust security methods without sacrificing user-friendliness, ease of use and efficiency.

1.5.5 Enhanced energy efficiency and autonomy

With the ever-growing scales at which IoT environments are being deployed nowadays in different sectors, there is a remnant question whether current IoT technologies underneath such deployments will meet the aggregate energy demands of IoT meshes. All processing around the IoT node adds up to its computational burden and ultimately, to its autonomy, which is an essential design factor for a number of critical applications (e.g., aerospace or health, among others). As a matter of fact, nowadays no IoT-related study should currently neglect its implications in terms of power consumed at the IoT node, as it has become a decisive factor for its practical adoption. Technologies inherently linked to the batteries themselves such as wireless charging, wake-up circuits, new materials and energy harvesting hardware are nowadays on the focus [38], but the importance of energy efficiency spreads further to the design of the wireless communication stacks, stimulating new wireless interfaces for data broadcasting (e.g., LiFi, Visible Light Communications, VLC), energy-efficient MAC techniques, routing protocols and data collection/compression schemes, among others [39]. This trend hints at a holistic design of the entire IoT application with energy as a strictly necessary design objective at all levels of the communication stack.

Paradoxically in a sense, an emerging market has sprung from the application of IoT networks to the improvement of the energy efficiency of other systems and processes, such as street lighting, home appliances, home automation and heating, ventilation and air conditioning (HVAC) control in buildings, among others [40]. The portfolio of applications under the same motto is increasing day by day, but they occur to be mostly composed of ad hoc solutions with limited interoperability and questionable scalability. A major research effort must be invested along this line, to avoid transforming the global IoT realm into myriads of isolated islands of disconnected functionality and limited exploitable knowledge. By ensuring cross-operations between such islands a competitive advantage can be gained by the simply richer data substrate from which to construct learning models for e.g., temperature prediction or energy consumption forecasting, but also spans business cases around the commercialization of the captured data.

1.5.6 (Big) Stream analytics

While this research area can be certainly covered by the aforementioned need for embeddable AI, stream data mining deserves its own space in the IoT technology roadmap. In a word stream mining models (also known as online learning) are subject to stringent storage and processing constraints, calling for customized, computationally lightweight, incremental learning algorithms that do not require the collected data to be stored anyhow [41]. The stream mining scenario may become even more involved when data samples themselves are modelled by nonstationary distributions, a fact that jeopardizes dramatically the discovery of patterns from data. When this is the case, models must not only incrementally build a mathematical model relating their inputs to the target outputs (e.g., IoT sensors inferring how wireless radio measurements relate to the occupancy of the hall), but they must also detect reliably when the discovered relationship between variables change to an extent so as to reactively rebuild the model and ultimately, capture the newly occurring pattern or concept. This paradigm, often referred to as concept drift or evolving concepts, is lately occurring more and more frequently in IoT environments, particularly when sensors and actuators are deployed in scenarios of inherently high dynamics (e.g., industrial plants, smart cities and large events), prone to undergo severe changes in the stationarity of the captured streams. A yet one more degree of complexity is posed by the eventual heterogeneity of such information flows, which has encouraged the adoption of information fusion schemes at different levels of IoT systems [42]. Although notable advances have been made in this area, there are still unaddressed questions in regards to the scalability of practical IoT deployments of stream mining models, since the aggregate demand of the whole IoT mesh is expected to surpass the processing capacity and memory of commodity servers even in cloud infrastructures without latency constraints [43]. To face this challenge, research must be steered towards the distribution of learning tasks all over the mesh, by resorting to distributed computation models as those allowed by software tools (e.g., Spark, Flink, Storm and Samza) or by leveraging concepts from the aforementioned Federated Learning approach.

1.5.7 Conclusions

While the above list does not exclude other topics of relevance for IoT systems, these open challenges should lie at the inner core of future research efforts conducted in this area. In essence, IoT systems acquire information about the sensed phenomena and deliver it to smart application and services, which prescribe (to users and/or actuators) how to react according to the captured data. Since IoT communications, protocols and platforms have evolved at a fast pace during the last years, it is not a matter of how the overarching goal of IoT systems will be accomplished, but a matter of when the society will enjoy the benefits of IoT environments as an elemental part of any daily routine.

References

[1] Ashton K. That 'Internet of things' thing: in the real world, things matter more than ideas. *RFID Journal.* 22 June 2009.

[2] Gartner Special Report 'Digital Business Technologies'. Available from https://www.gartner.com/en/newsroom/press-releases/2014-09-08-gartner-says-a-typical-family-home-could-contain-more-than-500-smart-devices-by-2022.

[3] That 'Internet of Things' thing. Available from https://www.rfidjournal.com/articles/view?4986

[4] Cassandras CG. Smart cities as cyber-physical social systems. *Engineering*. 2016;2(2):218–219.

[5] Poncela J, Vlacheas P, Giaffreda R, *et al.* Smart cities via data aggregation. *Wireless Personal Communications*. 2014;76(2):149–168.

[6] Kephart JO, Chess DM. The vision of autonomic computing. *Computer*. 2003;36(1):41–50.

[7] Menasce DA, Kephart JO. Guest editors' introduction: autonomic computing. *IEEE Internet Computing*. 2007;11(1):18–21.

[8] Horn P. Autonomic Computing: IBM's Perspective on the State of Information Technology. Available from https://scholar.google.it/scholar?cluster=15217490283224901802&hl=it&oi=scholarr.

[9] Leminen S, Westerlund M, Rajahonka M, *et al.* Towards IoT ecosystems and business models. In: Andreev, S., Balandin, S. and Koucheryavy, Y., Eds., *Internet of Things, Smart Spaces, and Next Generation Networking*, Vol. 7469. Springer, Heidelberg. 2012. p. 15–26.

[10] Palattella MR, Dohler M, Grieco A, *et al.* Internet of things in the 5G era: enablers, architecture, and business models. *IEEE Journal on Selected Areas in Communications*. 2016;34(3):510–527.

[11] Ray PP. A survey of iot cloud platforms. *Future Computing and Informatics Journal.* 2016;1(1–2):35–46.

[12] Lamarre E, May B. Making sense of Internet of things platforms. Available from https://www.mckinsey.com/business-functions/digital-mckinsey/our-insights/making-sense-of-internet-of-things-platforms.

[13] The Kaa Project Homepage. Available from https://www.kaaproject.org/ (Last Accessed 28 March 2019).

[14] The SiteWhere Project Homepage. Available from http://www.sitewhere.org/ (Last Accessed 28 March 2019).

[15] The Eclipse IoT Project Homepage. Available from https://iot.eclipse.org/ (Last Accessed 28 March 2019).

[16] The Paho Project Homepage. Available from https://www.eclipse.org/paho/ (Last Accessed 28 March 2019).

[17] The Californium Project Homepage. Available from https://www.eclipse.org/californium/ (Last Accessed 28 March 2019).

[18] The Hawkbit Project Homepage. Available from https://www.eclipse.org/hawkbit/ (Last Accessed 28 March 2019).

[19] The Kura Project Homepage. Available from https://www.eclipse.org/kura/ (Last Accessed 28 March 2019).

[20] The Agail Project Homepage. Available from https://projects.eclipse.org/projects/iot.agail (Last Accessed 28 March 2019).

[21] The Agile Project Homepage. Available from http://agile-iot.eu/ (Last Accessed 28 March 2019).

[22] The ThingSpeak Project Homepage. Available from https://thingspeak.com/ (Last Accessed 28 March 2019).

[23] The DeviceHive Project Homepage. Available from https://devicehive.com/ (Last Accessed 28 March 2019).

[24] Shi W, Cao J, Zhang Q, *et al.* Edge computing: vision and challenges. *IEEE Internet of Things Journal.* 2016;3(5):637–646.

[25] Satyanarayanan M. The emergence of edge computing. *Computer*. 2017; 50(1):30–39.

[26] ITU-R. 'Framework and Overall Objectives of the Future Development of IMT for 2020 and Beyond', Recommendation ITU-R M.2083. Available from https://www.itu.int/rec/R-REC-M.2083.

[27] Fettweis GP. The tactile Internet: applications and challenges. *IEEE Vehicular Technology Magazine*. 2014;9(1):64–70.

[28] Maier M, Chowdhury M, Rimal BP, *et al.* The tactile Internet: vision, recent progress, and open challenges. IEEE Communications Magazine. 2016; 54(5):138–145.

[29] Want R, Schilit BN, Jenson S. Enabling the Internet of things. *Computer*. 2015;48(1):28–35.

[30] LeCun Y, Bengio Y, Hinton G. Deep learning. *Nature*. 2015;521(7553): 436–444.

[31] Ditzler G, Roveri M, Alippi C, *et al.* Learning in nonstationary environments: a survey. *IEEE Computational Intelligence Magazine*. 2015;10(4):12–25.

[32] Microsoft. Microsoft, Editor. Embedded Learning Library, Microsoft. Available from https://github.com/Microsoft/ELL (Last Accessed 28 March 2019).

[33] McMahan B, Ramage D. Blog GR, Editor. Federated Learning: Collaborative Machine Learning without Centralized Training Data. Google Research Blog. Available from https://research.googleblog.com/2017/04/federated-learning-collaborative.html (Last Accessed 28 March 2019).

[34] Pan SJ, Yang Q. A survey on transfer learning. *IEEE Transactions on Knowledge and Data Engineering*. 2010;22(10):1345–1359.

[35] Zhang ZK, Cho MCY, Shieh S. Emerging security threats and countermeasures in IoT. In: Proceedings of the 10th ACM Symposium on Information, Computer and Communications Security – ASIA CCS '15. Singapore: Republic of Singapore. 2015. pp. 1–6.

[36] Babar S, Mahalle P, Stango A, *et al.* Proposed security model and threat taxonomy for the Internet of Things (IoT). *Communications in Computer and Information Science*. 2010;89 CCIS:420–429.

[37] Farooq MU, Waseem M. A critical analysis on the security concerns of Internet of things (IoT). *International Journal of Computer Applications*. 2015;111(7):1–6.

[38] Kamalinejad P, Mahapatra C, Sheng Z, *et al.* Wireless energy harvesting for the Internet of Things. *IEEE Communications Magazine*. 2015;53(6):102–108.

[39] Rault T, Bouabdallah A, Challal Y. Energy efficiency in wireless sensor networks: a top-down survey. *Computer Networks*. 2014;67:104–122.

[40] Xu LD, He W, Li S. Internet of things in industries: a survey. *IEEE Transactions on Industrial Informatics*. 2014;10(4):2233–2243.

[41] Bifet A, Holmes G, Kirkby R, *et al.* MOA massive online analysis. *Journal of Machine Learning Research*. 2011;11:1601–1604.

[42] Tsai CW, Lai CF, Chiang MC, *et al.* Data mining for Internet of things: a survey. *IEEE Communications Surveys and Tutorials*. 2014;16(1):77–97.

[43] De Francisci Morales G, Bifet A, Khan L, *et al.* IoT Big Data Stream Mining. In: Proceedings of the 22nd ACM SIGKDD International Conference on Knowledge Discovery and Data Mining – KDD '16. New York: ACM Press; 2016. pp. 2119–2120.

Chapter 2

Sensors for the Internet of things

Francisco Javier Ferrero Martín[1]

This chapter is focused on the basic concepts of sensors for the Internet of things (IoT). There is a wide variety of sensors, so to better manage them some type of classification will first be necessary. Then, the main characteristics of the sensors will be defined. Understanding these features will undoubtedly help in selecting the most suitable sensor for IoT applications. Finally, examples of commercial sensors particularly suitable for IoT applications are provided.

2.1 Sensor, smart sensor and IoT sensor

In the context of the IoT, sensors are the primary devices required to gather data from the environment and send them to the cloud (Figure 2.1). Basically, a sensor is a device that converts a physical variable (temperature, pressure, vibration, etc.) from the process or system to a representative electrical signal. Information provided by the sensor can be used to know the value of the stimulus or to control the process or system. The increasingly reduced price of these devices is enabling low-cost IoT developments and extending their application to a wide range of applications.

The concept of smart sensor is not new. In 1986 Joseph Giachino of the Ford Motor Company reported the first definition for a smart sensor: 'smart sensors are those that include some logic functions and/or make some type of decision' [1]. In that paper the concepts of self-calibration, computation, communication and multisensing were introduced. In 1998 the IEEE 1451.2 Standard defined a smart sensor as a sensor 'that provides functions beyond those necessary for generating a correct representation of a sensed or controlled quantity' [2].

It is widely accepted that a smart sensor should has at least the functions included in the block diagram of Figure 2.2. The sensing element provides an electrical signal with a sensor gain. For instance, a temperature sensor with an analog output provides a senor gain of mV/°C. An accelerometer with digital output offers a sensor gain of LSBs/g or codes/g. The signal conditioning block usually

[1]Department of Electrical, Electronic, Computers and Systems Engineering, University of Oviedo, Campus of Gijón, 33204 Gijón, Spain

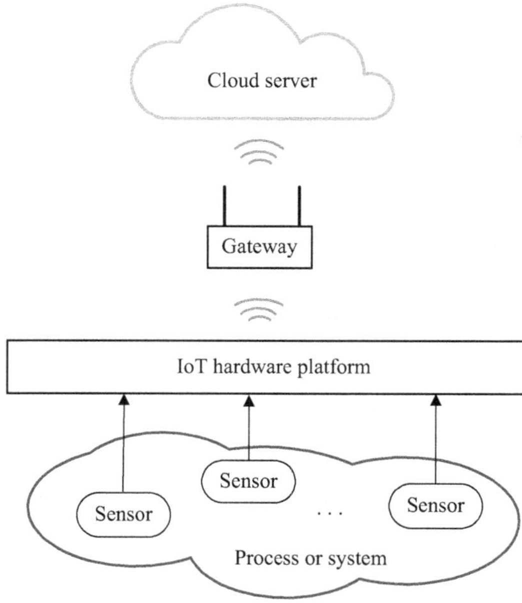

Figure 2.1 Basic architecture of an IoT system

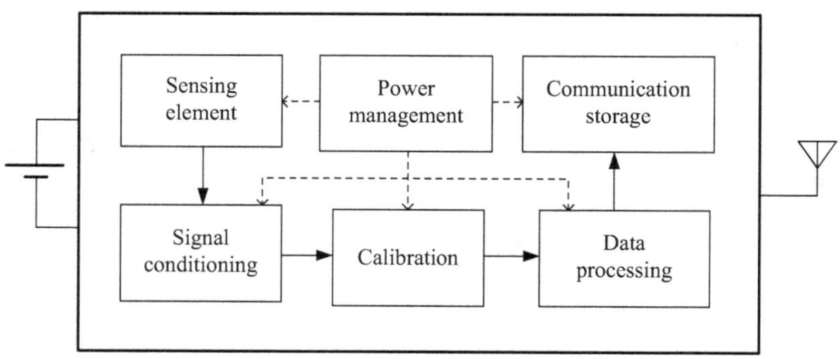

Figure 2.2 Functional block diagram of a smart sensor

includes amplification and filtering, enabling the sensor to be matched to the analog-to-digital converter (ADC) and the signal to noise ratio (SNR) to be improved.

Smart sensors are factory-calibrated for offset and gain. The trimming values are stored inside the device in a nonvolatile memory. Any time the device is turned on, the trimming parameters are downloaded into the registers to be used during the active operation. This enables the device to be used without further calibration. For example, in an accelerometer, the calibration process will involve observation of the MEMS accelerometer's output in several different orientations, with respect to gravity. Self-test enables the sensor functionality to be checked without moving it. If the output signals change within the amplitude specified, the sensor is working properly and the parameters of the interface chip are within the defined specifications.

The data processing function converts the calibrated sensor data into a measurement result. For example, in a tilt-sensing application, the smart sensor will convert acceleration-based responses to gravity into orientation angle estimates, through [3]

$$\theta = \tan^{-1}\left(A_{x,\mathrm{OUT}}/A_{y,\mathrm{OUT}}\right)$$

The power management (PM) function is devoted to manage the power sequencing requirements of all the components in the signal chain, converting the supply from the energy source into voltage(s) that support optimal operation in all of the components and finally, in systems that have regular measurement intervals, the PM system provides scheduling to trigger each measurement event [4]. The communication/storage function supports data staging and connectivity with all IoT cloud services (encryption/security, storage and analytics). Frank's Understanding Smart Sensors [5] is a very complete book about smart sensors.

IoT sensors are smart sensors specially designed to fulfil the requirements of IoT applications. The main features of IoT sensors are as follows:

- Ultra-low-power consumption,
- Low cost,
- Ultra-small size,
- Low maintenance,
- Practical and easy-to-use built-in features,
- High performance and
- Pinout compatibility.

The device that performs these functions can be built around a general purpose MCU IC for low-end applications or around a system-on-chip (SoC) integrating a specifically designed applications processor for high-end applications. These two approaches can be better understood by considering two examples.

Example 1: Holter monitor

A Holter monitor is a wearable battery-operated device that measures and records heart activity continuously for up to 96 hours on an AA battery. It contains electrodes that attach to the skin and helps doctors to determine the cause of symptoms such as dizziness, to detect pacemaker malfunction or assess the effectiveness of medications. Figure 2.3 shows the functional block diagram of a Holter monitor interface based on an MCU IC (MSP430FR5994), an ADC, an SD card and other peripherals.

The ECG electrodes are attached to the ADS1299-4. It is a four-channel, low-noise, 24-bit delta-sigma ($\Delta\Sigma$) ADC with an integrated ECG front end. The ECG

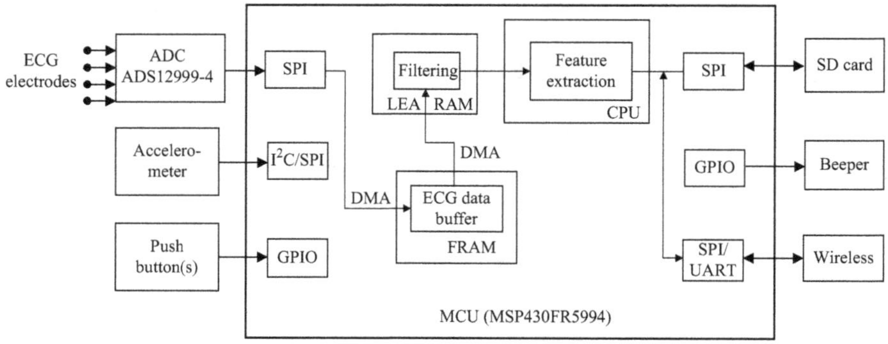

Figure 2.3 Functional block diagram of a Holter monitor based on an MCU

electrodes output a voltage from approximately 0.02 to 5 mV, which must be captured in the presence of much larger noise sources. These range from 50/60-Hz power-line to other sources in the patient's body. As a result, the front end includes a low-noise differential-input amplifier with a high common-mode rejection ratio (CMRR). The ADC interfaces with the MCU via SPI. The MCU saves the SPI data from multiple ECG channels to a data buffer in the on-chip ferroelectric random access memory (FRAM) using direct memory access (DMA). This improves processing efficiency by accumulating a block of data before requiring CPU input. At regular intervals – every second, for example – the CPU sends the data from a specific ECG channel to a buffer in the low energy accelerator (LEA) RAM for filtering and other processing.

Many other effects, including poor electrode placement, can affect the quality of the results. Excessive patient movement is another culprit. The design includes an accelerometer to capture such events. A complete description of this application is reported in [6,7].

Example 2: Fall detector

A fall detector detects changes in body position when moving by tracking acceleration changes in three orthogonal directions of an individual wearing a sensor. The data are then analysed algorithmically to determine whether the individual's body is falling. If an individual falls, the device works with a GPS module and a wireless transmitter module to determine the position and issues an alert for assistance.

Figure 2.4(a) shows the functional block diagram of the fall detector. It consists of a digital three-axis MEMS accelerometer (Analog Devices ADXL345) connected to a 32-bit microcontroller built-in the popular Wi-Fi SoC ESP8266 (Espressif Systems), through the I^2C interface. The low power and compact design of both devices make them ideal for the IoT developments.

The accelerometer features several special functions. The main one is that it detects activity and inactivity by comparing the acceleration in the axes with the

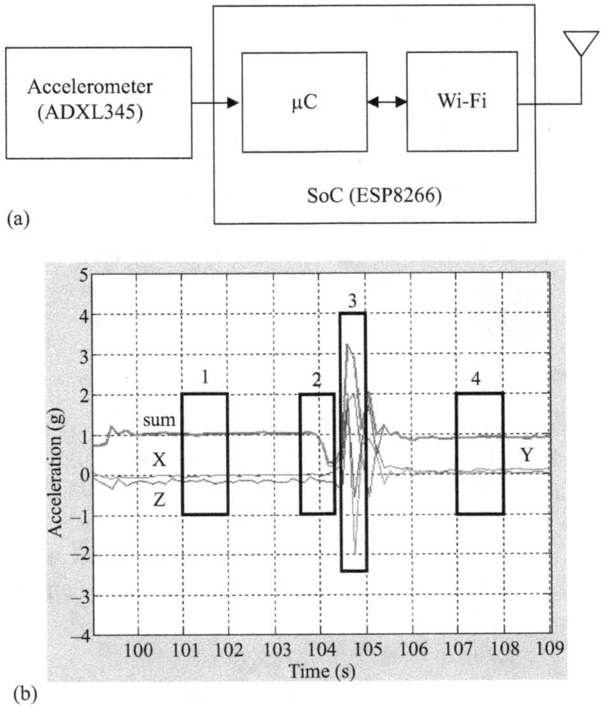

Figure 2.4 (a) Block diagram of a fall detector based on an SoC and (b) acceleration change curves during the process of falling human body (1: Initial status, 2: Weightlessness, 3: Impact, 4: Motionless)

thresholds set by the user. This sensor is also capable of detecting hits in any direction, and free falls when the device falls into a vacuum. Its high sensitivity enables inclinations of only 1° to be detected.

The ESP8266 can be configured and controlled via UART using AT command or by the integrated microcontroller. The programming was done under the environment and Arduino libraries using the compiler Arduino core for ESP8266 Wi-Fi chip. A complete description of this application is reported in [8]. Figure 2.4(b) shows the phases of a fall that always follow the same pattern: (1) rest, (2) fall, (3) impact and (4) inactivity. A more complete description of this application can be found in [8].

2.2 Sensor classification

There is a large variety of sensors which makes it difficult to proceed rationally to their study without some previous classification criteria. The most common criteria for classification are the following.

2.2.1 According to the output signal

- *Analog sensors*: the output varies, at macroscopic level, continuously. The information is in the amplitude, although the sensors with output in the time domain are usually included in this group. The IoT SoC needs to convert the sensor signal into the digital domain by means of an ADC.
- *Digital sensors*: the output varies in the form of discrete steps or jumps. They do not require A/D conversion and the transmission of their output is easier. They also have greater precision and often greater accuracy. IoT SoC needs to include a digital interface that complies with the one used by the sensor to drive the output sensor signals into the microcontroller in the SoC.

2.2.2 According to the excitation method

- *Active sensors*: require an external excitation source. Resistive sensors such as thermistors, RTDs (resistor-temperature detectors) and strain gauges are examples of active sensors.
- *Passive sensors*: do not require an external excitation source to generate the output signal. Examples of passive sensors are thermocouples and photodiodes. An IoT system needs to be able to support analog and digital sensors.

2.2.3 According to the variable parameter

- *Resistive sensors*: are based on the variation of the electrical resistance of a metallic wire or semiconductor element. There are many physical magnitudes that affect the value of electrical resistance. Examples of resistive sensors are strain gauges, RTDs and thermistors.
- *Inductive sensors*: are based on the variation of the inductance. The inductance can be expressed by

$$L = \frac{N^2}{R}$$

  where N is the number of turns and R the reluctance of the circuit. Most inductive sensors are of variable reluctance and it is a displacement that modifies it. R can also vary by Foucault currents [9]. There are inductive sensors to measure touch, presence, distance, angle and flow.
- *Capacitive sensors*: are based on the capacity variation of a capacitor. This capacity depends on the geometrical arrangement of the conductors and the dielectric material disposed between them. In its simplest form the capacity of a capacitor formed by two equal parallel flat plates, with area A, distance d between plates, and a material between them with relative dielectric constant ε_r, is given by

$$C = \varepsilon_0 \varepsilon_r \frac{A}{d}$$

Thus, any phenomenon or magnitude that produces a variation in ε_r, A or d, will cause a change in the C capacity and, in principle, could be detected by a capacitive sensor. The range of application of capacitive sensors is extraordinary [10]. They can directly sense a variety of variables – motion, chemical composition and electric field – and, indirectly, sense many other variables which can be converted into motion or dielectric constant, such as pressure, acceleration, fluid level and fluid composition.

2.2.4 According to the measured magnitude

There a lot of possible magnitudes capable of being measured but, from the point of view of IoT applications, the following sensors are the most common:

- Temperature sensors,
- Humidity sensors,
- Pressure sensors,
- Flow sensors,
- Level sensors,
- Proximity and displacement sensors,
- Gas and chemical sensors,
- Motion sensors and
- Optical sensors.

2.3 Sensor characteristics

It is necessary to understand the characteristics of the sensors because the behaviour of an IoT system is conditioned by such characteristics. Some of the most important characteristics of the sensors are the following:

- *Transfer function*: is the relation between the input to the sensor (x) and its output (y) in static regime. It can be a simple linear, logarithmic, exponential, power function, etc. relationship. Figure 2.5 shows a typical linear transfer function.
- *Operating range*: is the set of values between the upper (x_{max}) and lower (x_{min}) limits between which the measurement can be made. For example, a temperature sensor is designed to measure temperature between −50 °C and 150 °C, its operating range is −50 °C/150 °C.
- *Span* (input full-scale, FS): is the algebraic difference between the upper and lower limits of measurement ($x_{max}-x_{min}$). In the case of the previous temperature sensor, the span is 200 °C.
- *Full-scale output* (FSO): is the difference between the outputs for the ends of the operating range ($y_{max}-y_{min}$).
- *Sensitivity* (or sensor gain): is the slope of the transfer function. For a linear transfer function as $y = mx + b$, the coefficient m is the sensitivity. A high sensitivity is of interest and, if possible, constant. For a nonlinear transfer function, the slope is not a fixed value. At a particular point, x_0 of the input variable, the sensitivity is defined as

$$S(x_0) = \left.\frac{dy}{dx}\right|_{x=x_0}$$

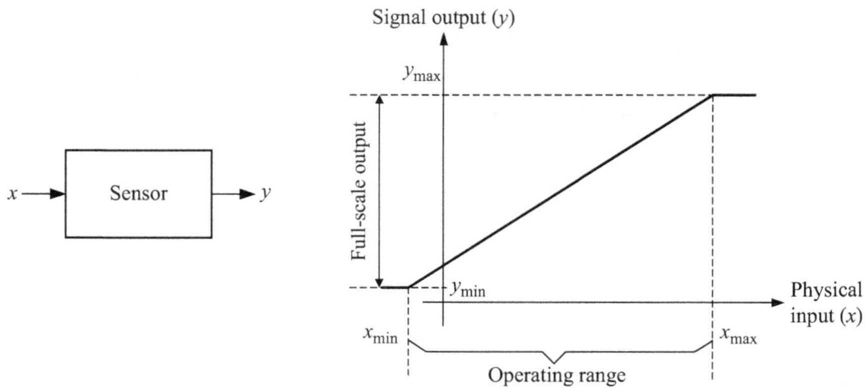

Figure 2.5 Example of a transfer function

- *Accuracy*: is a measure of the maximum deviation between the value measured by the sensor and the ideal or actual value presented at its input. In most applications, accuracy is one of the more critical specifications that a sensor needs to meet.

  Accuracy is usually specified in terms of inaccuracy or error. In general, sources of error include zero error, span error, nonlinearity, thermal effect, thermal effect and repeatability. Other system considerations that can affect accuracy include response time, long-term stability and more.
- *Precision*: is the degree of agreement of the results among themselves when measuring several times under the same determined conditions (environmental, operator, etc.). Note that the results of many measurements can be very similar but, nevertheless, they can be very different from the true value.

  Accuracy and precision are frequently used interchangeably; however, there are obvious differences between the two specifications. Figure 2.6 shows the relationship between accuracy and precision [11]. A more precise sensor has a narrower distribution and a more accurate sensor is closer to the actual value.
- *Nonlinearity*: is the maximum deviation of the calibration curve with respect to the straight line by which it has been approximated (see Figure 2.7). It is generally expressed in % with respect to the FSO. This characteristic is usually referred to as error of linearity and even as linearity.

  Keep in mind that this error depends on the approach that has been made. In many cases, the linearization is usually carried out according to the ends of the measuring field or taking into account the most usual measurement area. It is also usually done by the method of least squares so that the error will be the minimum possible over the whole field of measurement. This last method of linearization is typically the one used by sensor manufacturers to indicate their linearity.
- *Resolution*: is the smallest measurement a sensor can reliably indicate. In analog sensors, resolution can be expressed in units of the measured variable or in per cent on full-scale. If resolution is given as a per cent, that value must be

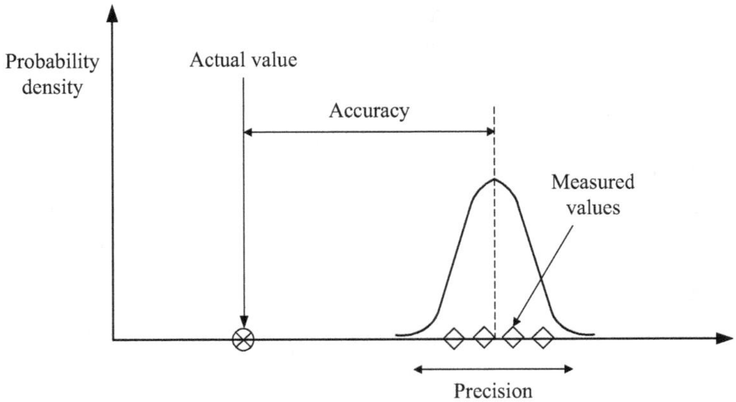

Figure 2.6 The relationship between accuracy and precision

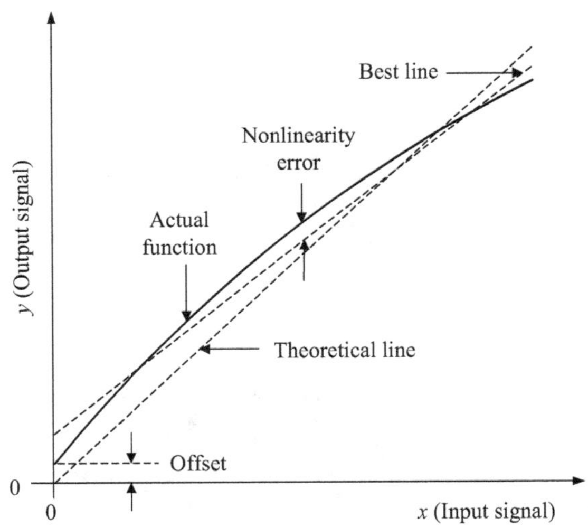

Figure 2.7 Nonlinearity error

multiplied by the sensor's range to determine the smallest possible measurement. Digital sensors usually specify the resolution of the ADC used in the device.

Resolution is one of the most frequently misunderstood and poorly defined specifications. Resolution is not accuracy. A very inaccurate sensor could have very high resolution, and a low-resolution sensor may be sufficiently accurate in some applications.

The electrical noise in a sensor's output is the primary factor limiting its smallest possible measurement. Resolution is only meaningful within the context of the system bandwidth, the application and the measurement method and unit of measure used by the sensor manufacturer [12].

- *Repeatability*: is the ability of a sensor to give the same value of the measured quantity, by measuring several times in the same determined conditions (environmental, operator, etc.) when the measurements are made in a short time interval. The repeatability error is not always specified separately but is usually combined in a total figure including linearity and hysteresis.
- *Reproducibility*: is the ability of a sensor to give the same value of the measured quantity when determining the same parameter with a specific method, but with a set of long-term measurements or made by different people or with different instruments in different laboratories.
- *Hysteresis*: is the maximum difference in the output for the same input, depending on the direction in which it is reached. Hysteresis may cause false inaccurate readings.

Example

Tables 2.1 and 2.2 show the general specifications of the Silicon Labs Si7005 digital humidity and temperature sensor, while Figure 2.8 shows its accuracy, including hysteresis.

Table 2.1 Specifications of the Silicon Labs Si7005 humidity sensor

Parameter	Symbol	Test condition	Min	Typical	Max	Unit
Operating range		Noncondensing	0	–	100	%RH
Accuracy		20–80% RH 0–100% RH	–	±3.0	±4.5	%RH
Repeatability/noise			–	0.05	–	%RH RMS
Resolution			–	–	12	Bits
Response time	$T_{63\%}$	1 m/s airflow	–	18	–	s
Drift versus temperature			–	0.05	–	%RH/°C
Hysteresis				±1	–	%RH
Long-term stability			–	<0.25	–	%RH/year

Table 2.2 Specifications of the Silicon Labs Si7005 temperature sensor

Parameter	Symbol	Test condition	Min	Typical	Max	Unit
Operating range			−40	–	85	°C
Accuracy		Typical at 25 °C maximum	–	±0.5	±1	°C
Repeatability/noise			–	0.04	–	°C RMS
Resolution			–	–	14	Bit
Response time	$T_{63\%}$	Time to reach 63% of final value	–	1.5	–	s
Long-term stability			–	<0.05	–	°C/year

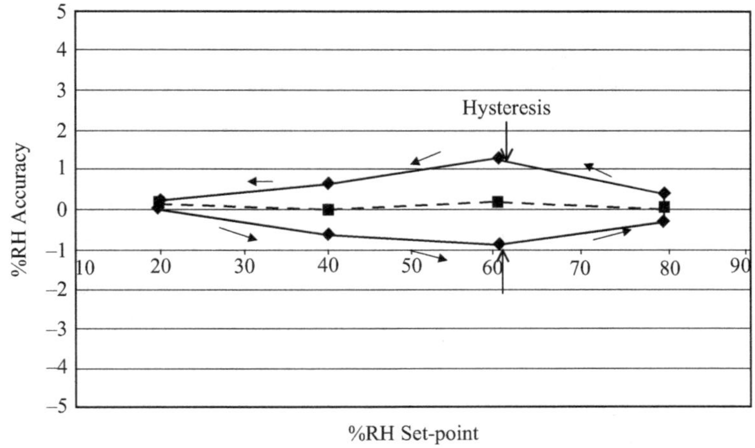

Figure 2.8 Silicon Labs Si7005 measuring sensor accuracy, including hysteresis

2.4 Applications and manufactures of IoT sensors

IoT sensors are quickly becoming ubiquitous. Table 2.3 shows some of the most important fields of applications together with the typical sensors involved.

The following is a list of the main manufacturers of IoT sensors:

- *ams* is a global leader in advanced sensor solutions. Its sensor portfolio includes environmental sensors, light sensors, magnetic position sensors, flow sensing, wireless connectivity, sensor interfaces and CMOS imaging sensors.
- *Analog Devices*' sensor technologies portfolio includes MEMS accelerometers, MEMS gyroscopes, MEMS inertial measurement units (IMUs), optical sensors, anisotropic magnetoresistive (AMR) sensors, temperature sensors and Hall-effect sensors.
- *Figaro* offers a wide range of gas sensor products for the detection of various gases, from explosive gases such as propane, toxic gases such as carbon monoxide, to air quality sensors for volatile organic compounds (VOCs) that are responsible for sick-house syndrome.
- *Hillcrest Labs*, a subsidiary of Interdigital, is a leading global supplier of software components and intellectual property to enable the intelligent use of sensors in consumer electronics and IoT devices. Hillcrest's products are used in the development of devices in the robotics, virtual reality (VR), augmented reality (AR), TV and IoT markets.
- *InvenSense* is a TDK Group company. It is a world-leading provider of MEMS sensor platforms found in mobile, wearables, smart home, industrial and automotive products.
- *Kionix Inc.* is a global MEMS inertial sensor manufacturer. Kionix offers high-performance, low-power accelerometers, gyroscopes and six-axis combination

Table 2.3 Applications and sensors for IoT

Smart cities	**Healthcare**	**Domotic and home automation**	**Transportation**
Temperature	Pressure	Gyroscope	Gyroscope
Light	Temperature	Accelerometer	Accelerometer
Chemical (CO)	Chemical	Magnetic	Magnetic
Noise	Light (IR, X-ray)	Pressure	Chemical
	Bio sensor	Temperature	Pressure
	Inertial	Chemical	Temperature
Industrial control	Smart environment	Security and emergencies	Retail
Pressure	Temperature	Gyroscope	Light
Light	Humidity	Accelerometer	Temperature
Chemical	Gases (CO, CO_2, NO_2)	Gases	Gases (CO, O_3, NO, NO_2)
		Light	
Temperature	Light (IR, visible)		Magnetic
Hall effect	Atmospheric pressure	Level	
Accelerometer	Relative humidity		
Smart metering	Logistic	Smart water	Smart animal farming
Temperature	Temperature	Temperature	Temperature
Conductivity	Pressure	pH	pH
Dissolved oxygen	Humidity	Dissolved oxygen	Dissolved oxygen
pH	Luminosity	Conductivity	Conductivity
Radiation	Gases (CO, CO_2)	Oxidation–reduction potential	Oxidation–reduction potential
Light	Vibration		

sensors plus comprehensive software libraries that support a full range of sensor combinations, operating systems and hardware platforms.

- *MEMSIC* designs, manufactures and markets CMOS MEMS IC products that have on-chip mixed signal.
- *Monnit* sells wireless sensors, wireless gateways, monitoring software and controllers.
- *NXP Semiconductors* (acquired by Qualcomm) provides motion sensors (accelerometers and gyroscopes), pressure sensors (gauge, differential, absolute and tire pressure monitors), magnetic sensors (angular sensors, speed sensors, magnetometers and eCompass) and other sensors (temperature, touch and capacitive).
- *Sensirion* is the world's leading manufacturer of digital microsensors and systems. The product range includes gas and liquid flow sensors, differential pressure, as well as environmental sensors for the measurement of humidity and temperature, VOCs, carbon dioxide (CO_2) and particulate matter (PM2.5).
- *Silicon Labs* is a leading supplier of intelligent sensor solutions characterized by superb reliability, compact size, high levels of integration and unmatched ease of use for a variety of applications. Its sensor product portfolio includes optical sensors, digital relative I^2C humidity and temperature sensor ICs, biometric sensors, magnetic Hall-effect sensors and capacitive touch sense microcontroller devices.

- *SPEC Sensors* designs and manufactures high-performance gas sensors for IoT. Carbon monoxide alarms, air pollution monitoring, indoor air quality and breath analysis are some of the gas sensing applications that demand high-performance measurement. Electrochemical gas sensing technology is the preferred solution for these applications due to the measurement performance and the ultra-low-power consumption needed for battery operation.
- *STMicroelectronics* offers the widest range of MEMS and sensors covering a full spectrum of applications from low-power devices for IoT and battery-operated applications to high-end devices for accurate navigation and positioning, industry 4.0, augmented virtual reality components and smartphones.
- *TE Connectivity* designs and manufactures connectivity and sensor products.
- *Texas Instruments Incorporated* designs, manufactures and sells processors, MCUs, power management circuits and all kind of sensing products.
- *Xsens* is the leading innovator in 3D motion tracking technology and products. Its sensor fusion technologies enable a seamless interaction between the physical and the digital world in consumer devices and professional applications.

2.5 Temperature sensors

Temperature sensors can be used in nearly every IoT environment, from the factory floor to agricultural fields. In manufacturing, these sensors can continually measure the temperature of a machine to ensure it stays within a secure threshold. On the farm, they are used to track the temperature of soil, water and plants to maximize output. It is used to control the performance of the IoT device at varying temperatures.

Table 2.4 shows the types of sensors for applications that require temperature sensing [13]. For basic temperature sensing where accuracy of ±2 °C to ±3 °C is adequate, the thermistors tend to be a popular choice due to their low cost and

Table 2.4 Temperature sensor types

Sensor type	Advantages	Disadvantages
Thermistors	Low cost Minimal BOM	Lower accuracy Requires external AC
Resistance temperature detectors (RTDs)	Highest accuracy Wide temperature range	High cost Large BOM Slow response time Require external ADC
Thermocouples	Widest temperature range	Large BOM (require cold junction compensation) Lower accuracy Requires external ADC
Integrated semiconductors	Best price/performance ratio for high accuracy Lowest power consumption Minimal BOM Integrated ADC	Limited temperature range More expensive than thermistors

minimal bill of materials (BOM). However, for MCU-based applications if an ADC input is not available, then semiconductor sensors are the preferred option. For designs that require higher accuracy better than ±2 °C, lower consumption and/or high-resolution measurement, semiconductor sensors provide the best performance for the price if the sensing is within −40 °C to +125 °C.

Figure 2.9 shows an example of a semiconductor temperature sensor. It is the Silicon Labs Si705x digital temperature sensor. It offers low power consumption and high accuracy across the entire operating voltage and temperature range.

These monolithic CMOS ICs feature a band-gap temperature sensor element, an ADC with up to 14-bit resolution, signal processing, calibration data and an I²C

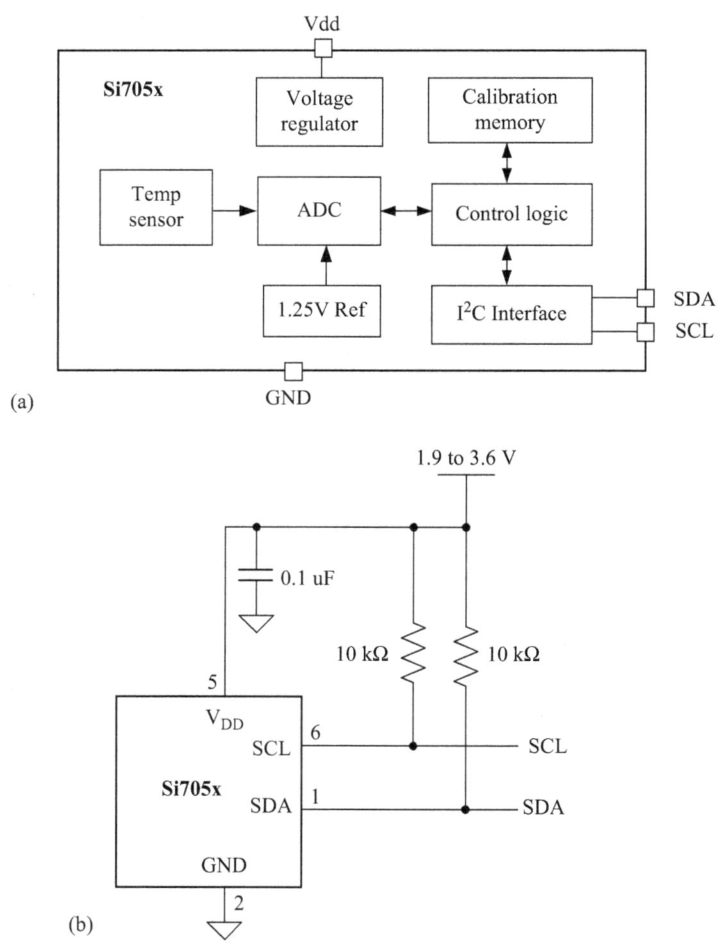

Figure 2.9 (a) Block diagram of the Silicon Labs Si705x semiconductor temperature sensor and (b) typical application circuit for temperature measurement

interface. The temperature sensors are factory-calibrated, and the calibration data are stored in the on-chip nonvolatile memory. This ensures that the sensors are fully interchangeable, with no recalibration or software changes required.

2.6 Humidity sensors

Humidity sensing technology is critical to a wide range of applications, including industrial HVAC/R, thermostats/humidistats, respiratory therapy, white goods, microenvironments/data centres, automotive climate control and de-fogging and asset, goods tracking and mobile computing devices.

The way to evaluate the humidity is by the relative humidity. This is defined as the ratio of actual water vapour present in air to the amount of water vapour that would be present in air at saturation when the air cannot absorb any more moisture. The saturation level is generally called the 'dew point' or 'frost point,' depending on the temperature. The relative humidity (RH) value can change significantly with even slight variations in temperature [14].

Discrete resistive and capacitive modules tend to exhibit high power consumption and demand a large amount of PCB real estate, which can make them difficult to integrate into devices. This is problematic for applications, such as asset tracking and for use in portable medical devices. Ideally, a humidity sensor should have a monolithic design to enhance reliability and minimize power consumption and size.

An example of monolithic CMOS IC is the Silicon Labs Si7005 digital relative humidity and temperature sensor. The sensor operates on the principle of measuring a change in capacitance; any changes to the dielectric constant of the polymer film will be detected as a change in relative humidity. Therefore, it is important to minimize the probability of contaminants coming into contact with the sensor. Dust and other particles as well as liquids can affect the RH reading.

Figure 2.10 shows the block diagram of the Silicon Labs Si7005. This IC integrates temperature and humidity sensor elements, an ADC, signal processing,

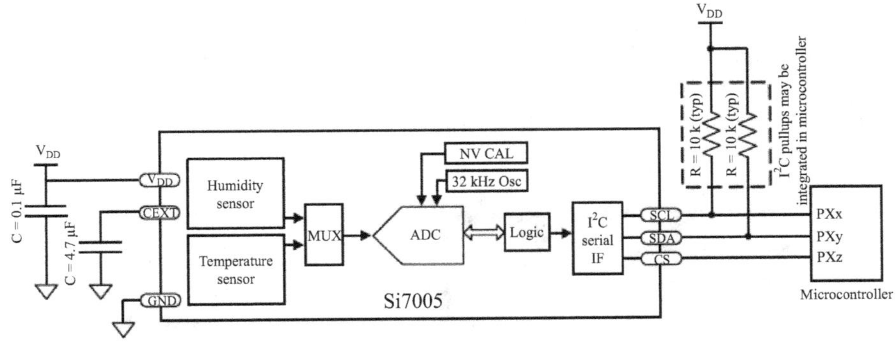

Figure 2.10 Functional block diagram of the Silicon Labs Si7005 relative humidity and temperature sensor

calibration data and an I^2C host interface. The patented use of industry-standard, low-K polymeric dielectrics for sensing humidity enables the construction of a low-power, monolithic CMOS sensor IC with low drift and hysteresis and excellent long-term stability. Both the temperature and humidity sensors are factory-calibrated, and the calibration data are stored in the on-chip nonvolatile memory.

2.7 Pressure sensors

Pressure sensors are generally used in many applications such as aircraft, automobiles, weather instrumentation and any other machinery that has pressure functionality implemented. Other applications include measuring flow and to calculate the level of a fluid or a loss of pressure due to a system leak.

There are three types of pressure measurements: gauge or gage pressure (the pressure is referenced to the atmosphere, absolute pressure (the pressure is measured against a reference vacuum sealed inside the chip assembly) and differential pressure (the pressure is measured as the difference between two pressures). Figure 2.11 shows these three types of pressure measurements.

MEMS pressure sensors have changed the way that system designers and application engineers measure pressure. The simplicity of use, small size, low cost and ruggedness allow these sensors to address applications in automobiles and industrial process control as well as medical and handheld portable products. For example, a highly accurate altitude measurement in a handheld navigation device such as a smartphone that has three-axis accelerometers, gyroscope and magnetometers adds a tenth degree of freedom. Pressure sensing allows the navigation device to locate the exact floor of a destination. MEMS pressure sensors typically measure the pressure difference across a silicon diaphragm.

Today, sensor experts estimate that MEMS technology is used in over 90% of the pressure sensors. Piezoresistive and capacitive sensing are commonly used in MEMS-based pressure sensors. For pressure measurements, piezoresistive bulk micromachined sensors are the most commonly used technology due to the ease of signal conditioning, broad selection and low cost. These MEMS sensors are used in most common pressure ranges from 0 to 4 into 0 to 100 psi.

An example is the STMicroelectronics LPS25HB. Figure 2.12 shows the block diagram of the LPS25HB pressure sensor. It is a piezoresistive absolute pressure sensor which functions as a digital output barometer. The device comprises a

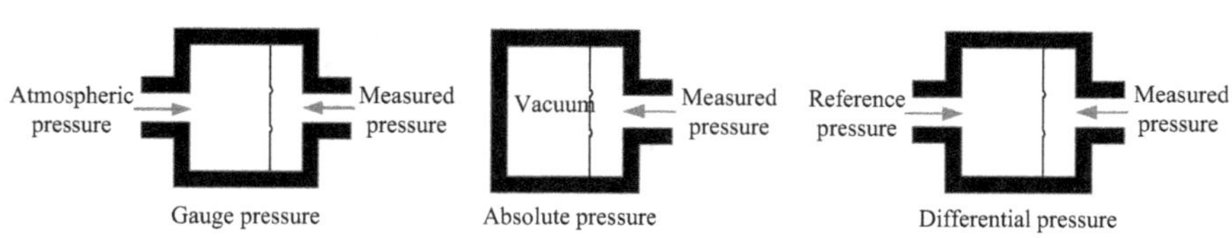

Figure 2.11 Types of pressure measurements

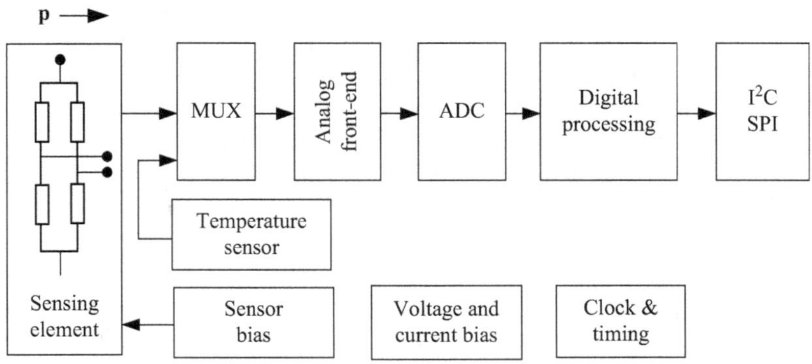

Figure 2.12 Block diagram of the STMicroelectronics LPS25HB pressure sensor

sensing element and an IC interface which communicates through I2C or SPI from the sensing element to the application.

The sensing element, which detects absolute pressure, consists of a suspended membrane manufactured using a dedicated process developed by ST. MEMS pressure sensors measure absolute pressure (*P*) and temperature (*t*) and store the values in two's complement registers which can be read via the I^2C or SPI host interface. The pressure sensor stores pressure as a 24-bit word. Both values are stored as two's complement integers.

When pressure is applied, the membrane deflection induces an imbalance in the Wheatstone bridge piezoresistances whose output signal is converted by the IC interface. The complete measurement chain is composed of a low-noise amplifier which converts the resistance unbalance of the MEMS sensors (pressure and temperature) into an analog voltage using an ADC. The pressure and temperature data may be accessed through an I^2C/SPI interface thus making the device particularly suitable for direct interfacing with a microcontroller.

2.8 Motion sensors

Motion is present in a very wide variety of IoT applications. Motion sensors include accelerometers, gyroscopes and inertial measurement units (IMUs). MEMS gyroscopes are used to measure angular rate. IMU sensors can integrate a combination of gyroscopes, accelerometers, magnetometers and pressure sensors to provide reliable position and motion discernment for stabilization and navigation applications.

Accelerometers can be used to sense not only acceleration but also vibration, shock and tilt [15]. Acceleration measures the change in velocity in a unit of time. Vibration can be thought of as acceleration and deceleration that happens quickly and in a periodic manner. Shock is an acceleration that occurs instantaneously but, unlike vibration, a shock is a nonperiodic function that typically happens once. When an object is moved to alter its tilt or inclination, some change in position with respect to gravity is involved. That movement tends to happen rather slowly compared to vibration and shock. The unit of these four modes of motion is g-force, that is, the gravity exerts on an object on the Earth (1g = 9.8 m/s^2).

MEMS accelerometers can be used to notice when a device is picked up and put down, which when detected can generate an interrupt that powers on and off functions automatically. Various combinations of functions can be kept active or put into the lowest power state possible. Movement-driven on/off features are human-friendly because they eliminate repetitive actions on the user's part. Another application is in automated external defibrillators (AED). Accelerometers embedded in the AED's chest pads can be used to give the rescuer feedback on the proper amount of compression by measuring the distance the pad is moved.

Slight changes in vibration serve as a leading indicator of worn bearings, misaligned mechanical components and other issues in machinery, including industrial equipment. Very small MEMS accelerometers with very wide bandwidth are ideal for monitoring vibration in motors, fans and compressors. Being able to perform predictive maintenance lets manufacturing companies avoid damage to expensive equipment and prevent breakdowns that cause costly productivity loss.

The disc drive protection found in many notebook PCs is among the most widely implemented applications of shock-sensing to date. An accelerometer detects tiny g-forces that indicate the notebook is falling or dropping, which is a precursor to a shock event: hitting the floor. Within milliseconds, the system orders the hard disc drive head to be parked. Parking the head stops contact with the disc platter during impact, preventing damage to the drive and the resulting data loss.

Tilt sensing has tremendous potential in gesture recognition interfaces as well. For instance, one-handed operation may be preferable in applications such as construction or industrial inspection equipment. The hand not operating the device remains free to control the bucket or platform where the operator stands, or perhaps to hold a tether for safety's sake. The operator could simply 'rotate' the probe or device to adjust its settings.

Compensating for the position of a device is another significant area where tilt measurement is useful. Take the electronic compass in a global positioning system (GPS) or mobile handset. A well-known problem here is the heading error that results when the compass is not positioned exactly parallel to the surface of the Earth. Industrial weigh scales are another example. In this application, the tilt of a loaded bucket relative to the Earth must be calculated to read the weight accurately. Pressure sensors, such as those used in automobiles and industrial machinery, are likewise subject to gravity's effects. These sensors contain diaphragms whose deflection changes depending upon the position in which the sensor is mounted. In all these situations, MEMS accelerometers perform the necessary tilt sensing to compensate for the error.

Figure 2.13 shows the functional block diagram of the Analog Devices ADXL345 accelerometer. It is a small, thin, ultra-low power, three-axis accelerometer with high-resolution (13-bit) measurement at up to $\pm$16g. Digital output data are formatted as 16-bit twos complement and is accessible through either an SPI (3- or 4-wire) or I2C digital interface.

The ADXL345 is well suited for mobile device applications. It measures the static acceleration of gravity in tilt-sensing applications, as well as dynamic

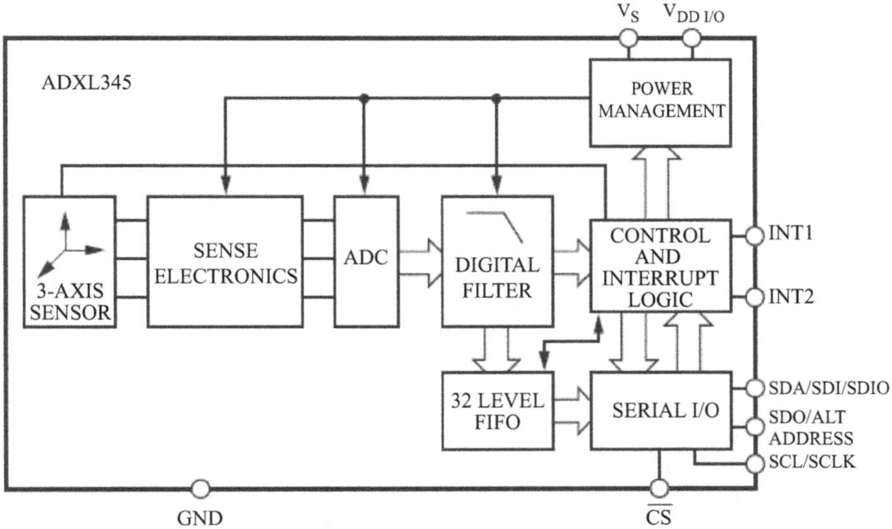

Figure 2.13 Block diagram of the Analog Devices ADXL345 accelerometer

acceleration resulting from motion or shock. Its high resolution (3.9 mg/LSB) enables measurement of inclination changes less than 1.0°. Several special sensing functions are provided. Activity and inactivity sensing detect the presence or lack of motion by comparing the acceleration on any axis with user-set thresholds. Tap sensing detects single and double taps in any direction. Freefall sensing detects if the device is falling. These functions can be mapped individually to either of two interrupt output pins. An integrated memory management system with a 32-level first in, first out (FIFO) buffer can be used to store data to minimize host processor activity and lower overall system power consumption.

Gyroscopes are used to measure angular rate, but unlike accelerometers, gyros are not affected by gravity, so they do not need anything external to make this measure. The angular rate is usually represented in degrees per second (°/s or dps). Typical applications of gyroscopes are navigation systems aid for dead reckoning, lift gate motion detections, accurate location for vehicle to vehicle and infrastructure, 360° view camera stabilization, car alarm, telematics and insurance vehicle tracking. An example of gyroscope is the InvenSense IAM-20380. Figure 2.14 shows the block diagram of this circuit. It consists of three independent vibratory MEMS rate gyroscopes, which detect rotation about the *X*-, *Y*- and *Z*-axes. When the gyros are rotated about any of the sense axes, the Coriolis Effect causes a vibration that is detected by a capacitive pickoff. The resulting signal is amplified, demodulated and filtered to produce a voltage that is proportional to the angular rate. This voltage is digitized using individual on-chip 16-bit ADCs to sample each axis. The full-scale range of the gyro sensors may be digitally programmed to ±250, ±500, ±1,000 or ±2,000 degrees per second (dps). The ADC sample rate is programmable from

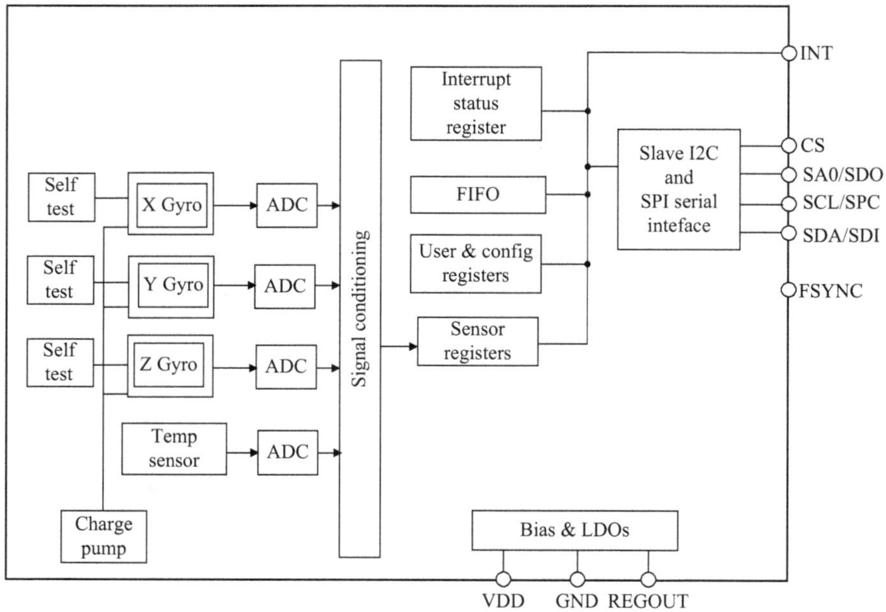

Figure 2.14 Block diagram of the InvenSense IAM-20380 gyroscope

8,000 samples per second, down to 3.9 samples per second, and user-selectable low-pass filters enable a wide range of cut-off frequencies.

Inertial measurement units include a multiaxis accelerometer, a multiaxis gyroscope and – to increase heading accuracy further – a multiaxis magnetometer. The IMU may, in addition, provide a full 6 degrees of freedom (6DoF). This brings ultra-high resolution to applications such as medical imaging equipment, surgical instrumentation, advanced prosthetics and automated guidance for industrial vehicles. Besides highly precise operation, another advantage of selecting an IMU is that its multiple functions can be pretested and precalibrated by the sensor's manufacturer [16].

An example of IMU is the BNO080. This circuit is manufactured by Bosch Sensortec and runs software provided by Hillcrest Labs. It integrates a triaxial 12-bit accelerometer with a range of ±8g, triaxial 16-bit gyroscope with a range of ±2,000 degrees per second, a triaxial geomagnetic sensor and a 32-bit ARM® Cortex™-M0+ microcontroller. The sensors are provided by Bosch Sensortec GmbH and the Cortex M0+ processor by Atmel Corporation. The BNO080 can be integrated as a co-processor in a mobile device such as a smartphone, tablet or ultrabook (Figure 2.15). This means the host processor can offload the sensor management and processing functions to the BNO080. The BNO080 constantly processes sensor data in the background while the host processor is in sleep state and can provide the host processor with real-time sensor data on demand.

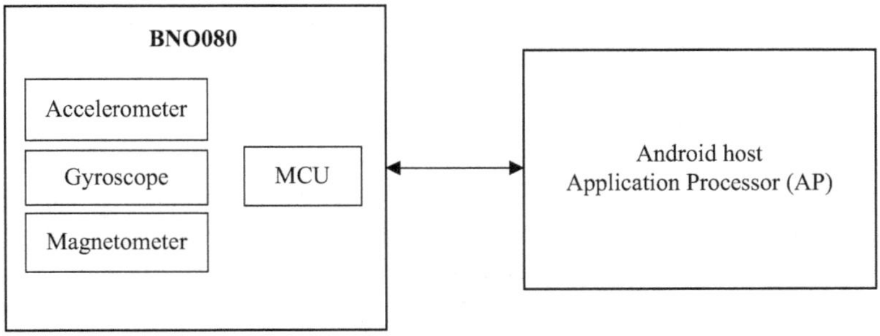

Figure 2.15 Bosh Sensortec BNO080 in a mobile device

2.9 Gas sensors

Environmental conditions have a major impact on our well-being, comfort and productivity. The key environmental parameters are ambient humidity and temperature, VOCs, particulate matter (PM2.5 and PM10) and CO_2.

The widespread use of new products and building materials has resulted in increased concentrations of indoor pollutants, in VOCs. Many of these compounds are known to impair well-being and productivity, for example, through headaches, eye irritation or dizziness. Sensors of VOCs are present in the development of new smart home and HVAC applications.

PM2.5 and PM10 refer to particulate matter with particle diameter up to 2.5 microns and 10 microns, respectively, and are among the most dangerous air pollutants. Due to their small size, PM2.5 particles can travel deep into the human lung and cause a variety of health issues, for instance, by triggering asthma attacks or contributing to cardiovascular disease.

Carbon dioxide is a key indicator of indoor air quality. Thanks to new energy standards and better insulation, houses have become increasingly energy-efficient, but the air quality can deteriorate rapidly. Active ventilation is needed to maintain a comfortable and healthy indoor environment and improves the well-being and productivity of the inhabitants.

An example of gas sensor for IoT applications is the ams CCS811. It is an ultra-low-power digital gas sensor solution which integrates a metal oxide gas sensor to detect a wide range of VOCs for indoor air quality monitoring with a microcontroller unit (MCU), which includes an ADC and an I^2C interface.

CCS811 is based on ams unique microhotplate technology which enables a highly reliable solution for gas sensors, very fast cycle times and a significant reduction in average power consumption. The integrated MCU manages the sensor drive modes and raw sensor data measured while detecting VOCs. The I^2C digital interface significantly simplifies the hardware and software design, enabling a faster time to market. CCS811 supports intelligent algorithms to process raw sensor measurements to output a TVOC value or equivalent CO_2 (eCO_2) levels, where the main cause of VOCs is from humans. CCS811 supports multiple measurement modes that have been optimized for low-power consumption during an active sensor measurement and idle mode extending battery life in portable applications. Figure 2.16 shows the recommended application circuit of this device.

Electrochemical gas sensing technology is the preferred solution for high-performance gas sensors due to its ultra-low-power consumption. SPEC sensors designs and manufactures amperometric gas sensors, that is, electrochemical sensors which generate a current proportional to the volumetric fraction of the gas. This manufacturer provides digital gas sensor development kit to measure the following gases: carbon monoxide (CO), hydrogen sulphide (H_2S), nitrogen dioxide (NO_2), sulphur dioxide (SO_2), ozone (O_3), ethanol (ETOH), indoor air quality (combustion) and respiratory irritants. Figure 2.17 shows the image of a 15 × 15 CO sensor 1,000 ppm.

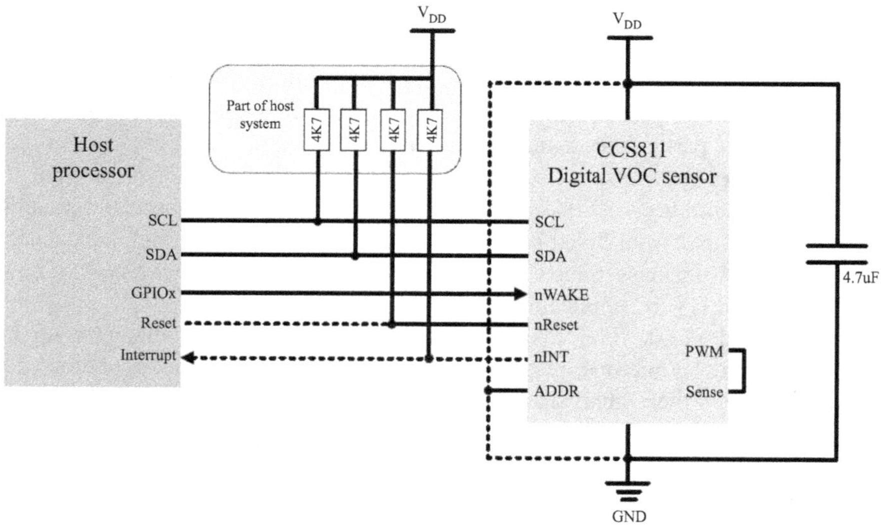

Figure 2.16 Recommended application circuit for ams CCS811 digital gas sensor

Figure 2.17 SPEC sensors screen-printed electrochemical CO sensor

2.10 Optical sensors

The light sensor products include a broad portfolio of digital ambient light, digital colour, proximity detection, light-to-digital (LTD), light-to-voltage (LTV) and light-to-frequency (LTF) sensors, in addition to linear sensor arrays, for intelligent light sensing in a broad range of applications including display management for display-based products (cell phones, TVs, tablets, computer monitors, etc.), medical diagnostic equipment, industrial processes and controls, LED solid-state lighting, consumer/commercial printer and health/fitness products.

Matching the sensor spectral response to that of the human eye response is vital because ambient light sensors are used to measure and help create ideal human lighting experiences. Strong rejection of infrared light, which a human does not see, is a crucial component of this matching.

An example of ambient light sensor is the Texas Instruments OPT3006 (Figure 2.18). The precision spectral response of the sensor tightly matches the photopic response of the human eye [17]. With strong infrared (IR) rejection, the OPT3006 accurately meters the intensity of light as seen by the human eye, regardless of the light source. The strong IR rejection also aids in maintaining high accuracy when design requires mounting the sensor under dark glass. The OPT3006, often in conjunction with backlight ICs or lighting control systems, creates light-based experiences for humans, and is an ideal replacement for photodiodes, photoresistors or lower-performing ambient light sensors.

There are optical sensors that include an ambient light and a proximity sensor in the same package. Typical applications of this kind of sensors are handsets, touchless switches, occupancy sensors, consumer electronics, notebooks/PCs, industrial automation, display backlighting control and photo-interrupters.

For example, the Silicon Labs Si1120 is a low-power, reflectance-based proximity and ambient light sensor with advanced analog signal processing and analog PWM

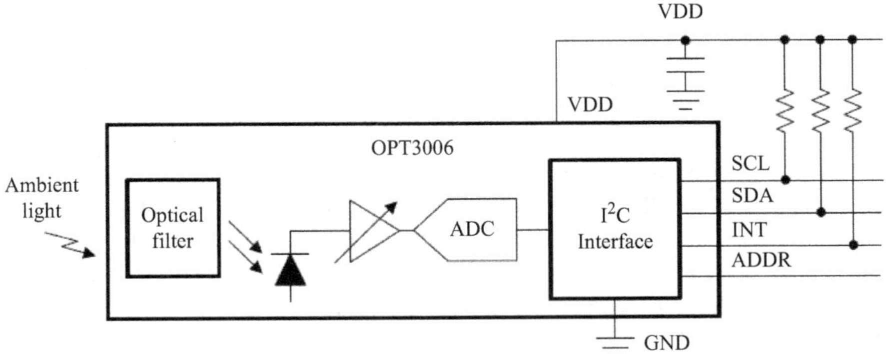

Figure 2.18 Block diagram of the Texas Instruments OPT3006 ambient light sensor

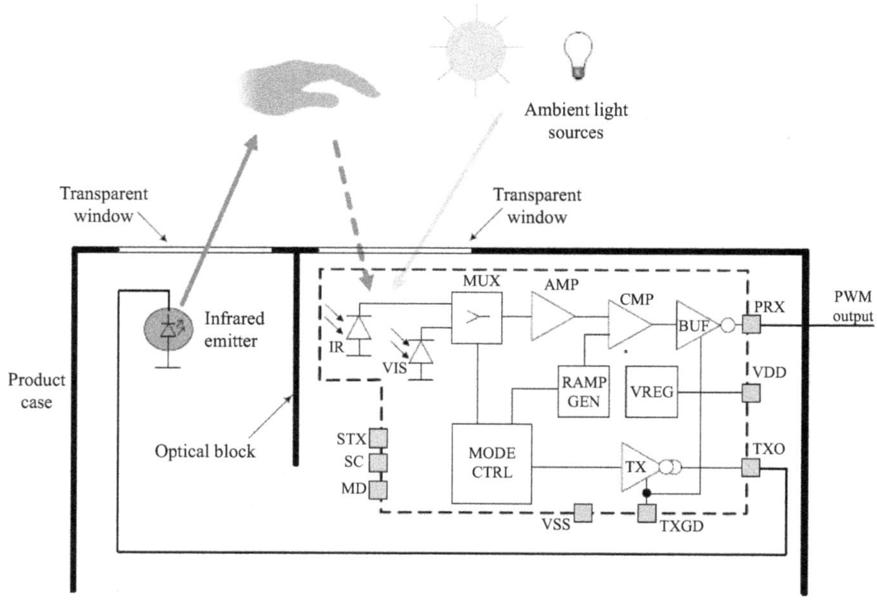

Figure 2.19 Block diagram of the Silicon Labs Si1120 proximity/ambient light sensor

output. It includes an integrated differential photodiode, signal processor and LED driver. Proximity sensing is based on the measurement of reflected light from an external, optically isolated, strobed LED. A separate visible light photodiode is used for ambient light sensing. Figure 2.19 shows a functional block diagram of this sensor.

An example of a proximity detection application is controlling the display and speaker of a cellular telephone. In this type of application, the cell phone turns off the power-consuming display and disables the loudspeaker when the device is next to the ear, then re-enables the display (and, optionally, the loudspeaker) when the phone moves more than a few inches away from the ear.

2.11 Ultrasonic sensors

One of the most effective areas of sensor technology is ultrasonic. In IoT applications ultrasonic sensors are used to measure the time interval between an ultrasonic signal that is sent and received, or what is commonly referred to as time-of-flight (TOF).

An example of ultrasonic circuit is the Texans Instruments' TDC1000 [18]. It is a fully integrated analog front-end (AFE) for ultrasonic sensing measurements of level, fluid identification concentration, flow and proximity/distance applications common in automotive, industrial, medical and consumer markets.

Figure 2.20 shows how ultrasonic works for a single transducer measurement. The TDC1000 AFE excites the sensor and detects the echo once it returns. The AFE excites the transducer by hitting it with a series of pulses and with frequencies in the

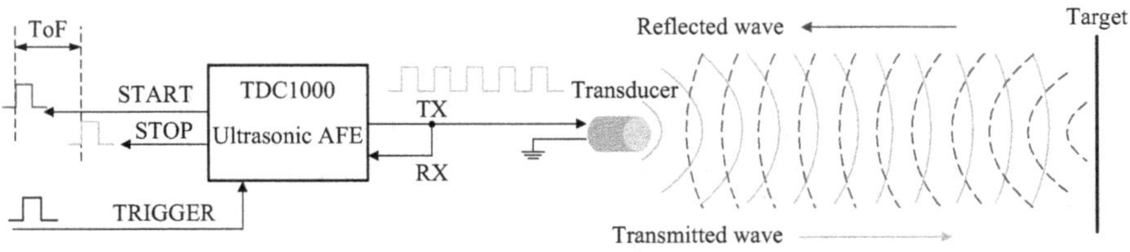

Figure 2.20 Functional block diagram of a single transducer measurement [19]

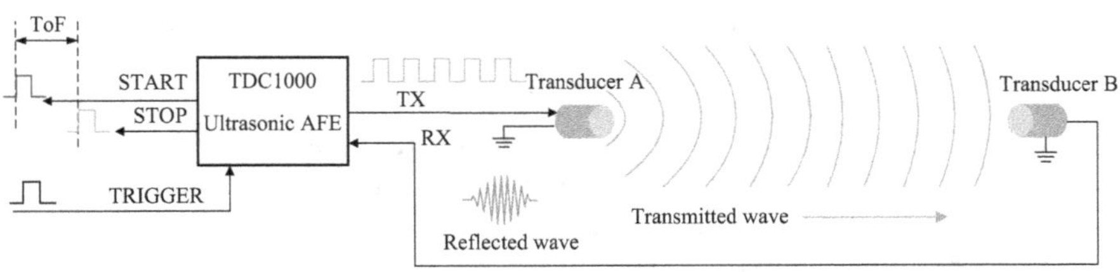

Figure 2.21 Functional block diagram of a dual transducer measurement [19]

range of kHz to MHz. The TDC1000 allows a maximum number of 31 pulses to excite the sensor, and its frequency can have a range from 31.25 kHz to 4 MHz.

The excitation of the sensor is marked with a START pulse, while the echo is denoted with a STOP pulse. The difference in time between START and STOP time-of-flight (TOF) indicates the fluid level, fluid ID/concentration and distance.

For dual transducers applications such as flow meter (Figure 2.21), the TDC1000 AFE utilizes the transducers in a pitch-and-catch fashion. In this method, transducer A is excited by the TDC1000 while transducer B acts as a receiver to generate STOP pulses. The time-of-flight (TOFAB) between excitation and STOP pulses indicates the distance between the transducers. A differential TOF, TOFAB – TOFBA, needs to be done to find the flow of medium (water or gas) between the two transducers.

The proximity, distance and level of the fluid can be determined by means of

$$d = \frac{\text{TOF} \times c}{2}$$

where d is the distance or the fluid level in metres (m), TOF is the time-of-flight in seconds (s) and c is the speed of sound in the fluid in metres per second (m/s).

2.12 Conclusions

The IoT couldn't exist without smart sensors. They are particularly useful in plants or warehouses because they can keep track of temperature and humidity, log data for historical records and quality management, or be used as triggers for alarms or process management.

When we consider IoT sensors, the following challenges have to be taken into account [19]:

- *The technology itself*: Manufactures of IoT sensors are pushed to leverage their core MEMS and system know-how to achieve the almost impossible. There are physical constraints challenging the engineer, package sizes cannot shrink forever, whilst demands for lower power and higher performance continue to rise incessantly.
- *New requirements*: These new requirements include lower power and longer battery-life requirements, improved networking capability through embedded antennas and the ability to process or store short bursts or long streams of real-time data.
- *The industry's fragmentation*: The IoT world is characterized by a highly fragmented structure of competing technological platforms. The demands placed on the sensor subsystem comprising of the sensors, microcontrollers and actuators vary widely across the IoT space.
- *The increasing complexity*: IoT systems are inherently complex, and OEMs often require turnkey solutions or reference designs. Techniques such as miniaturization and hybridization will come into play to drive advances.
- *The sensor fusion*: Individual sensors have inherent limitations and can err, which can be corrected or compensated for by complementary sensing nodes. For instance, gyroscopes suffer from offset drifts over time, which can be compensated for using a companion accelerometer. Therefore, integrating inputs from multiple sensors for more accurate and reliable sensing, sensor fusion can produce much higher levels of recognition and offer new responses.

References

[1] Giachino J. Smart Sensors. *Sensors and Actuators*. 1986;10:239–248.

[2] IEEE 1451.2 Standard. A Smart Transducer Interface for Sensors and Actuators. *Transducer to Microprocessor Communication Protocols and Transducer Electronic Data Sheet (TEDS) Formats*. Piscataway, NJ: IEEE Standards Department. 1998.

[3] Fisher C. *Using an Accelerometer for Inclination Sensing*. Available from: https://www.analog.com/media/en/technical-documentation/application-notes/AN-1057.pdf [Accesed 30 May 2019].

[4] Looney M. *The Power of Low Noise in IoT Smart Sensors*. Available from: https://www.analog.com/en/technical-articles/the-power-of-low-noise-in-iot-smart-sensors.html [Accesed 30 May 2019].

[5] Frank R. *Understanding Smart Sensors*. 3rd Edition. Norwood, MA: Artech House. 2013.

[6] Pickering P., *How to Design an IoT Smart Sensor with an Ultra-Low-Power MSP430FR MCU*. Available from: https://www.electronicdesign.com/embedded-revolution/how-design-iot-smart-sensor-ultra-low-power-msp430fr-mcu [Accesed 30 May 2019].

[7] Zhang Z. *Smart Sensing with Ultra-low-power MCUs – Part 4: Holter Monitor*. Available from: https://e2e.ti.com/blogs_/b/industrial_strength/archive/2017/05/08/smart-sensing-with-ultra-low-power-mcus-part-4-holter-monitor. [Accesed 30 May 2019].

[8] López A., Pérez D., Ferrero F., Postolache O. A Real-Time Algorithm to Detect Falls in the Elderly. *Proceedings of the 13th Annual IEEE International Symposium on Medical Measurements & Applications*; Rome, Italy, June 2018. pp. 1–5.

[9] Kasemsadeh B., LaPointe L. *Inductive Sensing Touch-On-Metal Buttons Design Guide. Texas Instruments*. Available from: http://www.ti.com/lit/an/snoa951/snoa951.pdf. [Accesed 30 May 2019].

[10] Baxter L. K. *Capacitive Sensors: Design and Applications*. Piscataway, NY: Wiley-IEEE Press. 1996.

[11] All Sensors Corporation. *Understanding Accuracy and Precision for MEMS Pressure Sensors*. Available from: https://allsensors.com/engineering-resources/white-papers/accuracy-and-precision-for-mems-pressure-sensors. [Accessed 30 May 2019].

[12] Lion Precision. *Understanding Sensor Resolution Specifications and Performance*. Available from: http://www.lionprecision.com/tech-library/technotes/article-0010-sensor-resolution.html. [Accessed 30 May 2019].

[13] Silicon Labs. *Measuring Temperature Accurately with Semiconductor Sensors*. Available from: https://www.silabs.com/documents/public/white-papers/measuring-temperature-accurately-with-semiconductor-sensors.pdf. [Accessed 30 May 2019].

[14] Silicon Labs. *Simplifying Humidity Measurement with Single-Chip Relative Humidity and Temperature Sensor ICs*. Available from: https://www.silabs.com/documents/public/white-papers/cmos-solution-for-humidity-sensing.pdf. [Accessed 30 May 2019].

[15] Analog Devices. *The Five Motion Senses: Using Mems Inertial Sensing to Transform Applications*. Available from: https://www.analog.com/media/en/technical-documentation/white-papers/The_Five_Motion_Senses.pdf [Accessed 30 May 2019].

[16] Scannell B. *High Performance Inertial Sensors Propelling the Internet of Moving Things*. Available from: https://www.analog.com/en/technical-articles/high-perf-inertial-sensors-propelling-internet-of-moving-things.html. [Accessed 30 May 2019].

[17] Harmon D. *Why Does Accuracy Matter So Much with an Ambient Light Sensor*. Available from: https://www.electronicdesign.com/optical/why-does-accuracy-matter-so-much-ambient-light-sensor. [Accessed 30 May 2019].

[18] Le A. *Ultrasonic Sensing Basics for Liquid Level Sensing, Flow Sensing, and Fluid Identification Applications*. Available from: http://www.ti.com/lit/an/snaa220a/snaa220a.pdf. [Accessed 30 May 2019].

[19] Gemelli M. *Smart Sensors Fulfilling the Promise of the IoT*. Available from: https://www.sensorsmag.com/components/smart-sensors-fulfilling-promise-iot. [Accessed 30 May 2019].

Chapter 3

Sensor communication interfaces and standards

Vítor Viegas[1,3] *and José Miguel Dias Pereira*[2,3]

The chapter talks about the interfaces used to communicate with sensors and the standards available to channel data to higher logical levels. These communication interfaces and standards aim at abstracting the sensor layer in order to achieve interoperability at the layers above. Two standards are covered in detail: the IEEE 1451 standard that defines a smart transducer interface for sensors and actuators, and the Open Process Control (OPC) platform. The first one is an ongoing work with academic roots, while the second one is an industry-born platform. Both solutions are described in terms of architecture and internal structure, and are criticized in terms of pros and cons.

3.1 Introduction

The global market for smart transducers is growing fast pushed by the Internet of Things (IoT), Industrial IoT (IIoT), smart cities, smart buildings, smart health and many other 'smart' buzzwords. The market is expected to increase steadily reaching $60 billion by 2022 globally [1,2]. The arrival of more smart transducers, from so many manufacturers, in so diverse applications, demand the creation of standards and guidelines to make possible all these devices to work together. Only the definition of basic functionalities, open communication channels and common programming interfaces will open the door to interoperability. This job is in the hand of all players in the market, including researchers, manufacturers, developers, end users and regulatory authorities.

This chapter presents two initiatives to streamline the information flow in measurement and control systems: the IEEE 1451 standard and the OPC platform. The first one is a standard created specifically for smart transducers, proposing hardware and software interfaces from the primary transducer up to the network; the second one is a well-established 20-year-old platform that followed the evolution of the technology.

[1]CINAV – Escola Naval, Base Naval de Lisboa, Alfeite, Almada, Portugal
[2]ESTSetúbal/IPS, Setúbal, Portugal
[3]Instituto de Telecomunicações, Lisboa, Portugal

3.2 IEEE 1451 standard

The IEEE 1451 standard [3], or 1451 std in short, defines a set of normalized hardware and software interfaces to connect transducers, microprocessors and communication networks. Using these interfaces, any compliant sensor may be attached to any network-enabled microprocessor and start working immediately, without human intervention, following the principle of Plug and Play. At present, the 1451 std includes a total of seven clauses. The first six handle different connections between the sensor and the microprocessor:

- 1451.0: Introduces the concept of Transducer Interface Module (TIM) as a kind of a 'super' data acquisition board that bridges transducers to the microprocessor. The clause explains how the TIM shall work by defining common functionalities, internal data structures and software interfaces to access it;
- 1451.2: Materializes the concept of TIM when it connects to the microprocessor through a wired, point-to-point communication interface;
- 1451.3: Materializes the concept of TIM when it connects to the microprocessor through a wired, multidrop communication interface;
- 1451.4: Defines functionalities, data structures and communication interfaces to add to analog legacy transducers in order to make them smarter (minimally compliant with the 1451 std);
- 1451.5: Materializes the concept of TIM when it connects to the microprocessor through a wireless communication interface and
- 1451.7: Defines data structures and communication protocols to access transducers through a Radio Frequency IDentification (RFID) link.

The seventh clause, numbered 1451.1, introduces the concept of Network Capable Application Processor (NCAP) as a bridge between the network and the transducers (usually connected through a TIM). It explains how the NCAP shall work by defining behaviours, information models and communication services.

Figure 3.1 presents a diagram explaining the architecture of the 1451 std. The NCAP connects to one or more TIM using plug-and-play 1451.X interfaces. Third-party interfaces (non-1451.X) are allowed but discouraged. The 1451.0 interface is strongly recommended but not mandatory. At the core, the NCAP runs the end-user application, which normally consists of a loop that reads data from sensors, processes it according to a predefined control strategy and writes data to actuators. Meanwhile, the application publishes relevant data and events on the network and responds to requests from remote peers.

The following sections explain in more detail each one of the 1451 clauses.

3.2.1 The 1451.0 clause

The 1451.0 clause [4] presents the TIM as a microprocessor peripheral intended to add intelligence to the attached transducers. As shown in Figure 3.2, the TIM includes all the circuitry to condition and digitize the signals coming from the transducers, as well as extra logic and data structures to support advanced features such as Plug and Play, self-diagnostics and self-calibration.

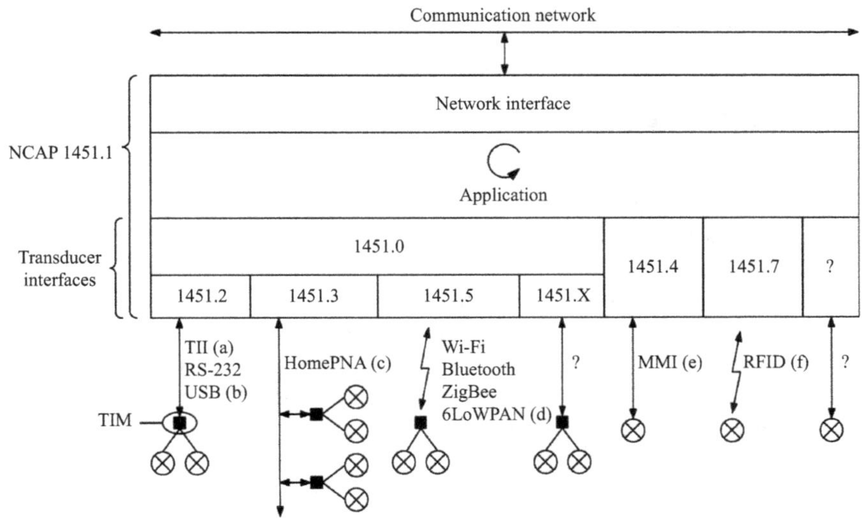

Legend: (a) Transducer Independent Interface; (b) Universal Serial Bus; (c) Home Phoneline Networking Alliance; (d) IPv6 over Low power Wireless Personal Area Networks; (e) Mixed Mode Interface; (f) Radio Frequency IDentification.

Figure 3.1 Architecture of the 1451 std

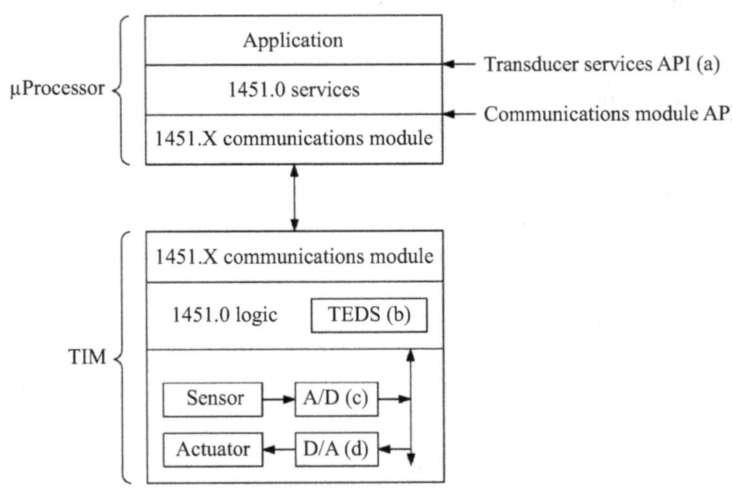

Legend: (a) Application Programmable Interface; (b) Transducer Electronic Data Sheet; (c) Analogue to Digital; (d) Digital to Analogue.

Figure 3.2 TIM structure and context

The 1451.0 clause describes the TIM in very generic terms to make it independent from third-party or 1451.X interfaces. The idea is to promote interoperability by introducing an abstraction layer right above the transducer interfaces. The clause defines common functionalities, electronic data sheets and programming interfaces for the TIM, which will be detailed in the following paragraphs.

3.2.1.1 Transducer types

The TIM shall support three types of transducers:

- Sensor: The sensor acquires one or more samples from a physical quantity and returns the result as a digital word. The acquisition takes place after a trigger order.
- Actuator: The actuator changes the state of an output according to a digital word. The action takes place after a trigger order.
- Event sensor: The event sensor serves to synchronize a given task with the occurrence of a given event. The event can be a state change on a digital input, or a threshold crossing of an analog input. Once activated, the event sensor waits to be triggered by the event.

3.2.1.2 Sampling modes

Transducers shall admit the following sampling modes depending on the reaction to the trigger:

- Trigger-ordered: The sampling process is usually stopped and only starts after receiving a trigger order. In the case of sensors, a predefined number of n samples are read from the input and transferred to memory; in the case of actuators, n samples are transferred from memory and written to the output. Once the transference is complete, the sampling process stops again.
- Continuous: Works like trigger-ordered with the difference that the sampling process only stops after an explicit order from the microprocessor.
- Immediate: Works like trigger-ordered with the difference that the trigger order is implicit in the read and write functions.
- Free-running with pre-trigger: The sampling process runs continuously and autonomously, with the acquired samples being transferred to a circular buffer. After receiving the trigger order, n samples are transferred to memory, from which a predefined number p was acquired before trigger and $n-p$ was acquired after trigger.

3.2.1.3 Data structures

Each transducer has a data structure that keeps record of its working condition. The transducer together with its data structure makes a so-called 'transducer channel'. The channel 0 is predefined, a virtual channel that represents the TIM as a whole. As shown in Figure 3.3, the data structure is composed of a memory (to store the samples associated with the transducer, if any) and three 32-bit registers:

- Condition register: Keeps track of the working condition of the transducer channel as described in Table 3.1;

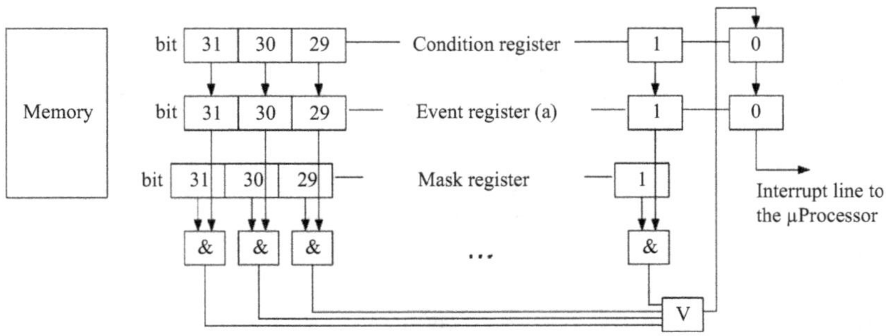

Legend: (a) The event register memorizes logical values equal to 1.

Figure 3.3 Data structure of a transducer channel

Table 3.1 Status bits

Bit	TIM (channel 0)	Transducer channel ($\neq$ 0)
0	Service request	Service request
1	TEDS changed	TEDS changed
2	Invalid command	Reserved
3	Command rejected	Command rejected
4	Missed data or event	Missed data or event
5	Data/event	Data/event
6	Hardware error	Hardware error
7	Not operational	Not operational
8	Protocol error	Reserved
9	Data available/data processed	Data available/data processed
10	Busy	Busy
11	Failed calibration	Failed calibration
12	Failed self-test	Failed self-test
13	Data over range or under range	Data over range or under range
14	Corrections disabled	Corrections disabled
15	Consumables exhausted	Consumables exhausted
16	Reserved	Not-the-first-read-of-this-data-set
17–23	Reserved	Reserved
24–31	Open to manufacturers	Open to manufacturers

- Event register: Memorizes the activation of the condition bits. If a bit condition passes from 1 to 0 (because the condition is no longer verified), the corresponding event bit keeps equal to 1 until it is explicitly reset by the microprocessor and
- Mask register: Defines the events that can generate an interrupt request to the microprocessor. The mask bits are AND-wired with event bits so that a mask equal to 0 disables the corresponding event. The interrupt requests are all OR-wired to generate a single interrupt line to the microprocessor. The microprocessor can later interrogate the condition and event registers to know the cause of the interruption.

3.2.1.4 Addressing scheme

The 1451.0 clause defines a high-level addressing scheme to identify individual and co-related transducer channels. The addresses are 16-bit wide and are divided into the following ranges:

- Address 0x0000[1]: Identifies the TIM on a point-to-point connection with the microprocessor (used in the clause 1451.2 only);
- Addresses between 0×0001 and 0×7FFF: Used to identify an individual transducer channel on a given TIM and
- Addresses between 0×8000 and 0xFFFF: Used to identify a group of co-related transducer channels located on a single TIM or in several TIMs. The actions addressed to a group affect all the channels of that group.

3.2.1.5 Messages

According to the 1451.0 clause, the data exchange between the TIM and the microprocessor is done using three types of messages (see Figure 3.4):

- Command messages (from the microprocessor to the TIM): Used to invoke a given command on the TIM:
- Reply messages (from the TIM to the microprocessor): Reply to an invoked command and
- Service request message (from the TIM to the microprocessor): Used to interrupt the microprocessor asking for attention on a given issue.

3.2.1.6 Transducer electronic data sheets

The transducer electronic data sheet (TEDS) is a descriptive memory of the TIM. As soon as the microprocessor detects the presence of a TIM, it tries to read its TEDS to become aware of its characteristics and capabilities. This information is all the microprocessor needs to start working properly with the TIM. This is done automatically, without human intervention, respecting the principle of transducer Plug and Play.

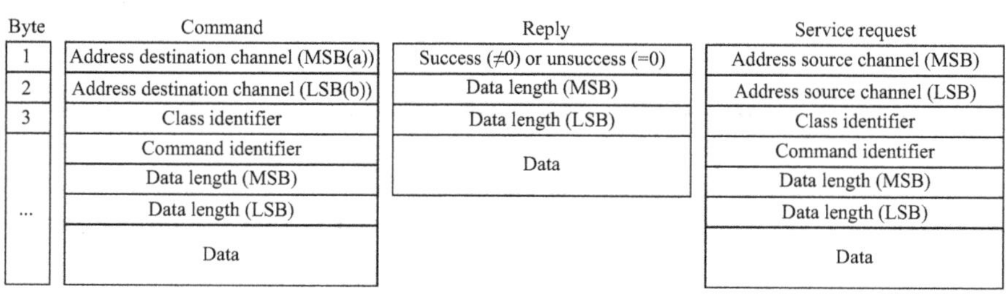

Legend: (a) Most Significant Byte; (b) Least Significant Byte.

Figure 3.4 Messages between the TIM and the microprocessor. All bytes are transmitted starting by the most significant bit

[1]0x means hexadecimal notation.

According to the specifications of the 1451.0 clause, the TEDS shall be organized in ten tables described as follows:

- Meta TEDS (mandatory, binary, read-only): Describes the general characteristics of the TIM, including identification, number of transducer channels, bindings to channel groups and timing parameters (such as operational and self-test time outs);
- Channel TEDS (mandatory, binary, read-only): Describes the characteristics of a given transducer channel, including type of transducer, measurement parameters (units, range, accuracy, sensitivity, etc.), data model (number of significant bits, bit alignment, number of samples, sampling period, etc.) and timing parameters. There is one channel TEDS for each channel, except for channel 0 which is described by the meta TEDS;
- Calibration TEDS (optional, binary, read and write): Describes the mathematical equation that converts raw data into engineering units. It also accommodates the calibration period of the transducer and the date of the last calibration. The calibration TEDS is associated with a transducer channel;
- Frequency response TEDS (optional, binary, read and write): Describes the frequency response, magnitude and phase, of a transducer channel;
- Transfer function TEDS (optional, binary, read and write): Describes the transfer function, poles and zeros, of a transducer channel;
- Text TEDS (optional, text, read and write): Used to store text messages in several languages;
- End-user application specific TEDS (optional, binary, read and write): Used to store binary data with no specific format;
- User's transducer name TEDS (mandatory, binary, read and write): Stores the tags associated with the transducer channels;
- Manufacturer-defined TEDS (optional, binary, read and write): Stores data needed to interact with manufacturer's software tools and
- PHY TEDS (mandatory, binary, read-only): Describes the communication interface that connects the TIM to the microprocessor.

3.2.1.7 Commands

The 1451.0 clause defines dozens of commands to interact with the TIM. The commands have a 16-bit identifier, where the most significant byte identifies the class of the command, and the least significant byte identifies the command itself. Here are some examples of commands covering several functionalities:

- Commands related to the sampling mode: read sampling mode, set transducer channel data repetition count (*n*), set transducer channel pre-trigger count (*p*), trigger command, abort trigger, etc.
- Commands related to the data structures: read condition register, read/clear event register, read/write service request mask register, read/write transducer channel data-set segment, etc.
- Commands related to addressing: address group definition.

- Commands related to TEDS tables: query TEDS, read/write TEDS segment, update TEDS, etc.
- Miscellaneous commands: reset, run self-test, wakeup, read 1451.0 version, etc.

3.2.1.8 Application programming interfaces

The 1451.0 clause defines three programming interfaces to communicate with the TIM:

- Communications module application programming interfaces (API): This API makes the connection to the communications module 1451.X (see Figure 3.5). The API includes low-level functions to search, register and unregister modules, control the operating state of registered modules and perform read and write operations.
- Transducer services API: This API makes the interface between the application (running in the NCAP) and the TIM. The API includes high-level functions to report the presence of TIMs and transducer channels, access TEDS structures, send commands to channels and trigger measurements.
- HTTP API: The goal of this API is to make the TIM visible on the web by exposing a subset of its functionalities through the HTTP protocol. The API, listed in Table 3.2, is hosted on a web server that tunnels the HTTP requests/responses to/from the TIM. The requests are made according to the following format: *http://<host>:<port>/<path>?<parameters>*, where *<host>* represents the internet address of the web server, *<port>* represents the communication port (80 by default), *<path>* identifies the function to be invoked on the TIM (see Table 3.2), and *<parameters>* represents the parameters needed for the remote call organized in pairs *name=value* separated by the *&* character. The HTTP responses can be given as web pages or plain text files.

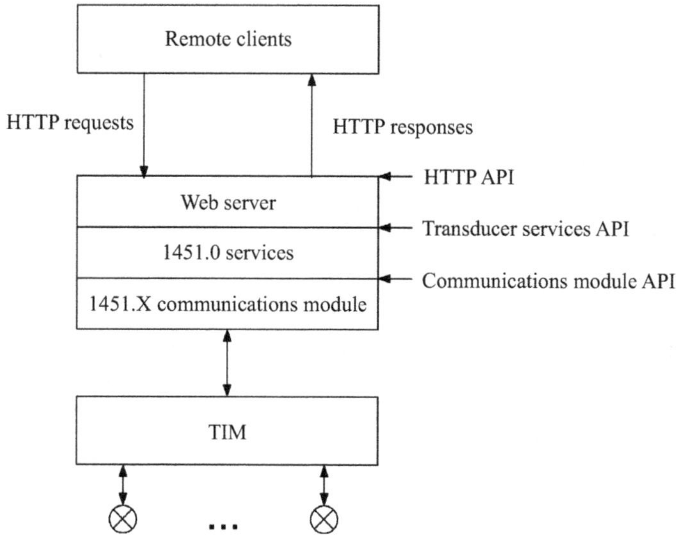

Figure 3.5 Tunnelling of HTTP requests/responses to/from the TIM

Table 3.2 Functions of the HTTP API

Interface	Function	Path
TIMDiscovery	*reportTims*	*1451/Discovery/TIMDiscovery*
	reportChannels	*1451/Discovery/TransducerDiscovery*
TransducerAccess	*readData*	*1451/TransducerAccess/ReadData*
	startReadData	*1451/TransducerAccess/StartReadData*
	measurementUpdate	*1451/TransducerAccess/MeasurementUpdate*
	writeData	*1451/TransducerAccess/WriteData*
	startWriteData	*1451/TransducerAccess/StartWriteData*
TransducerManager	*sendCommand*	*1451/TransducerManager/SendCommand*
	startCommand	*1451/TransducerManager/StartCommand*
	commandComplete	*1451/TransducerManager/CommandComplete*
	trigger	*1451/TransducerManager/Trigger*
	startTrigger	*1451/TransducerManager/StartTrigger*
TedsManager	*readTeds*	*1451/TEDSManager/ReadTeds*
	readRawTeds	*1451/TEDSManager/ReadRawTeds*
	writeTeds	*1451/TEDSManager/Write Teds*
	writeRawTeds	*1451/TEDSManager/WriteRawTeds*
	updateTedsCache	*1451/TEDSManager/UpdateTedsCache*

Table 3.3 Lines of the TII interface

Line	Actuated by	Description
DIN	Microprocessor	Data from the microprocessor to the TIM
DOUT	TIM	Data from the TIM to the microprocessor
DCLK	Microprocessor	Clock
NIOE	Microprocessor	Delimiter of data frames (active low)
NTRIG	Microprocessor	Trigger line (active low)
NACK	TIM	Acknowledgement of data bytes
NINT	TIM	Interrupt line (active low)
NSDET	TIM	Line to detect the presence of the TIM (active low)
POWER	Microprocessor	Power supply
COMMON	Microprocessor	Ground

3.2.2 The 1451.2 clause

The 1451.2 clause [5] is the embodiment of the 1451.0 clause for the case where the microprocessor connects to the TIM through a point-to-point, wired communication interface, known as Transducer Independent Interface (TII) (see Figure 3.1). The TII is a synchronous, bit serial connection, similar to the SPI interface. The data and clock lines are complemented by self-detection, trigger, service request and power lines, making a total of ten lines, as described in Table 3.3. The communication is of type master/slave, where the microprocessor is the master and the TIM is the slave. The TIM only interrupts the microprocessor when it has a pending service request.

The TIM shall implement the specifications of the 1451.0 clause, including data structures, addressing scheme, TEDS tables and commands. The PHY TEDS must be customized for the TII interface.

The high number of lines of the TII and the absence of a standardized plug have been much criticized. This problem can be solved in future revisions of the clause with the adoption of pure-serial, commercial interfaces such as RS-232 and USB.

3.2.3 The 1451.3 clause

The 1451.3 clause [6] is the embodiment of the 1451.0 clause for the case where the microprocessor connects to one or several TIMs through a multidrop bus (see Figure 3.6). The bus is a transmission line (twisted pair or coaxial cable) that transports power and data in different frequency bands:

- Power: Each TIM shall receive a minimum voltage of +12 VDC on any point of the bus. Each TIM shall consume no more than 250 mA from the bus, otherwise it shall be supplied by an external power supply;
- Synchronization: The clocks of all TIMs can be synchronized (if needed) by a dedicated (and optional) signal at 2 MHz and
- Data: Data communications shall be made at frequencies between 4.75 MHz and 9.25 MHz adopting the HomePNA technology.

Each TIM shall implement the specifications of the 1451.0 clause, including data structures, addressing scheme, TEDS tables and commands. The PHY TEDS must be customized for the HomePNA interface.

3.2.4 The 1451.4 clause

The 1451.4 clause [7] defines digital wiring, communication protocols and TEDS structures to make analog transducers smarter (minimally compatible with the 1451 std). The idea is to embed the TEDS on the transducer (using an EEPROM

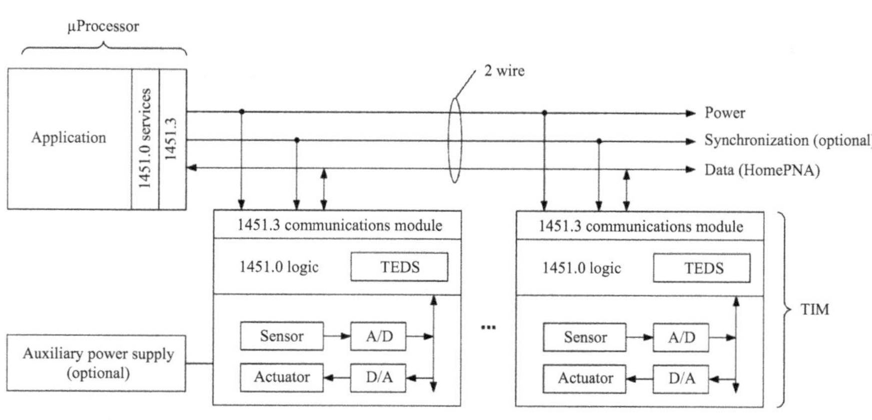

Figure 3.6 TIM in the context of the 1451.3 clause

memory or similar) and add a digital communication interface to access it. The upgraded transducer shall support, at least, the advanced features of self-identification and self-calibration.

The (new) digital communication interface, together with the (old) analog interface, form the so-called 'Mixed Mode Interface' (MMI). The MMI can be of class 1 if both interfaces share the same wiring or be of class 2 if the digital interface has its own wiring. The digital interface makes use of two wires and is based on the 1-Wire technology. A single communication port on the micro-processor can read the TEDS tables of multiple 1451.4 transducers (see Figure 3.7).

The TEDS tables are deliberately compact to occupy the smallest amount of memory, still they contain all the information needed for the microprocessor to use the transducer properly. This includes data to identify the transducer, describe its capabilities and reflect its calibration condition. These data are organized into two main parts:

- Basic TEDS (mandatory, binary, 64 bits length, read-only): Contains the transducer ID composed by the following fields: manufacturer ID (14 bits), model number (15 bits), version letter (5 bits), version number (6 bits) and serial number (24 bits). The emission, maintenance and revocation of manufacturer ID is regulated by the IEEE [8].
- User and Manufacturer TEDS (optional, binary, variable size): Describes the technical characteristics of the transducer and provides information about its calibration. These data are organized according to standard templates that cover the most common types of transducers (accelerometers, voltage output sensors, current output transmitters, thermocouples, RTD and many others). Templates 25–39 are used to store technical specifications, while templates 40–42 are specific for calibration data. The user can add free-style data at the end of the TEDS.

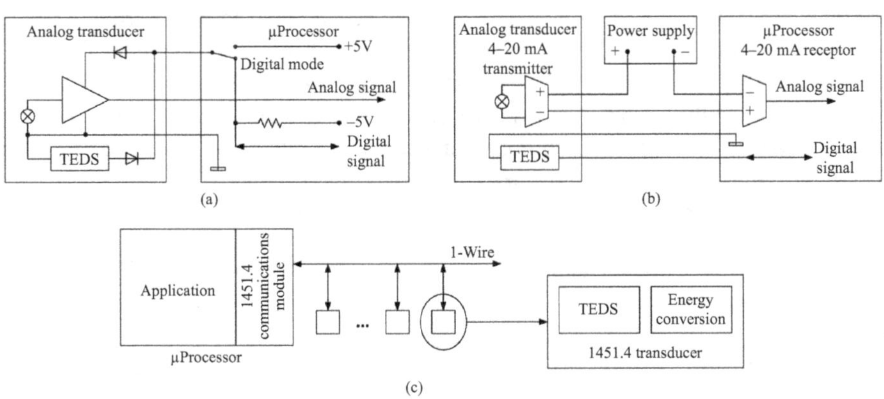

Figure 3.7 Several arrangements of MMI: (a) MMI class 1 added to a voltage output transducer (the digital interface is activated by inverting the supply voltage); (b) MMI class 2 added to a current output transducer and (c) the digital interface can reach multiple 1451.4 transducers

The 1451.4 clause is a kind of transition standard that makes the bridge between legacy transducers and the 1451 std. The 1451.4 clause is not compliant with the 1451.0 clause and it does not support the concept of TIM.

3.2.5 The 1451.5 clause

The 1451.5 clause [9] is the embodiment of the 1451.0 clause for the case where the microprocessor connects to one or several TIMs through a wireless communication interface (see Figure 3.1). Four wireless interfaces are currently supported: Wi-Fi, Bluetooth, ZigBee and 6LoWPAN. The 1451.5 clause delegates on these technologies the implementation of all OSI layers except the application layer; the same is to say that these technologies handle all communication details, including physical specifications, medium access control and management of radio channels.

At the application layer, the TIM shall implement the specifications of the 1451.0 clause, including data structures, addressing scheme, TEDS tables and commands. The PHY TEDS must be customized for the wireless interface used. If the TIM is powered by batteries, it shall implement energy saving mechanisms to maximize autonomy.

3.2.6 The 1451.7 clause

The 1451.7 clause [10] defines data structures and communication protocols for sensors attached to RFID tags. The 1451.7 clause is not compliant with the 1451.0 clause and it does not support the concept of TIM.

The 1451.7 clause can be implemented in any microprocessor equipped with an RFID tag reader. The microprocessor sends a command through the RFID interface and waits for the reply from the sensor. No new command is sent before the previous one is answered or until a time out expires. The sensor receives the command, executes it and replies. The command and reply messages are bit strings tunnelled through the RFID interface. The command messages consist of a command code (five bits), followed by input data (of variable length); the reply messages consist of the command code (repeated), the response code (three bits that signal error conditions) and output data (of variable length). As an option, all messages can be encrypted to ensure their authenticity.

The 1451.7 clause requires the sensor to implement five data structures:

- Sensor identification (read-only): 64-bit word, compliant with EUI-64 [11], that identifies uniquely and universally the sensor;
- Sensor characteristics TEDS (binary, read-only): Describes the technical characteristics of the sensor, including sensor type, measurement type (instantaneous value, average value, peak values, standard deviation, etc.), scaling (gain and offset) and security policy (RFID level or sensor level);
- Sample and configuration record (read and write): Allows the configuration of the sensor in terms of sampling period, alarm thresholds and security policy;

- Event administration record (read and write): Provides additional information about the data acquisition task (running or stopped, active alarms, sample count at a predetermined time or events, etc.) and
- Event record (read and write): Record of measurement values and event counts.

The 1451.7 clause defines 21 commands to talk with RFID sensors. There are commands to read and write data structures, to open and close secure communication sessions, and to run and stop data acquisition tasks.

3.2.7 The 1451.1 clause

The 1451.1 clause [12] presents the NCAP as a bridge between the network and the transducers (see Figure 3.8). On the field side, NCAP implements high-level functions to communicate with transducers; on the network side, the NCAP implements high-level functions to communicate with other processors; inside, the NCAP runs the application that processes data from both sides (field and network) and generates actions that decide the next state of the system.

The 1451.1 clause defines two models for inter-NCAP communications: the client/server for one-to-one communications, and the publisher/subscriber model for one-to-many communications. The client/server model follows the remote procedure call paradigm, according to which a given NCAP (the client) can invoke remote function from another NCAP (the server) and consume the returned results. Figure 3.9 describes this model step by step:

1. The client port in the client's process space is bound to the remote (dispatch) address of the server object;

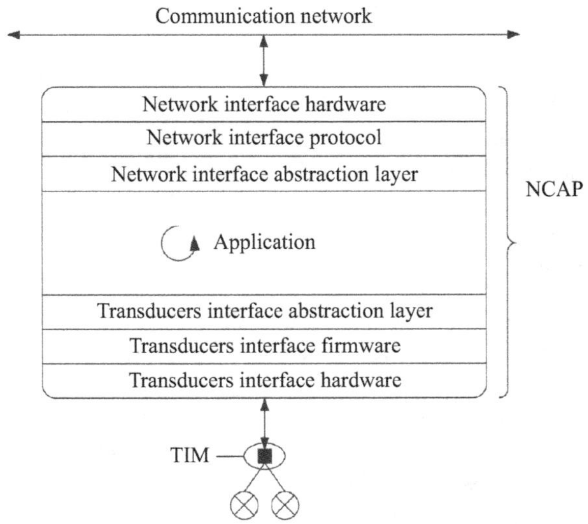

Figure 3.8 NCAP structure and context

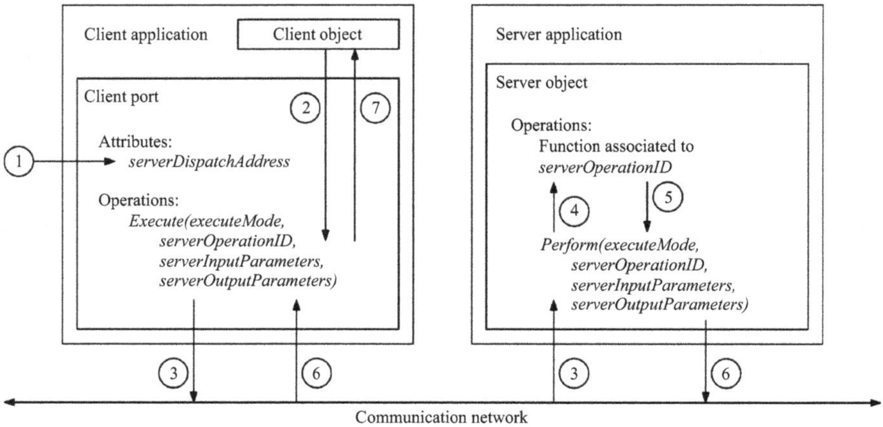

Figure 3.9 Client/server communication model (the numbered circles have correspondence with the main text)

2. The client invokes the Execute function on the client port indicating which operation of the remote server object is to be executed. It provides input arguments for that operation and references to its output arguments;
3. The network infrastructure invokes the Perform function on the target server object;
4. The Perform function invokes the desired server object operation;
5. When the operation completes, it returns to Perform with its output arguments filled in;
6. The network infrastructure returns the operation results to the client port and
7. The Execute function returns the operation results to the invoking client.

On the other hand, the publish/subscribe model follows the message paradigm, according to which a given NCAP (the publisher) can send messages to the network whenever it has some announcement to make. These messages, called publications, contain metadata describing the domain, semantics and syntax of the information published. Based on these metadata, any remote NCAP (subscriber) can filter the publications of its interest and consume their content. Figure 3.10 describes this model step by step:

1. The publisher configures the metadata of the publication in the publisher port.
2. The subscriber configures the metadata of the subscription in the subscriber port.
3. The subscriber invokes the function *AddSubscriber* in the subscriber port to register the callback function to be called if a valid publication is received.
4. The publisher invokes the *Publish* function in the publisher port, passing it the contents of the publication as an input argument.

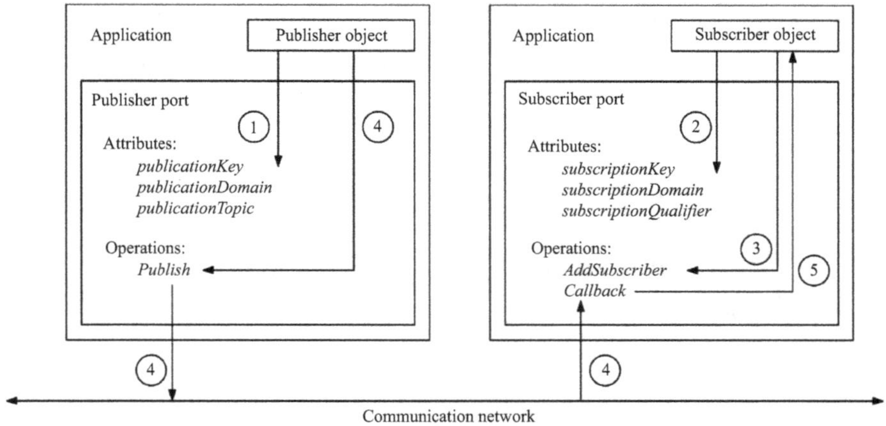

Figure 3.10 Publish/subscribe communication model (the numbered circles have correspondence with the main text)

5. The subscriber port filters incoming publications by comparing publication's metadata with port's metadata. If there is a match, the subscriber port invokes the registered callback operation passing it the contents of the publication.

The 1451.1 clause also defines an information model composed by the hierarchy of classes shown in Table 3.4. These classes are the bricks used to build applications: there are classes intended to process data (blocks), to store data (components) and to communicate across the network (services). The hierarchy is extensible by adding non-1451.1 classes to satisfy the requirements of a given application.

3.2.7.1 Blocks

Block classes, which include the NCAP Block, transducer blocks and function blocks, are intended to acquire and process data. Each block may own one or more non-block objects and all blocks are owned by the top-level NCAP Block. The activation or inactivation of a block affects all its child objects.

The NCAP Block is the head of the application and its main duties are as follows:

- System start-up: The NCAP Block is responsible for creating the initial set of objects that guarantee minimum system operability. The start-up information shall be kept in a nonvolatile storage device, such as the hard disc or an EEPROM memory;
- Keep track of all network-visible objects: To accomplish this task, it is necessary that all NCAP Blocks announce their presence and broadcast

Table 3.4 1451.1 information model (the classes in bold are non-abstract)

- *Root*
  - *Entity*
    - *Block*
      - ***NCAPBlock***
      - *FunctionBlock*
      - *BaseTransducerBlock*
        - *TransducerBlock*
          - ***DotXTransducerBlock***
    - *Component*
      - ***Parameter***
        - ***ParameterWithUpdate***
          - *PhysicalParameter*
            - ***ScalarParameter***
              - ***ScalarSeriesParameter***
            - ***VectorParameter***
              - ***VectorSeriesParameter***
        - ***TimeParameter***
      - ***Action***
      - ***File***
        - ***PartitionedFile***
      - ***ComponentGroup***
    - *Service*
      - *BasePort*
        - *BaseClientPort*
          - ***ClientPort***
          - ***AsynchronousClientPort***
        - *BasePublisherPort*
          - ***PublisherPort***
          - ***SelfIdentifyingPublisherPort***
            - ***EventGeneratorPublisherPort***
      - ***SubscriberPort***
      - ***MutexService***
      - ***ConditionVariableService***

updated information about all network-visible objects they own. To do this, the 1451.1 std defines two publications:

1. PSK_NCAPBLOCK_ANNOUNCEMET: After completing the start-up process, the NCAP Block issues this publication to notify its presence across the network and
2. PSK_NETWORK_VISIBLE_SERVER_OBJECT_PROPERTIES: Whenever a network-visible object suffers a profound change (as it happens when a parameter gets its metadata changed or its tag modified), the corresponding NCAP Block communicates that occurrence to the entire network by issuing this publication.

Transducer blocks, in turn, provide high-level functions to interact with transducers, such as read data from sensors, write data to actuators, read status bits, clear mask registers, access TEDS structures and many others. The 1451.1 clause is natively prepared to include transducer blocks from 1451.X interfaces, but additional transducer blocks can be included to support other interfaces as well.

As the name suggests, function blocks behave like software functions: they receive input data, process it and return the results. Examples of typical function blocks include PID controllers, square root extractors, comparators and multiplexers/demultiplexers.

3.2.7.2 Components

Component classes include parameters, actions and files. Parameter classes are data repositories used to store information related to blocks. The 1451.1 std distinguishes four types of parameter classes, as follows:

- Parameter: Represents a simple variable that can be read and written;
- Parameter with an update: Adds additional support for data synchronization between the parameter and its owning block. For example, the function UpdateAndRead instructs the owning block to acquire fresh data (from a sensor) and store it in the parameter;
- Physical parameters: Provide additional metadata structures to describe measuring quantities. Typical metadata fields include units, maximum and minimum values, data type, uncertainty, sampling period, among other attributes. Whenever the information held by the physical parameter changes, in terms of either value or metadata, it broadcasts the publication PSK_PHYSICAL_ PARAMETRIC_DATA to inform the network and
- Time parameter: Behaves as a timer by providing time representations of some event. Whenever the timer ticks, it issues the publication PSK_TIMER_TICK to share its timestamp with all subscribers across the network.

The diversity of data types and metadata fields supported by physical parameters makes possible the representation of (almost) every conceivable physical quantity, including scalars, vectors, waveforms and matrices.

3.2.7.3 Services

Service classes include communication ports and mutexes. All communication ports provide functions to get and set the timeout and the priority of the on-wire messages. Two priority values are possible: MP_NORMAL and MP_EXPEDITE.

Client ports implement the client/server communication model presented in Figure 3.9. Besides the basic functionality described earlier, client ports offer two advanced modes of operation:

- Asynchronous mode: This is a nonblocking operation mode in which the Execute function returns to the invoking client as soon as the message is transmitted. Later, the client can query the port as to whether the remote server operation has completed and, if so, request from the port any returned data and
- Send-and-Forget mode: This is also a nonblocking operation mode like the previous one. The difference is that the client does not care about any returned data at all.

Publisher/Subscriber ports implement the publish/subscribe communication model presented in Figure 3.10. They provide functions to issue publications, to edit the metadata of publications and subscriptions and to register subscription callbacks.

Finally, mutexes are semaphores that guarantee exclusive access to network-visible NCAP resources. They provide atomic functions, such as Lock and Unlock, to allow one and only one remote client to execute a given NCAP operation at a time.

3.2.8 Opinion about the IEEE 1451 standard

After having an overview of the 1451 std it is time to evaluate its weaknesses and strengths. Here is a list of strengths:

- It is ambitious because it covers all types of transducers (digital and analog, new and legacy) and a wide variety of transducer interfaces (1451.X or not, point-to-point or multidrop, wired or wireless);
- It is neutral in terms of hardware and software. It doesn't depend on any proprietary technology;
- It adds 'intelligence' to the systems in two ways: at the transducer level it defines logic to support self-identification, self-diagnostics and self-calibration; at the application level it defines function blocks to implement data processing algorithms and
- It is well designed and makes a coherent whole.

The major weakness of the 1451 std is its complexity, which prevents a wider adoption by the market. The (a few) application examples that can be found are mostly academic.

3.3 OPC platform

The Open Process Control Foundation [13], or OPC Foundation for short, is an industry consortium created to promote the secure, reliable and open exchange of data in industrial automation applications. The consortium was formed in 1995 by five top companies in the area of process control – Fisher-Rosemount (now Emerson Process Management), Rockwell Automation, Opto 22, Intellution (now part of GE Fanuc) and Intuitive Technology (now part of Wizcon Systems) – and Microsoft, and emergent player in the area of software and operating systems. The primary goal was the development of a software model to facilitate communications between field controllers and supervision applications. The model was finished in 1996 under the name 'OPC Data Access' (OPC-DA), and it was heavily based on two Microsoft technologies, Common Object Model (COM) and Distributed COM (DCOM).

The OPC-DA model [14], represented in Figure 3.11, is based on a client/server architecture. The OPC server is a Windows® application that acts as a bridge between the field devices and the supervision layer. On the field side, it interacts with field controllers – such as Programmable Logic Controllers (PLC), Distributed Control Systems (DCS) and fieldbuses masters – using dedicated device drivers; on the supervision side, it exposes variables that can be accessed remotely by OPC

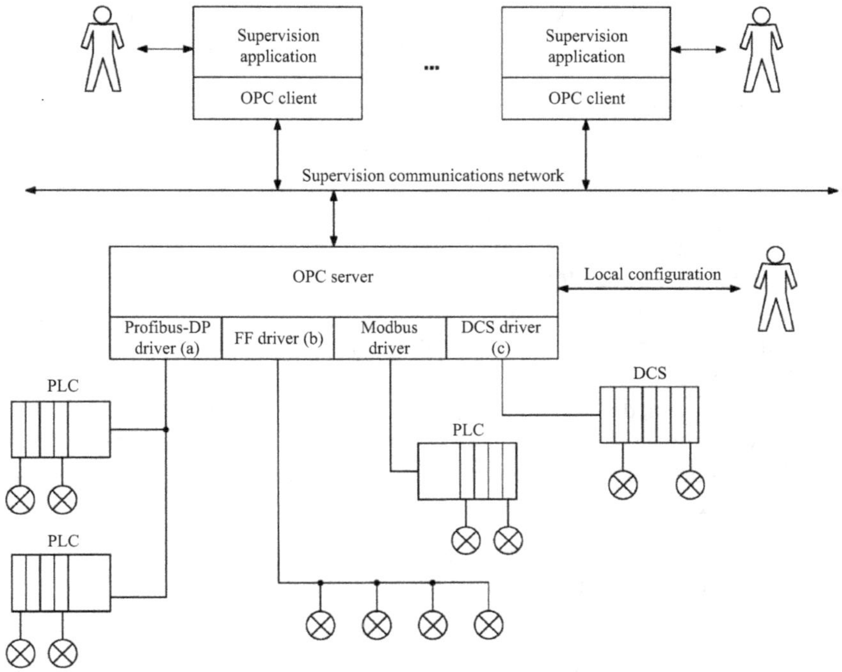

Legend: (a) Profibus Decentralized Periphery; (b) Foundation Fieldbus; (c) Proprietary Distributed Control System.

Figure 3.11 Architecture of the OPC platform

clients. The mapping between the network-visible variables and the field variables, as well as the configuration of the device drivers, is done through the local interface of the server. The OPC-DA model defines software interfaces to gather data between the server and its clients. The clients talk to the server without worrying about the details, the intrinsics, the tricks of the field devices.

The OPC-DA model was well accepted by the market and soon appear the first suppliers of OPC servers, clients and device drivers. Today, the OPC Foundation includes more than 300 members among equipment manufacturers, software producers, system integrators, research institutions and end users. During these years, the OPC Foundation has released other data models of which stand out:

- OPC-AE (Alarms and Events): Defines rules and software interfaces to notify the occurrence of alarms and events;
- OPC-HAD (Historical Data Access): Defines rules and software interfaces to access historical data and
- OPC-UA (Unified Architecture): Defines a common architecture that covers the functionalities of all data models.

Although these models were designed for industrial automation, they can be employed in all applications that require high density of transducers, high volume of data and low downtimes, which are requisites that are familiar to IoT in general. The following sections talk about each one of these models in a little bit more detail.

3.3.1 OPC data access

The OPC-DA model defines mechanisms to transfer data between the OPC server and one or more clients. As shown in Figure 3.12, the server acquires data from the field through dedicated drivers, designed to communicate with specific fieldbuses (Modbus, Profibus, Foundation Fieldbus or other). The acquired data are organized as field variables, known as 'OPC items', which are visible across the network. The OPC-DA model defines software interfaces that allow a remote client to establish a connection with a server, browse the OPC items, organize them into groups and access their data.

The OPC-DA model supports connections between servers and clients only. One client can connect to one or more servers, and one server can attend several clients, but connections between servers are not allowed. The OPC-DA model defines three main communication mechanisms:

- Synchronous calls: The client invokes a remote read/write operation on the server and waits for the results;

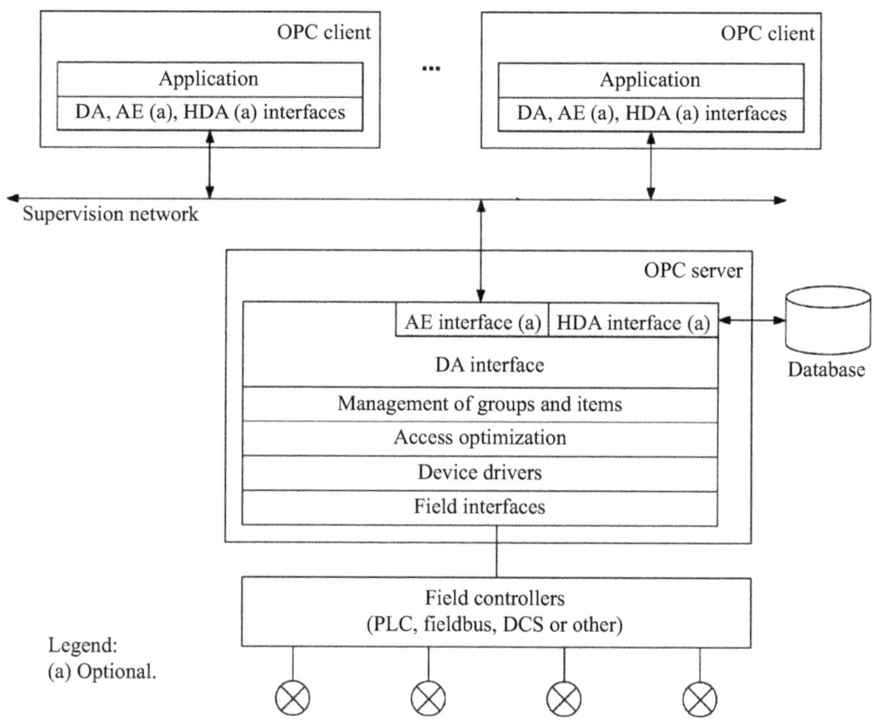

Figure 3.12 Details of the OPC server, including the software interfaces DA, AE and HDA

- Asynchronous calls: The client makes a remote call to the server but does not wait for the response. Later, it receives a notification with the results and
- Event subscription: The client registers a callback function on the server to be notified when one or more OPC items get their value changed.

The DA interface is implemented by three main objects:

- OPC Item: Represents a network-visible variable. The OPC Item is characterized by three attributes: value, quality (good, bad or unknown) and timestamp. A quality of good means that the value is valid, otherwise the value may not reflect the actual content of the variable. The timestamp contains the instant when the value was last updated. OPC Items are always accessed in group to optimize network communications;
- OPC Group: Represents a set of correlated OPC items. The OPC Group implements software interfaces through which a remote client can create groups, invoke read/write operations (synchronously or asynchronously), register callback functions, set the update rate, etc. and
- OPC Server: Represents the software process that supports the DA interface.

The client has great flexibility in its relation to the server: it can create the OPC groups it wants and populate them with the desired OPC items. The client can also define the behaviour of each OPC group by editing the following attributes:

- Name: Each OPC group has a name. The name must be unique among the other groups that belong to the client;
- Active: In the active state, the OPC group is updated with fresh data from the field, otherwise it is not;
- Cached data: The server keeps an image of the field in a fast-access local memory known as 'cache'. Through this attribute, the client can tell if it wants to receive data from the cache or from the devices in the field. In the first case, data are gathered more quickly; in the second case, data are gathered in live;
- Per cent dead band: The client can specify the minimal variation (in percentage) above which value changed notifications are fired. This adjustment affects all the analog variables in the OPC group and
- Update rate: Rate at which the client wants to receive value changed notifications. The server considers the update rates of all clients to adjust the refresh rate of its cache.

After this brief presentation of the OPC-DA model, it is time to see how the client and the server talk to each other. In the case of synchronous calls, things go this way:

1. The client establishes a remote connection with the server using TCP/IP sockets hidden behind DCOM;
2. The client creates one or more OPC groups and populates them with one or more OPC items;
3. The client activates the OPC groups;

4. For a given OPC group, the client invokes synchronous read/write operations and waits until the results arrive or until a time out expires. In the case of read operations, data can come from the server cache or from the devices in the field (further prolonging the wait) and
5. The server executes the remote call and returns the results.

In the case of asynchronous readings, things go this way:

1. The client establishes a remote connection with the server;
2. The client creates an OPC group and populates it with OPC items;
3. For that OPC group, the client registers a callback function to return results;
4. The client activates the OPC group;
5. The client invokes an asynchronous read operation against the OPC group. Data can be provided by the server cache or by the devices in the field;
6. The client repeats the asynchronous call;
7. The server invokes the registered callback function passing it the results of the first reading operation and
8. The server invokes again the callback function passing it the results of the second reading operation.

The delivery of notifications to the client is guaranteed because communications are supported over TCP/IP. If data are channelled through the same socket (which is common), notifications arrive by the same order they were being emitted.

Finally, when the client subscribes value changed notifications, things work this way:

1. The client establishes a remote connection with the server;
2. The client creates an OPC group and populates it with OPC items;
3. For that OPC group, the client registers a callback function to receive value changed notifications;
4. The client configures the OPC group in terms of dead band and update rate;
5. The client activates the OPC group;
6. The server invokes the registered callback function passing it the values of OPC items that were changed over its dead band;
7. The server waits, at least, the time corresponding to the update rate and
8. Step 6 is repeated.

OPC servers and clients are software applications running on a Windows® operating system. All software interfaces are COM-enabled, and all network communications must rely on DCOM over TCP/IP.

3.3.2 OPC alarms and events

The OPC-AE model [15] defines mechanisms to notify clients about the occurrence of alarms and events in the domain of the OPC server (see Figure 3.12). A client can connect to the server, browse through the alarms and events available, and subscribe those of its interest. Alarms and events are usually associated with OPC items or to the server process.

The term 'alarm' refers to an abnormal working situation that deserves attention at the supervision layer. The server is responsible for maintaining a list of subscriptions and for monitoring the occurrence of alarms in its domain. If an alarm is detected, all clients subscribing that alarm shall be notified. The alarm is described by a top-level condition and one or more sub-conditions. Alarm conditions support the following attributes:

- Name: Name of the alarm (e.g., LevelAlarm). The name of a condition must be unique within the server;
- Active: The alarm condition becomes active if one of its sub-conditions becomes active;
- Active sub-condition: If the active attribute is true, this is the name of the sub-condition which is currently active (e.g., HighAlarm in the context of the LevelAlarm). For single-state conditions, the value of this attribute is the condition name;
- Quality: Quality of the underlying OPC item;
- Enabled: The client can enable or disable the alarm condition. In the disabled state, the client ceases to receive alarm notifications;
- Acknowledged: The client shall acknowledge the alarm condition after understanding it;
- Last acknowledged time: Time of the most recent acknowledgement (of any sub-condition);
- Subcondition last active: Time of the most recent transition into the currently active sub-condition;
- Condition last active: Time of the most recent transition into this condition;
- Last inactive: Time of most recent transition out of this condition;
- Acknowledger ID: Identification of the client who last acknowledged this condition and
- Comment: Comment string passed in by the client during acknowledgement.

The alarm sub-conditions are disjoint, that is, only one can be active at a given time. They are described by the following attributes:

- Name: Name of the sub-condition (e.g., HighAlarm). The name of the sub-condition must be unique within the alarm condition;
- Definition: Textual, mathematical or logical expression that objectively describes the sub-condition (e.g., LT.PV > 80);
- Severity: Priority of the sub-condition, ranging from 1 to 1,000, where 1 is the highest priority and 1,000 is the lowest priority. An alarm condition can have several sub-conditions with different severities and
- Description: Brief message, intended to be read and understood by the supervision staff, that describes the sub-condition (e.g., level of water too high in the main boiler).

The term 'event' refers to a happening, not necessarily bad or abnormal, that deserves to be seen by the supervision layer. If an event is detected, all clients subscribing that event shall receive a callback (notification) containing the following information:

- Type (condition-related, tracking-related or simple): Condition-related events are associated with state transitions in alarm conditions (like when the alarm LevelAlarm becomes inactive). Tracking-related events are associated with actions from clients over OPC items (like when a client changes the value of an OPC item). Simple events are all the rest, such as a failure on a device driver or a warning of excessive temperature on an electronic board;
- Source: A reference to the object which generated the event notification. In the case of a condition-related event, it can be the tag name (e.g., FIC101) if the event pertains to a tag entering the LevelAlarm condition. It can also be the tag name for a tracking-related event such as the operator changing the set point value for FIC101. For a simple event such as system error, the source value might be the string system;
- Time: Time of the most recent occurrence of the event;
- Event category: The category to which the event belongs. Event categories are groupings of events supported by the server, such as process events, batch events or system events. Categories can only include events of only one type;
- Severity: Priority of the event (similar to the severity of alarm sub-conditions) and
- Message: Brief message, intended to be read and understood by the supervision staff, that describes the event (e.g., communications error).

The AE interface is implemented by two main objects:

- OPC Event Subscription: Represents the subscription of an alarm or event and
- OPC Event Server: Represents the software process that supports the AE interface. It handles the registering and deregistering of alarm and event subscriptions.

These objects implement software interfaces through which remote clients can browse alarms and events, register callback functions, enable or disable notifications, acknowledge alarms, etc. These software interfaces are complemented by others, implemented on the client side, intended to sink events. All software interfaces, both on server and client sides, must be compatible with COM and DCOM technologies.

3.3.3 OPC historical data access

The OPC-HDA model [16] defines mechanisms for an OPC server to provide historical data for its clients (see Figure 3.12). The provided data are normally used to build graphics showing the evolution of one or more field variables over a time window. If the window is static, the graphics are known as 'trend graphics' or simply 'trends'; if the time window is refreshed automatically, the graphics are known as 'playback graphics'. Some servers can also apply mathematical operations to the data contained in the time window and return the results.

The server is responsible for creating and maintaining a database to store historical data. The OPC-HDA model says nothing about the structure of the database or the mechanisms used to register data into the database. It just assumes that the database exists, and that the data logging is robust, compact and uninterrupted.

The field variables logged into the database are usually OPC items. A client can connect to the server, browse the available OPC items and request historical data from those of its interest. All read operations cover a time window defined by the client, which is characterized by three main parameters:

- Start time: Time of the first sample of the historical record;
- End time: Time of the last sample of the historical record. If the end time is less than the start time, the time window looks backward. The historical data include the first sample of the record and exclude the last one and
- Resample interval: The client can ask the server to resample the historical data. The server will use linear interpolation (or other) to fill the voids.

Read operations can be

- Synchronous: The client makes a remote call to the server asking for historical data and waits for the results;
- Asynchronous: The client asks for historical data but does not wait for the response. Later, it receives a notification with the results and
- Playback: The client asks for historical data but does not wait for the response. Instead, it starts receiving periodic notifications, carrying fresh data intended to build a playback graphic.

The object OPC-HDA Server implements software interfaces through which remote clients can browse OPC items, define time windows, read trends, request playback graphics, etc. These software interfaces are complemented by others, implemented on the client side, intended to sink callbacks. All software interfaces, both on server and client sides, must be compatible with COM and DCOM technologies.

3.3.4 OPC unified architecture

The OPC-DA, -AE and -HDA models, also referred to as 'classical OPC', were well accepted by the market, with thousands of applications spread all over the world. Despite their success, they have some limitations that must be underlined:

- Limitations in terms of information modelling: The information provided by classical OPC models is somewhat poor, has no context and is disaggregated. It is poor because it is limited to elementary types (such as Booleans, integers, reals and timestamps), which distorts the representation of more complex structures such as software (virtual) and real objects. It has no context because the exposed data are not complemented by metadata to describe its meaning. And finally, it is disaggregated because the DA, AE and HDA interfaces work separately, meaning that relations between co-related variables must be established on the client side;

- Dependence on proprietary technologies: Classical OPC relies on Microsoft resources, namely the Windows® operating system and the COM and DCOM technologies and
- Limitations in terms of security and robustness: Communications based on DCOM are difficult to configure and hard to pass through firewalls. Moreover, they are not safe because they do not support encryption mechanisms.

In 2001, the OPC Foundation started the development of a new software model to address these issues and also to unify the classical models under a common architecture. The work was completed five years later, in 2006, with the release of the OPC Unified Architecture (OPC-UA) [17–28].

The OPC-UA model inherits the client/server architecture of classical OPC as shown in Figure 3.13. Clients connect to the server and make remote calls to read, write and subscribe data. The UA communication stack converts the procedural calls into binary or text messages that travel through the network. On the server side, the UA communication stack converts back the binary messages into procedural calls ready to be executed. The replies and notifications from the server are treated the same way but in the opposite direction.

The all set of information exposed by the UA server is called 'address space'. This virtual space is organized in nodes related to each other through references (similar to pointers in C language). The nodes are created from seven elementary classes:

- Base Node: Is the root of all classes. It includes some attributes common to all nodes, namely: node ID (unique identifier that distinguishes the node universally), node class (enumeration that classifies the node as object node, variable node, method node, reference node or data type node), browse name (path of the node in the space address of the UA server), display name (character string containing the name of the node), description (character string explaining the meaning of the node) and references (list of pointers that connect the node to other nodes);
- Object Type Node: Represents software (virtual) or real objects;
- Variable Type Node: Represents an attribute, property, field or state of a given object;
- Method Type Node: Represents an action over the object;
- Reference Type Node: Represents a relation between two nodes (e.g., organizes, has component, has property, has type definition, has notifier, etc.);
- Data Type Node: Used to represent complex, structured data types and
- View Node: Used to group co-related nodes in the space address of the UA server. Views act as filters of information because they show only the nodes of interest (according to a given criterion) and hide all the rest.

The specialization of objects, variables and data types is not made by inheritance (since there are not classes derived from the elementary types); instead, it is made by reference to custom classes. Nevertheless, the elementary classes and their

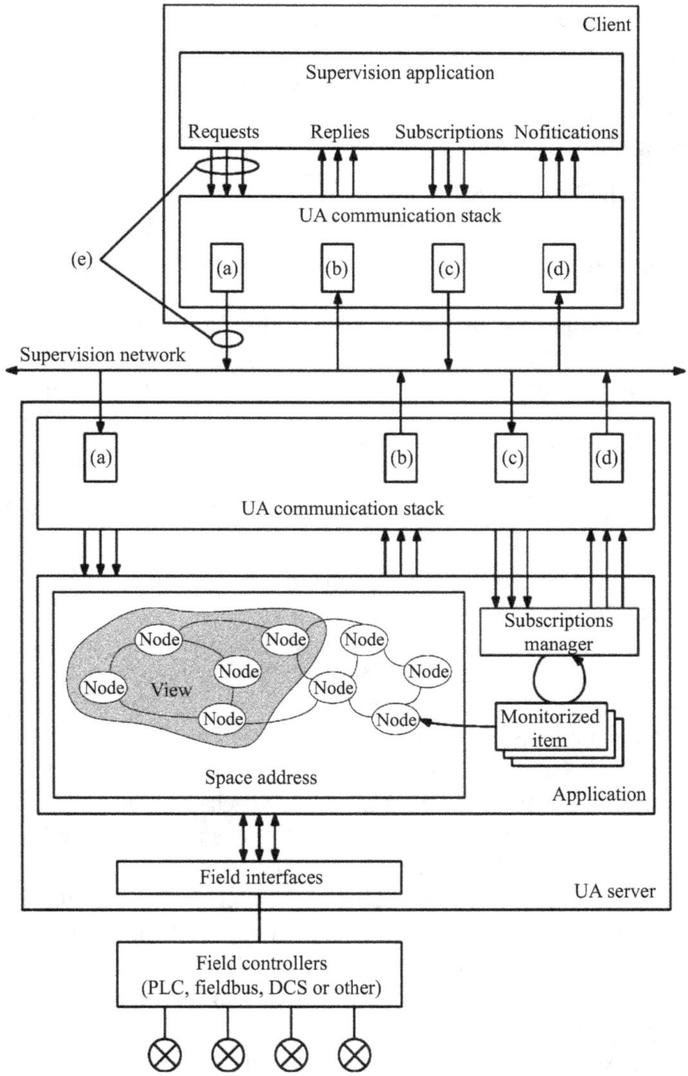

Legend: (a) Request message; (b) Repy message; (c) Subscription message; (d) Notification message (callback); (e) A different number of arrows reflect the presence of buffering mechanisms.

Figure 3.13 Architecture of the OPC-UA model

attributes are enough to implement all the features of classical OPC. Custom classes can be added to support other information models (including that of the 1451.1 clause, represented in Table 3.4).

The OPC-UA model does not define software interfaces neither is dependent on proprietary technologies (as it happens in classical OPC). It just defines the content and format of the messages exchanged between the client and the server.

In terms of content, the messages are composed by a header, input parameters (if it is a request from the client to the server or a notification from the server to client) and output parameters (if it is a reply from the server to client). In all cases, the order, type and meaning of the parameters are explained in the specifications. In terms of format, two options are available:

- UA binary: The message fields are coded directly in binary and sent over TCP/IP. Issues such as byte order, null representation and string encoding are properly documented. The binary format is very compact and is best choice to maximize the transmission speed and
- UA XML: The message fields are coded in text, according to the SOAP protocol, and sent over HTTP. This format is fully compatible with web services [29–31], making it the best option in terms of interoperability with third-party applications.

The request message and the corresponding reply message make an 'operation'. The OPC-UA model defines dozens of operations divided in the following groupings:

- Node management: Operations to add and delete nodes and references from the address space of the UA server;
- Attribute: Operations to read and write attributes. In the case of the value attribute, exclusive of variable objects, it is also possible to read historical data if available;
- Method: Operations to invoke methods;
- Monitored item: Operations to register, renew, edit and deregister subscriptions. Every client can ask to be notified about changes on every attribute of every node in the space address. The subscriptions remain active for a limited leasing time, after which they are eliminated;
- Subscription: Operations to notify clients about changes on attributes;
- View: Operations to browse the nodes of a view. The default view is the whole space address of the UA server. Views are defined a priori, using the local interface of the UA server, and cannot be changed by the clients;
- Query: Operations to filter attributes and nodes in the context of a view. The filter can be based on current or past (historical) values of an attribute;
- Discovery: Operations to discover a given UA server on the network and
- Security channel and session: Operations to establish a secure communication session with a UA server using an encrypted channel. The encryption mechanisms guarantee that the messages remain private and consistent (no one can read them without corrupting them).

Figure 3.14 presents the address space of a possible UA server. The address space models the information related to the electrical motor *m1*, which belongs to the view area1. The motor exposes two variables and two methods. The variables provide information about the velocity and torque of the motor; the methods allow to switch the motor on or off. Remote clients can connect to the server, browse its nodes, filter the view area1, change the state of the motor *m1* and read its variables

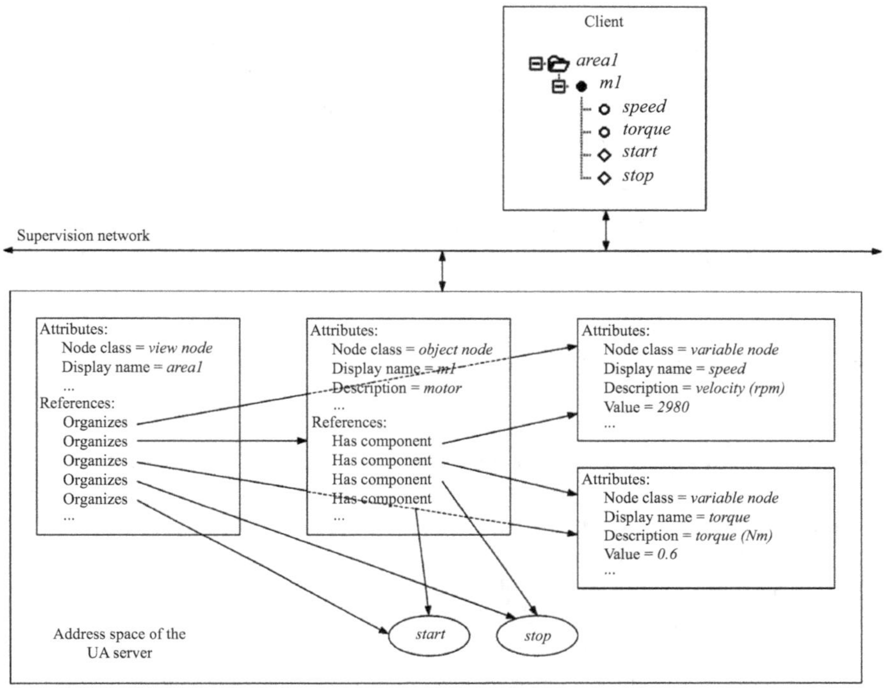

Figure 3.14 Example of a space address exposed by an UA server

(including past values). Reading operations can be made synchronously (blocking) or asynchronously (through callback functions). Clients can add nodes to the view and exclude nodes from the view, but they cannot delete the view.

3.3.5 Opinion about the OPC platform

After knowing the OPC platform in a little more detail, it is time to analyse its weaknesses and strengths. The main strengths are as follows:

- The OPC platform is comprehensive because it includes several data models covering live data, historical data, alarms and events. This range of features addresses the needs of most applications and
- The OPC platform is a mature and robust solution. It has good commercial support and counts with thousands of applications spread all over the world. The adoption of the OPC-UA model still is moderate but is increasing.

On the other hand, some weaknesses must be underlined:

- Classical OPC models depend on Microsoft technologies, namely the Windows® operating system and the technologies COM and DCOM. This limitation doesn't apply to OPC-UA, which is agnostic in terms of operating system and network communications and

- The OPC platform does not make transducers 'smarter' since it does not add support for transducer self-identification, self-diagnostics and self-calibration. The OPC platform does not care too much about the low-level layers; it just picks data from the field, organizes it in groups, hierarchies and views, and shares it with the upper layers.

Although the OPC platform is designed for industrial automation, it can be easily adapted for IoT applications in general. It is a robust solution for all applications that require high density of transducers, high volume of data and low downtimes.

3.4 Conclusion

The chapter talked about hardware and software interfaces to communicate with sensors and share data with higher logical levels. The 'Holy Grail' of a universal interface that connects any sensor to any microprocessor to any network, automatically and transparently, does not exist. Instead, there are platforms that connect to multiple sensor interfaces and aggregate data under a common information model, visible across the network. Two examples of such platforms were covered in some detail: the 1451 std and the OPC platform.

In the case of the 1451 std, the sensor interfaces are merged into the TIM where they are enriched with logic to support advanced features such as Plug and Play, self-diagnostics and self-calibration. The TIM is then connected to the microprocessor where data are processed and exposed through the 1451.1 information model.

In the case of the OPC platform, data are gathered using dedicated field interfaces and is exposed as OPC items organized in OPC groups. Support for alarms and events, as well as for historical data, can be added on the fly. The OPC-UA model allows the building of 'personalized' information models composed by nodes of different types connected through references (pointers).

The 1451 std is an ambitious work that tries to add plug-and-play capabilities and network visibility to any kind of transducer connected through any kind of interface. However, the industry insists in ignoring it as there are no real application examples and the commercial support is absent.

The OPC platform, in turn, is a mature and robust solution with good commercial support. Its main drawback, which was the dependence from Microsoft technologies, was solved with the release of the OPC-UA model. Being born in the context of industrial automation, the OPC platform fits perfectly in all IIoT applications and can easily be adapted to general purpose IoT applications.

Abbreviations and symbols

6LoWPAN	IPv6 over Low-power Wireless Personal Area Networks
AD	Analog-to-Digital
AE	Alarms and Events

API	Application Programming Interface
CAN	Controller Area Network
COM	Component Object Model
DA	Digital-to-Analog
DA	Data Access
DCOM	Distributed Component Object Model
DCS	Distributed Control System
DP	Decentralized Periphery
EEPROM	Electrically Erasable Programmable Read-Only Memory
FF	Foundation Fieldbus
HDA	Historical Data Access
HomePNA	Home Phoneline Network Alliance
HTTP	HyperText Transfer Protocol
ID	IDentification, Identifier
IEC	International Electrotechnical Commission
IEEE	Institute of Electrical and Electronics Engineers
IP	Internet Protocol
IoT	Internet of Things
IIoT	Industrial Internet of Things
ISO	International Organization for Standardization
LSB	Least Significant Byte
MMI	Mixed Mode Interface
MSB	Most Significant Byte
NCAP	Network Capable Application Processor
OPC	Open Process Control
OSI	Open Systems Interconnection
PID	Proportional Integral Derivative
PLC	Programmable Logic Controller
RFID	Radio Frequency IDentification
RS	Recommended Standard
RTD	Resistance Temperature Detector
SOAP	Simple Object Access Protocol
SPI	Serial Peripheral Interface
TCP	Transmission Control Protocol
TEDS	Transducer Electronic Data Sheet
TII	Transducer Independent Interface
TIM	Transducer Interface Module
UA	Unified Architecture
USB	Universal Serial Bus

References

[1] *Smart sensors market witnessing high growth* [online]. 2016. Available from http://www.sensorsmag.com/components/smart-sensors-market-witnessing-high-growth [Accessed 1 April 2018]

[2] *Smart sensor market expected to reach $60 billion, by 2022, globally* [online]. 2017. Available from http://www.sensorsmag.com/components/smart-sensor-market-expected-to-reach-60-billion-by-2022-globally [Accessed 1 April 2018]

[3] Song E., Lee K., 'Understanding IEEE 1451 - Networked smart transducer interface standard'. *IEEE Instrumentation & Measurement Magazine*. 2008; 11(2):11–17

[4] *IEEE 1451.0-2007 – IEEE standard for a smart transducer interface for sensors and actuators – Common functions, communication protocols, and Transducer Electronic Data Sheet (TEDS) formats*. New York: IEEE Instrumentation and Measurement Society, 2007

[5] *IEEE 1451.2-1997 – IEEE Standard for information technology – Smart transducer interface for sensors and actuators – Part 2: Transducer to microprocessor communication protocols and Transducer Electronic Data Sheet (TEDS) formats*. New York: IEEE Instrumentation and Measurement Society, 1997

[6] *IEEE 1451.3-2003 – IEEE standard for a smart transducer interface for sensors and actuators - Digital communication and Transducer Electronic Data Sheet (TEDS) formats for distributed multidrop systems*. New York: IEEE Instrumentation and Measurement Society, 2004

[7] *IEEE 1451.4-2004 – IEEE standard for a smart transducer interface for sensors and actuators – Mixed-mode communication protocols and Transducer Electronic Data Sheet (TEDS) formats*. New York: IEEE Instrumentation and Measurement Society, 2004

[8] *IEEE 1451.4™ - 2004 MANUFACTURER ID* [Online]. 2018. Available from https://standards.ieee.org/develop/regauth/manid/index.html [Accessed 1 April 2018]

[9] *IEEE 1451.5-2007 – IEEE standard for a smart transducer interface for sensors and actuators – Wireless communication protocols and Transducer Electronic Data Sheet (TEDS) formats*. New York: IEEE Instrumentation and Measurement Society, 2007

[10] *IEEE 1451.7-2010 – IEEE standard for a smart transducer interface for sensors and actuators – Transducers to Radio Frequency Identification (RFID) systems communication protocols and transducer electronic data sheet formats*. New York: IEEE Instrumentation and Measurement Society, 2010

[11] *Guidelines for use of Extended Unique Identifier (EUI), Organizationally Unique Identifier (OUI), and Company ID (CID)* [online]. 2017. Available from http://standards.ieee.org/develop/regauth/tut/eui.pdf [Accessed 1 April 2018]

[12] *IEEE 1451.1-1999 – IEEE standard for a smart transducer interface for sensors and actuators – Network Capable Application Processor (NCAP) information model.* New York: IEEE Instrumentation and Measurement Society, 1999

[13] *OPC Foundation – The industrial interoperability standard* [online]. 2018. Available from https://opcfoundation.org [Accessed 1 April 2018]

[14] *OPC Data Access specification version 3.00.* USA, OPC Foundation, 2003

[15] *OPC Alarms and Events specification version 1.10.* USA, OPC Foundation, 2002

[16] *OPC Historical Data Access specification version 1.20.* USA, OPC Foundation, 2003

[17] Mahnke W., Leitner S. H., Damm M. *OPC Unified Architecture.* Germany, Springer, 2009

[18] *OPC Unified Architecture specification, part 1 – overview and concepts version 1.01.* USA, OPC Foundation, 2009

[19] *OPC Unified Architecture specification, part 2 – security model version 1.01.* USA, OPC Foundation, 2009

[20] *OPC Unified Architecture specification, part 3 – address space model version 1.01.* USA, OPC Foundation, 2009

[21] *OPC Unified Architecture specification, part 4 – services version 1.01.* USA, OPC Foundation, 2009

[22] *OPC Unified Architecture specification, part 5 – information model version 1.01.* USA, OPC Foundation, 2009

[23] *OPC Unified Architecture specification, part 6 – mappings version 1.00.* USA, OPC Foundation, 2009

[24] *OPC Unified Architecture specification, part 7 – profiles version 1.00.* USA, OPC Foundation, 2009

[25] *OPC Unified Architecture specification, part 8 – data access version 1.01.* USA, OPC Foundation, 2009

[26] *OPC Unified Architecture specification, part 9 – alarms and conditions version 1.00.* USA, OPC Foundation, 2010

[27] *OPC Unified Architecture specification, part 10 – programs version 1.00.* USA, OPC Foundation, 2007

[28] *OPC Unified Architecture Specification, part 11 – historical access version 1.00.* USA, OPC Foundation, 2007

[29] *XML Web Services* [online]. 2018, Available from https://www.w3schools.com/xml/xml_services.asp [Accessed 1 April 2018]

[30] Alonso G., Casati F., Kuno H., Machiraju V. *Web Services – concepts, architectures and applications.* Berlin, Germany, Springer, 2004

[31] Freeman A., Jones A. *Microsoft .NET XML Web Services step by step.* USA, Microsoft Press, 2003

Chapter 4

Multisensor IoT interface with Bluetooth Low Energy

Nagaraj Hegde[1], *Matthew Bries*[1], *and Edward Sazonov*[1]

Bluetooth Low Energy (BLE), also referred to as Bluetooth Smart, is targeted towards low-power wireless sensor applications, connecting the world of Internet of Things (IoT). With BLE support in the Android OS being relatively new, the literature lacks methodologies for implementing a BLE multisensor interface with Android hub. In this article, we try to fill this gap, as we describe and address the implementation challenges in the development of a multisensor BLE system. In doing this, we propose the 'sequential actions following successful callback (SASC)' methodology, which helps in achieving reliable BLE connection establishment, data exchange and disconnection processes in a BLE network with no manual intervention. SASC methodology was used in a real-life use case for gait data collection using SmartStep sensors and 19 days of human subject tests resulted in 100% data retrieval. Profiling tests were conducted to benchmark the throughput, CPU loading and power consumption in the smartphone, for varying BLE connection intervals and a number of simultaneous sensor connections. The results of Android application throughput evaluation suggest that the maximum achievable throughput is ~20 kbps for a single sensor connection and ~70 kbps for a seven-sensor connection under SASC methodology. Based on the results of the CPU loading and power consumption tests, it may be advisable to use BLE connection interval of 13.75 ms or greater to keep the CPU loading and power consumption figures low. The presented results may help designers choose a trade-off between the number of sensors, connection interval, CPU loading, power consumption and throughput when working with multisensor BLE systems connected to IoT infrastructure.

4.1 Introduction

Bluetooth Low Energy (BLE, v 4.0) is a wireless personal area network technology designed by the Bluetooth Special Interest Group. BLE is aimed at applications such as fitness, healthcare, security and home entertainment control [1], playing a big role

[1]Department of Electrical and Computer Engineering, The University of Alabama, Tuscaloosa, AL 35487, USA

in connecting the world of IoT. BLE is intended to provide significantly lower power consumption compared to previous Bluetooth versions. In our previous work, we have achieved an improvement of two orders of magnitude in the power consumption of a wireless sensor with BLE, compared to Bluetooth v 2.1 [2,3].

Compared to other mobile operating systems (OSs), the BLE support in the Android OS is relatively new [4]. The implementation of an Android application that deals with single BLE sensor system was addressed in our previous work [2,3], but BLE multisensor scenarios involving an Android device are not described in the literature. There are many applications that may benefit from the use of multiple wireless sensors, such as physiological monitoring [5,6], biofeedback [7–9], gait and activity monitoring [10,11], body sensor networks [12,13] and others.

Let us consider a BLE body sensor network scenario, in which several wireless sensors are located on different body parts and there is an Android IoT Hub system (stationary or mobile) to collect the data from the sensors. The IoT Hub system interacts with all the sensors and is also connected to the Internet. We consider the scenario wherein several BLE sensors communicate with the IoT Hub simultaneously without manual intervention during operation. In such a situation where the sensors can come into and out of range from the IoT Hub at different times, there is a need for a systematic methodology to handle the communication between the sensors and the IoT Hub. This situation is different from the case wherein a smartphone user manually interacts with two or more BLE sensors.

In general, BLE communication between a sensor and an IoT Hub offers the following challenges:

- The application running in the IoT Hub needs a reliable BLE connection establishment procedure with the sensor.
- The information exchange mechanism between the IoT Hub and the sensor should enable the sensor to reliably communicate with the Android application running on the IoT Hub. Also, there should be a minimal loss in the sensor data collected.
- The BLE disconnection process needs to be handled correctly, making sure that the necessary commands are all conveyed from the IoT Hub to the sensor, and that the sensor's microcontroller peripherals (such as notification event timers) are turned off to save the battery power.

In a multisensor scenario, the above challenges are compounded, as the actions of connection establishment, data exchange and disconnection need to happen in a reliable manner with each of the sensors. In such scenarios, it was observed (through our laboratory experiments) that lack of a systematic design methodology results in many problems. These problems can lead to malfunctioning in some sensor nodes or throughout the whole network. One of the potential problems is that in an Android activity or a service handling a particular sensor, BLE actions are initiated without waiting for the successful completion of the previous actions. Another example is when the Android services or activities, each handling different simultaneous BLE connections, are blind to each other (no interservice communication). Different services/activities may initiate BLE actions independently of each other which may potentially collide. Such collisions in a multisensor network may result in some nodes failing to register the data writes from the IoT Hub application or others failing to be disconnected properly. This may also result in periodic data transmission timers to be unnecessarily active in some sensor nodes, draining the sensor's battery (while BLE is all about saving battery power).

Two challenges are addressed in this work. First, in implementing a BLE multisensor application that functions without user intervention, we propose, describe and evaluate the 'sequential actions following successful callback (SASC)' methodology. The SASC methodology will be utilized for handling BLE connection, data collection, event notification and disconnection processes in a BLE multisensor scenario. Second, we conduct profiling tests to evaluate the SASC methodology in terms of per cent data retrieval/throughput, CPU loading and power consumption for varying BLE connection intervals and numbers of sensors. Previously published research in [14] and [15] evaluated the performance of BLE in terms of throughput versus connection interval utilizing a single BLE sensor. Authors in [16] had presented a simulation tool that is capable of simulating multinodal BLE communication. The work in [17] described an accelerated connection establishment mechanism between the nodes in multinode BLE networks and validated the approach utilizing network simulation. The research in [18] described a new methodology for service mediation in multihop BLE networks based on named data networking approach. However, performance evaluation of BLE – implemented under an OS framework, such as Android, and in a multisensor scenario, remains open till date. To the best of our knowledge we believe that the evaluation of BLE in a real-world application scenario, involving multiple sensors and an Android client is being described for the first time in this article.

As an example platform for testing the multisensor BLE system, SmartStep [2,3,16] (a wearable monitor that is fully integrated into shoe insoles with BLE connectivity, that we have developed over the years), was used. Sections 4.2.1 details the application scenario under consideration and Section 4.2.2 describes the SASC methodology used to achieve Android multisensor BLE application along with the profiling tests conducted to evaluate the performance of the methodology.

4.2 Materials and methods

4.2.1 Application scenario

An application scenario in which several BLE sensors are worn on the body and there is an IoT Hub (Android device) that is interacting with these sensors, is considered. The sensors act as BLE server (or BLE peripheral device) and the IoT Hub acts as a BLE client (or BLE central device), similar to the single sensor case that we had presented earlier [3]. In the present context, the sensors are expected to come in and go out of the IoT Hub's wireless range together (as they are all worn on the body at the same time). When the sensors are worn on the body they collect data and whenever they are connected to the IoT Hub over BLE they transfer the collected data to the IoT Hub. The implementation of events that trigger the sensors to stop data collection and start advertising over BLE is left up to the application. One example of a scenario that can trigger this transition is when the users place their shoes with the SmartStep device on the charger. Other possible scenarios may include the usage of the inertial sensors to capture the time periods of sensor wear and decide on the required BLE actions (e.g., start advertising) accordingly.

4.2.2 Proposed SASC methodology

The following description relies on BLE terminology that is described in detail in [3]. In the proposed SASC methodology, every BLE action by the IoT Hub in a multisensor scenario is handled in a sequential manner to make sure that BLE actions across different services or within a particular service would not collide in time. Before initiating another BLE action, the SASC methodology makes sure that the previous action completes successfully. Successful completion is verified by waiting for the success flag in the callback methods for the particular BLE action.

For the sake of simplicity and clarity, we consider a scenario involving two sensors. Figure 4.1 shows the SASC system architecture, detailing the roles and responsibilities of the major components. As shown, UI_Activity implements functions such as initial scanning for sensors and selection of sensors that should automatically communicate with the IoT Hub. SASC comes into play after these specific sensors were selected. UI_Activity and the services that handle the sensors communicate via message broadcasting utilizing Android's 'sendBroadcast' method [19]. Sections 4.2.2.1–4.2.2.3 explain the implementation of SASC.

4.2.2.1 Connection establishment

After the desired sensors are selected by the user, the UI_Activity initializes the Android services to handle each of the sensors, and passes them the BLE address of their particular sensor. In this case of two sensors, let's call the two services Service1 and Service2, and the two sensors Sensor1 and Sensor2. Once initiated, these services start to scan for the particular BLE sensors as shown in Figure 4.2.

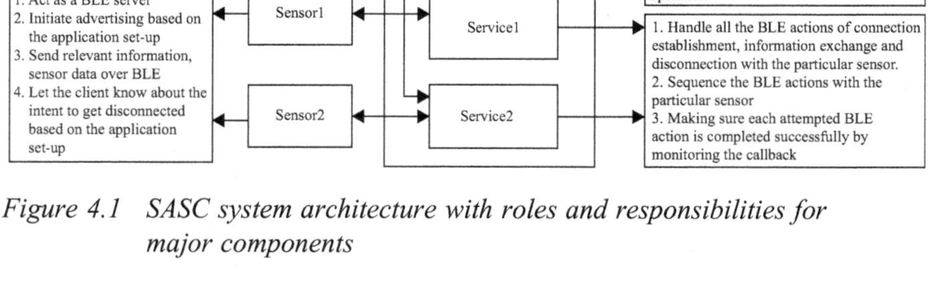

Figure 4.1 SASC system architecture with roles and responsibilities for major components

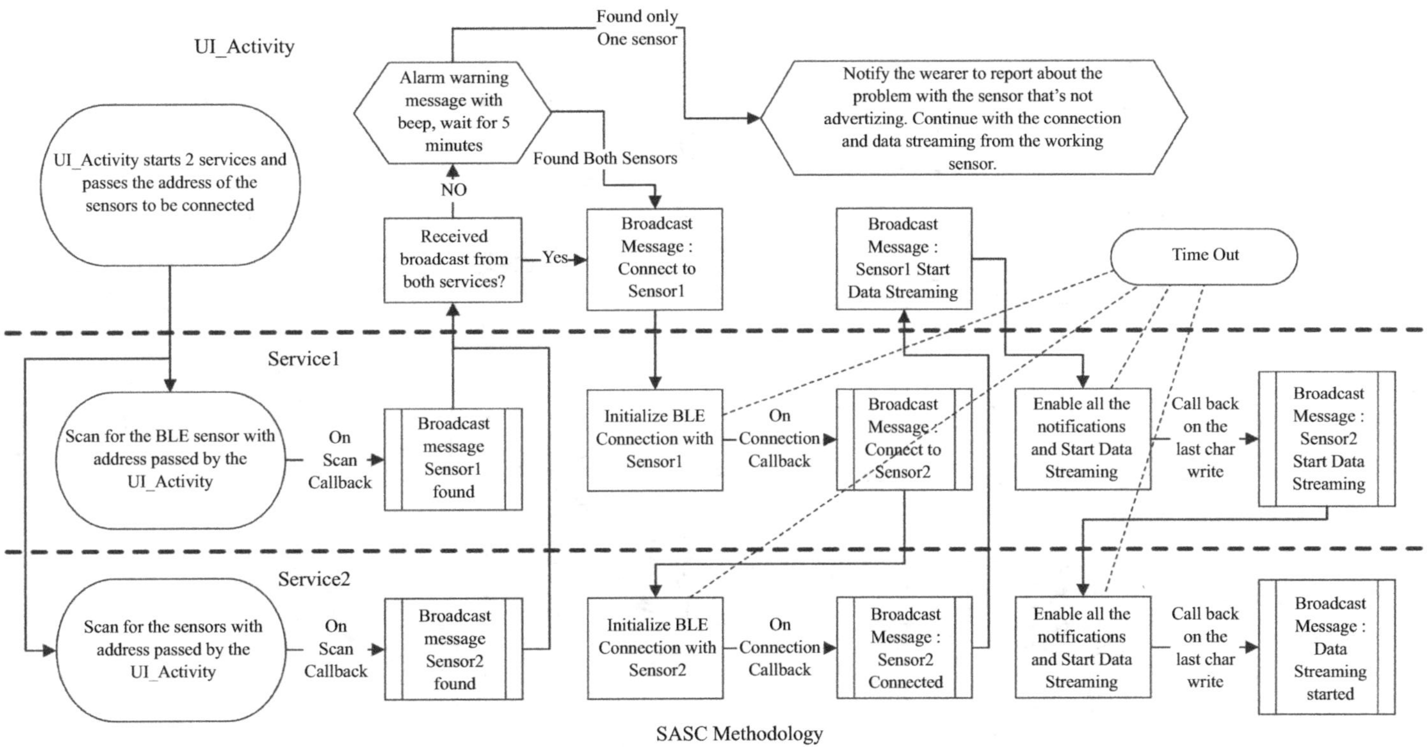

Figure 4.2 SASC – multisensor BLE Android methodology

When the scanning services find the appropriate sensors the scan callback function is executed. Inside the callback method, the services initiate a broadcast message that they have found the appropriate sensors. The connection establishment procedure will not begin until all the sensors are discovered by their respective services or the 5-minute time out occurs after one of the sensors is found. If all the sensors are discovered before the time out, the UI_Activity informs Service1 to connect to Sensor1. If for any reason, one of the sensors is not found within 5 minutes of finding the other sensors, the IoT Hub application will continue with the connection establishment and data streaming procedures with the sensor that is found. In such cases the UI informs the user about the failure to find a sensor by displaying appropriate messages.

As shown in Figure 4.2, inside the successful callback function for the connection event for Sensor1, the connection event for Sensor2 is triggered. Since there are two sensors in this scenario, successful connection with Sensor2 triggers the broadcast message to initiate the information exchange mechanism (in cases with more sensors, actions of connection establishment are carried out with each of the sensors before the characteristic writes are attempted). Unsuccessful connection establishment or notification enabling sequences are re attempted until a 5-minute time out. The time out paths are depicted by dotted lines in Figure 4.2.

4.2.2.2 Information exchange mechanism between the sensors and the IoT Hub

The information exchange between the IoT Hub and the sensors can happen via characteristic reads/writes or notification events. A characteristic write is performed by the IoT Hub to write a value to a characteristic on the sensor (e.g., current time). The IoT Hub waits for successful completion of the characteristic write operation before proceeding with any further BLE operations. As per SASC, if for any reason, a characteristic write is not completed, it is then performed again before attempting the next BLE action as detailed in Figure 4.3.

Sensors' operational status and data from the sensors are received in the IoT Hub via notification events. In the present application scenario, the data from the

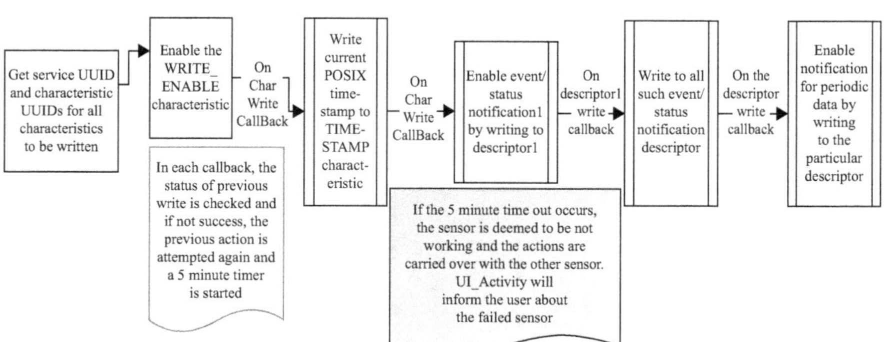

Figure 4.3 Example initialization sequence under SASC

sensors are sent via periodic notification events, handled by the notification event timers in the sensor's BLE stack. A descriptor write is performed by the IoT Hub to enable or disable notification events. While enabling or disabling notifications from the characteristics on the sensors, each service waits for the successful completion of the descriptor write operation in the corresponding callback method. Once again, as per SASC, if a descriptor write is not successfully completed, then it is attempted again before performing the next BLE action.

As shown in Figure 4.3, in the SASC method, all the special event notifications (which are not periodic in nature) are enabled before enabling the periodic data stream notification from any of the sensors. Also, all the necessary characteristic and descriptor writes are completed for all the sensors before starting periodic data stream notifications from any of the sensors.

4.2.2.3 Disconnection process

In the scenario where the IoT Hub and the sensors are 'always ON', the disconnection from the IoT Hub triggers the sensors to turn off BLE communication and start sensor data collection. The SASC methodology requires the sensors to request to be disconnected from the IoT Hub. Once the service handling a specific sensor receives the event notification notifying the sensor's wish to be disconnected, the service lets the UI_Activity know about it. When the UI_Activity approves the disconnection event, the service initiates descriptor writes to disable all the event notifications from the particular sensor. These descriptor writes are carried out in the same way as described in Section 2.2.2. After all the notification events of the particular sensor are disabled successfully following the SASC method, then the particular service triggers the disconnection event. After a successful disconnection event the UI_Activity is notified about the success. The service then enters a periodic scanning phase. Even if multiple disconnect notifications arrive at the same time, the disconnection events are handled in a sequential manner, as per the SASC methodology, and are sequenced by UI_Activity.

The method of disabling all notifying characteristics before a disconnection event with SASC ensures that the notification event timers are all halted and will not drain battery power during the sensor data collection period. When the sensors advertise the next time (as described in Section 4.2.1), the scanning services follow the same exact procedure of connection establishment and information exchange as shown in Figure 4.2.

4.3 Testing scenarios

4.3.1 Test sensor description

Our lab has been involved in the development of BLE-based wearable sensors for activity monitoring since some years now. These efforts have resulted in SmartStep, an insole-based activity monitoring system [2,3,16]. In this paper SmartStep is used as an example BLE sensor for testing of the SASC methodology. Each SmartStep sensor has an accelerometer and a gyroscope on the board as well as flash memory for optional storage of sampled sensor data. The BLE functionality on the device is implemented around Texas Instruments CC2540, using the Texas Instruments BLE stack 1.4.0 [20]. The sensor's BLE connection parameters and sampling frequency are configurable. Also the SmartStep's sensors can be used in real-time data streaming or in real-time data collection and offline data streaming as detailed in [21].

4.3.2 IoT Hub description

The IoT Hub was implemented as a 3D printed enclosure (Figure 4.4) that contained wireless Qi chargers for the SmartStep sensors and an Android smartphone that handled data collection from the SmartStep sensors. A smartphone was used because of the convenience of use, a functionally complete Android system containing the BLE, storage and a display screen. In general, the IoT Hub may utilize any functionally appropriate Android computer. The smartphone was a Google Nexus 5 with Qualcomm Snapdragon 800 chipset and quad-core 2.3 GHz Krait 400 CPU running under Android 6.0. The smartphone was powered via USB supply connected to a power outlet. The smartphone and wireless chargers operated independently, without direct communication; however, they interacted indirectly because SmartStep's firmware responded to wireless charging by starting the BLE advertising process.

4.3.3 Free-living validation of the proposed SASC methodology

The first test utilized the SASC methodology on two SmartStep systems in a real-world scenario. In this usage scenario, the user wears his/her shoes with the SmartSteps at the start of the day, taking the shoes from the IoT Hub, and performs all their daily activities. During wear, the SmartSteps' sensors are sampled at 50 Hz and the sampled data are logged in the SmartStep's flash memory. The BLE radio is turned off during this stage. At the end of the day, the user places the shoes on the IoT Hub for a wireless recharge of the SmartSteps' battery. When SmartStep detects the wireless charging it turns on BLE advertising, which allows the IoT Hub to discover

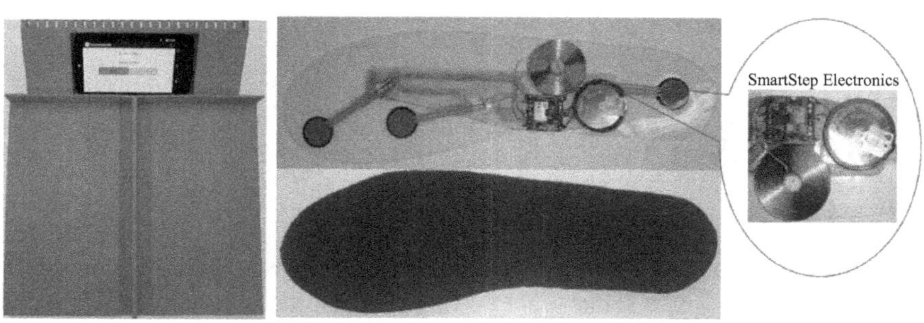

Figure 4.4 3D printed IoT Hub housing the smartphone and the SmartStep's sensor system

and connect to the SmartStep sensors. Once the connection is established, the SmartStep sensors transmit the data from their internal storage to the IoT Hub. After data transmission is complete, the IoT Hub transfers the data to a web server for further processing. To synchronize the data collected from the two SmartSteps for further signal processing, each SmartStep maintains time in POSIX format. The internal clock is corrected from the IoT Hub upon every BLE connection event.

The UI_Activity allows the user to configure the IoT Hub's settings for the first time. After the initial configuration, the IoT Hub works autonomously without user intervention. The UI_Activity displays relevant information and warning messages during operation. Figure 4.3 shows the different hardware system components.

The following experiment was conducted to verify the correct functionality of SASC methodology in handling multiple BLE sensors in a completely autonomous scenario, with no manual intervention during the course of action. The SmartSteps were initialized from the IoT Hub and were worn for a whole day by three subjects (two males with shoe size US M9 and one female with shoe size US W9). At the end of the day the shoes were placed on the IoT Hub, after which the data collected by the SmartStep insoles during the day were transferred to the IoT Hub. This procedure was repeated for 7 days by each of the subjects. At the end of the experiment, the data retrieved by the IoT Hub were analysed to determine the per cent data retrieval using the information from the POSIX timestamps in the SmartSteps. A quantitative measure of the per cent data retrieved (DR) by the IoT Hub was calculated for both sensors on each day using a MATLAB script. The metric was quantified as follows:

$$\mathrm{DRi} = (\mathrm{NTSi}/(\mathrm{LTSi} - \mathrm{FTSi} + 1)) \times 100 \tag{4.1}$$

where NTSi is the total number of timestamps received, LTSi is the last timestamp received and FTSi is the first timestamp received for the *i*th sensor ($i \in \{\text{left, right}\}$).

4.3.4 Benchmarking tests of the proposed SASC methodology

Even though modern Android devices are equipped with high-end CPUs and gigabytes of RAM, their wireless network interfaces are limited in the number of simultaneous connections that they can maintain, or the data rates that they can handle. The profiling tests were run to benchmark the SASC methodology for a varying number of simultaneous connections and BLE connection intervals for real-time streaming of sensor data. The throughput, the CPU loading and power consumption of the Android IoT Hub in real-time streaming mode (to be referred as RT_IoT Hub) were evaluated.

Theoretically, there is no set upper limit to the number of simultaneous connections a master can serve in BLE network [14], but the present day Android OS allows only seven [22]. Hence, we have carried out benchmarking tests that involve one to seven sensors, with connection intervals that range from 7.5 ms to 1,000 ms. The choice for the upper end of the connection interval range has been derived from the fact that, in general, BLE is meant for low-frequency data transmission, and 1 Hz is quite common in BLE applications (e.g., in a BLE heart rate monitor). The choice for the lower end of the connection interval is chosen since the shortest connection interval possible is 7.5 ms. The RT_IoT Hub's UI_Activity is designed to allow the user to select the desired number of sensors. Also, to help with running profiling tests for a particular amount of time, the UI_Activity, in this case, allows manual connect and disconnect functionalities.

4.3.4.1 Data retrieval and throughput evaluation

To evaluate the data retrieval and throughput of SASC methodology, RT_IoT Hub was used with different numbers of sensors and connection intervals. The connection intervals (frequencies) of 1,000 ms (1 Hz), 100 ms (10 Hz), 50 ms (20 Hz), 40 ms (25 Hz), 33 ms (33 Hz), 20 ms (50 Hz), 13.75 ms (72.5 Hz), 10 ms (100 Hz), 8.75 ms (114.3 Hz) and 7.5 ms (133.3 Hz) were used to test the throughput and the number of sensors varied from one to seven. The SmartStep sensors were used as standalone sensor modules and were loaded with real-time streaming firmware [21]. Each sensor sampled the accelerometer and gyroscope data at the specified sampling rate, the same as connection frequency, and along with the 4-byte timestamp, each sensor communicated a 16-byte data notification packet to the IoT Hub during every connection interval. Each test case was run for 10 minutes. Figure 4.5 shows the laboratory set-up of seven sensors with an Android phone used for the tests.

The data received in each of the cases were saved as a comma separated value file in the IoT Hub and was later used to determine the per cent data retrieval metric (DRi ($1<=i<=7$)), as described earlier. Average per cent data retrieved (PDR) for each case was obtained by averaging across DRi. The throughput for each case is computed using (2)

$$\text{Throughput} = (\text{PDR} * \text{NumCon} * \text{NBCI})/100 * \text{ConInt} \tag{4.2}$$

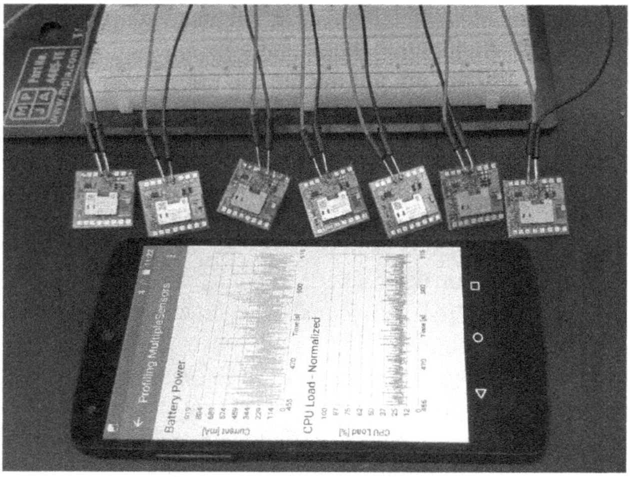

Figure 4.5 Laboratory set-up for profiling tests

where PDR is the average per cent data retrieved, NumCon is the number of simultaneous connections, NBCI is the number of bits in the user data per connection interval and ConInt is the connection interval used for the test.

Throughput calculations were made for two different scenarios. In the first, NBCI was kept at 128 (16 bytes of user data per connection interval), sent using one notification packet in every connection interval. These tests are run to emulate the real-world scenario of how SmartSteps are used in the field. However, the maximum allowed size of a notification packet is 20 bytes. Also, in TI CC2540 implementation of BLE, each connection event can carry up to 80 bytes of user data [23]. This means that each connection event can carry four notification packets when each packet is 20 bytes in length. Evaluating the performance of SASC at these limiting conditions can also help in comparing SASC's performance with other existing research for single sensor connections. Hence, a second set of throughput experiments was conducted across several sensors, with a connection interval of 7.5 ms and sending 80 bytes of user data per connection interval instead of 16.

4.3.4.2 CPU loading and power consumption profiling

The BLE stack at the BLE central side can be dealt either by the central device's main CPU or by the BLE transceiver on the central device as per the BLE specifications [24]. Since the central devices such as Smartphones are having high-end processing and computing abilities, the complete BLE host and application software stack can be implemented in Smartphone's CPU [25]. The Google nexus 5 utilizes BCM4339 (CYW4339) chipset which implements the BLE controller alone and not the host stack [26], and the Smartphone's CPU is involved in BLE communication directly. Hence it is beneficial to understand the CPU loading of the SASC methodology in a multisensor scenario. On the other hand, if the BLE host is implemented at the transceiver level, then profiling of CPU loading for multisensor BLE applications might not provide any useful information.

The profiling experiments were conducted to evaluate the CPU loading and power consumption in the Android device for different numbers of simultaneous BLE connections and varying BLE connection intervals. The power consumption by the BLE sensors themselves has been described in [3,21]. The Trepn Profiler from Qualcomm® [27] was used to profile both the CPU loading and power consumption on the smartphone while running the RT_IoT Hub application. The connection interval varied from 7.5 ms to 1,000 ms and the number of sensors varied from one to seven. The CPU loading and the power consumption by the RT_IoT Hub were recorded for each test case, which ran for 3 minutes (the time needed to obtain a stable CPU loading measure). During the profiling tests, it was made sure that no additional applications were running in the background which could potentially affect the results. Choice of expressing the CPU power consumption in mA was made because this gives a chance to calculate the battery run time immediately (mAh/mA).

4.4 Results

4.4.1 Free-living validation of SASC methodology

We had to exclude the data from 2 days of experiments from one of the subjects, as the left insole sensor had stopped functioning because of a sensor failure. For the remainder of the data (total of 19 days) the timestamp analysis showed that 100% of the data collected by the SmartStep sensors from all three subjects was successfully retrieved by the IoT Hub.

4.4.2 Per cent data retrieval and throughput evaluation

Average per cent data retrieved for the range of simultaneous connections and varying connection intervals are reported in Figure 4.6. With any connection intervals greater than or equal to 13.75 ms, 100% of the data were retrieved from the sensors by the application. The number of simultaneous sensor connections did not have an impact on the data retrieval for these connection intervals. However, for intervals shorter than 13.75 ms, a drop in the per cent data retrieval was observed. Computed throughputs are shown in Table 4.1.

4.4.3 CPU loading and power consumption profiling

Per cent CPU loading for different numbers of simultaneous connections and varying connection intervals are reported in Figure 4.7. The results show that as the number of sensors increases or as the connection interval decreases, the CPU loading of the RT_IoT Hub increases, which is to be expected.

Power consumption of the multisensor application for a different number of simultaneous connections and varying connection intervals are reported in Figure 4.8. The results suggest that when the number of sensors increased or when the connection interval decreased, the power consumption of the RT_IoT Hub increases.

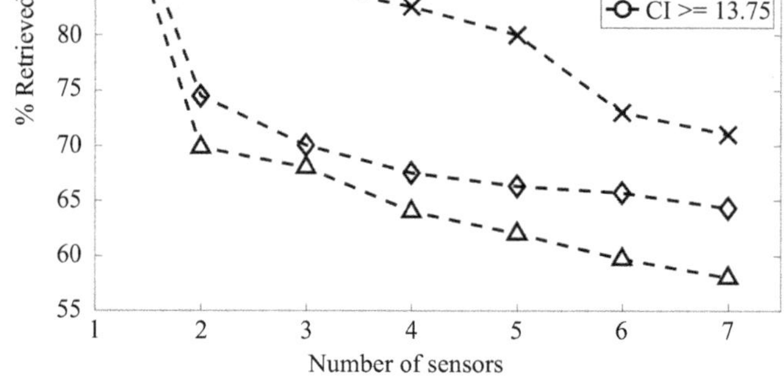

Figure 4.6 Per cent data retrieved by the IoT Hub's smartphone

Table 4.1 Computed BLE throughput (kilobits per second) with 16-byte application data per connection event

NS CI (ms)	1	2	3	4	5	6	7
1,000	0.12	0.25	0.38	0.51	0.64	0.76	0.89
100	1.28	2.56	3.84	5.12	6.40	7.68	8.96
50	2.56	5.12	7.68	10.2	12.8	15.3	17.9
40	3.20	6.40	9.60	12.8	16	19.2	22.4
33	3.87	7.75	11.63	15.5	19.3	23.2	27.1
20	6.40	12.80	19.20	25.6	32	38.4	44.8
13.75	9.30	18.61	27.92	37.2	46.5	55.8	58.5
10	12.80	21.65	32.44	42.2	51.2	56.0	63.6
8.75	14.57	21.79	30.72	39.4	48.4	57.6	65.8
7.5	16.96	23.82	34.81	43.6	52.9	61.1	69.2

NS – number of simultaneous connections; CI – connection interval.

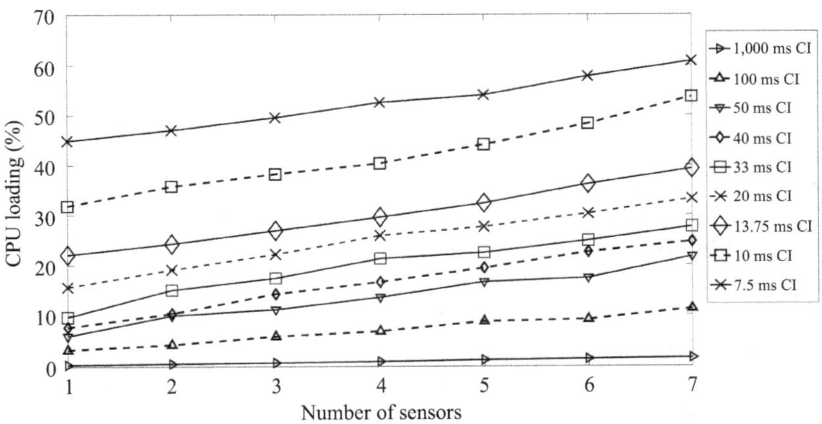

Figure 4.7 CPU loading on the IoT Hub's smartphone

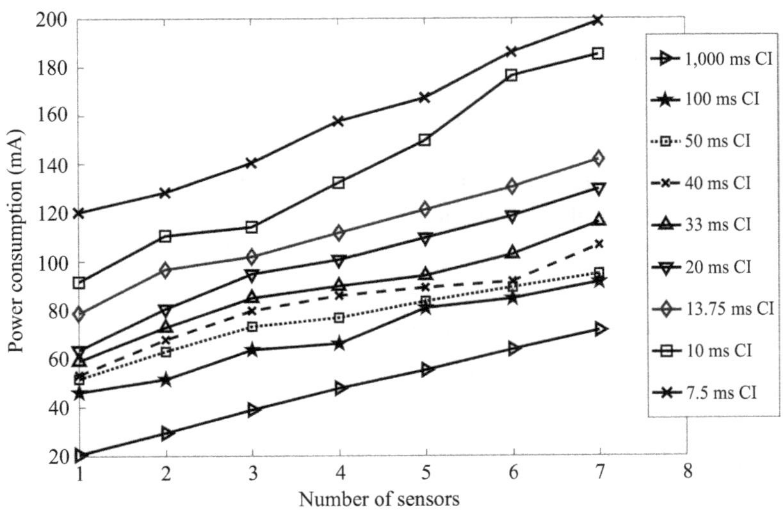

Figure 4.8 Power consumption in the IoT Hub's smartphone

4.5 Discussion

This work introduced and evaluated an Android application methodology that dealt with multiple BLE sensors. The human subject tests were run in real-life environments and the IoT Hub was able to retrieve all the data collected during a total of 19 days of experiments with three test subjects. The IoT Hub also could act upon the special event notifications and sensor status information correctly. These results suggest that the use of the proposed SASC methodology produces a reliable and robust multisensor Android application.

The results of the data retrieval tests (Figure 4.6), under general application scenario (16 byte of application data per connection interval) showed that in the cases with connection intervals less than 13.75 ms, for single sensor the retrieval rate is 100% but drops to 71% gradually as the number of simultaneous connections increases to seven sensors. At these connection intervals, the IoT Hub might be facing a lack of resources to deal with the many simultaneous connections. A closer look at the data received from each of the sensors revealed that for some of the sensors the retrieved data were close to 100% and for some others much less (sometimes around 30%) bringing the average down. Between a connection interval of 13.75 ms and a connection interval of 7.5 ms, there was a difference of 42% in the amount of retrieved data for the seven-sensor case. The data retrieval may be also dependent on the specific Android device (Google Nexus 5), or the BLE chipset used in the sensor (Texas Instruments CC2540). However, based on these results, it may be advisable to use a connection interval of 13.75 ms or greater if the applications intend to retrieve all the data.

As per the specifications of BLE v 4.0, the maximum data throughput is 270 kbps [24]. However, [14] reported 58.48 kbps in a realistic scenario involving the same CC2540 BLE solution from TI with 7.5 ms connection interval. However, under the Android OS framework and SASC methodology, our conducted tests demonstrated a maximum throughput of 19.1 kbps for single sensor connection (Table 4.2). The limitation may be because of the current implementation of BLE in the Android OS, the OS overhead, the application overhead or the Android device that was used. This result is in line with what is predicted in [28], under a smartphone framework. For a multisensor connection, the maximum achieved throughput from the SASC was ~70 kbps. To the best of our knowledge, there is no other published state-of-the-art research, dealing with BLE multisensor application running in Android. Hence, we have used the BLE specification in [24] and the research in [14] that deals with single sensor connection for comparison purposes.

Table 4.2 BLE throughput (kilobits per second) obtained with 80 bytes of application data per connection interval

NS 7.5 ms CI	1	2	3	4	5	6	7
Achieved throughput	19.1	32.5	47.4	56.7	62.6	67.5	73.4

In the TI BLE solution that we have used, given one notification packet per connection event, each connection event lasts approximately 3 ms [29]. Considering a case with 1,000 ms connection interval and a single sensor, there would be one connection event every second that the smartphone CPU needs to handle. In this case, smartphone CPU should be active for ~3 ms in each second, resulting in 0.3% CPU loading. In support of this argument, the observed loading for this case was 0.3%. This argument is valid for all the connection intervals tested, for single sensor case. However, for multiple sensors, this argument doesn't hold, and the reason can be that the Smartphone's CPU is a multicore system and is managed by an OS. The focus of this work was to propose and evaluate the performance of SASC and a more detailed study and analysis is required to identify the reasoning behind the CPU loading in a BLE multisensor scenario. Based on these results, it may be advisable to use a connection interval of 13.75 ms or greater to keep the CPU loading figures low. These results can aide the application designers in choosing the optimum number of sensors and connection intervals for specific applications, based on the CPU loading that they can afford for such applications.

The results of the power consumption tests (Figure 4.8) suggest that a connection interval of 10 ms or 7.5 ms causes the IoT Hub to consume considerably higher power. Looking at Figure 4.8, for single sensor case, at 7.5 ms, the power consumption is at 120 mA, while the same at 13.75 ms is 80 mA. This makes the power consumption at 7.5 ms case to be ~1.5 times that of 13.75 ms case. Across any number of sensors, the power consumption for the 7.5 ms case is approximately 1.5 times of that for the 13.75 ms case. High power consumption at short connection intervals (less than 13.75 ms) may be attributed to the BLE stack overhead. Based on these results, it may be advisable to use a connection interval of 13.75 ms or greater to keep the power consumption figures low.

We are utilizing notification methodology for data retrieval from SmartSteps, as was detailed in this chapter. As per this method, there will not be any application-level acknowledgement from the client (IoT Hub) to the server (SmartSteps). There is 'Indication' methodology in BLE, which when used will deal with application-level acknowledgement, while potentially reducing the throughput. As evident from the reported results, the packets get lost at the client side as the number of sensors is increased at the connection intervals less than 13.75 ms (Figure 4.6). Since we are more concerned with application-level throughput from the client's side, the PDR computation as performed is illustrative.

The presented methodology of computing per cent data retrieval had a limitation. In a case wherein only L last timestamps out of total N are received ($N-L$ being the loss), the described method of computing per cent data retrieval might still compute 100% data retrieval as it relies entirely on First and Last timestamps received. However, in real-life use cases, unless there is at least one connection event occurring before the time out (10 s in our case), BLE connections can't be alive. Hence, at the max, initial 9.99 s of data loss could be missing from consideration, out of the life cycle of the connection (which could span across many hours in real-life usage scenarios), which is tolerable. With our experimental knowledge, the above method of DRi has worked well in realistic situations.

The OS of the smartphone device has a major impact on the testing and operation of the system. Present day Android OS's limitation of maximum seven simultaneous BLE connections might be changed in future versions of the OS. Also, the reported SASC methodology has been validated only for Android OS and may need modifications for iOS or Windows mobile.

During all the conducted tests, the sensors and IoT Hub were placed statically without any movement during BLE connections and the sensors would never go out of IoT Hub's BLE range during data streaming session. If the target application demands BLE streaming functionality with moving sensors and/or IoT Hub, the system needs to account for intermittent connectivity. A small amount of data loss can be anticipated in that case during intermittent connections. The sensor and the Android application need to be tracking a possible disconnection due to packet losses (leading to eventual time outs) in such situations. Though SASC methodology doesn't deal with this scenario at present, a small amount of data loss can be anticipated in that case during intermittent connections. A similar scenario dealing with a single sensor was addressed in our previous work [3] and the methodology for such a system in a multisensor scenario can be explored in the future. Also, the proposed SASC methodology dealt with connection establishment, information exchange and disconnection processes in multisensor BLE scenarios. Efficient device discovery during the scanning is a valid problem on its own in a multisensor scenario that can be explored in future works.

We had used a CC2540-based solution from TI as an example platform in our experiments, as the SmartStep devices are built around this solution. However, there are new solutions from TI, namely CC2640, built around an ARM cortex M3 core [30]. These are supposed to be lower power and more robust, compared to the 8051-core-based CC2540. However, from SASC implementation perspective, which is meant for the Android BLE client, changing the BLE server chip won't demand a change in the presented methodology. Also, specifications for a newer version of Bluetooth, Bluetooth 5.0, were recently released [31], which could open up other possibilities, such as longer range for BLE in the future.

4.6 Conclusion

The article detailed the challenges in implementing a multisensor Android BLE application and a new methodology – SASC – was proposed to address these challenges. Benchmarking tests and free-living tests were conducted to validate the SASC implementation schemes. The benchmarking tests gave quantified results on metrics such as throughput, CPU loading on the smartphone and power consumption by the Android application for various number of sensors and varying connection intervals. The results of Android application throughput evaluation suggest that the maximum achievable throughput is ~20 kbps for a single sensor connection and ~70 kbps for a seven-sensor connection under SASC methodology. The results of the CPU loading and power consumption tests suggest that CPU loading and power consumption figures can be kept low by using a connection interval of 13.75 ms or greater. The proposed SASC methodology was also used in a real-life use case for gait data collection using SmartStep sensors, and a total of 19 days of human subject tests resulted in 100% data retrieval. These results suggest that the proposed methodology can serve as a foundation for applications that require managing multiple BLE connections in the world of IoT.

Authors' contributions

NH and ES designed the research. NH and ES contributed in designing the hardware. NH contributed in the firmware design and Android application implementation. NH and MB performed the testing. NH, MB and ES contributed in writing and editing of the manuscript.

Conflicts of interest

The authors declare no conflict of interest.

References

[1] Home. Bluetooth Technology Special Interest Group [Internet]. [cited 14 April 2015]. Available from https://www.bluetooth.org/en-us

[2] Sazonov ES, Hegde N, Tang W. Development of SmartStep: an insole-based physical activity monitor. *Conf Proc IEEE Eng Med Biol Soc*. 2013;2013: 7209–7212.

[3] Hegde N, Sazonov E. SmartStep: A Fully Integrated, Low-Power Insole Monitor. *Electronics*. 2014; 3(2):381–397.

[4] Bluetooth Low Energy. Android Developers [Internet]. [cited 1 May 2015]. Available from https://developer.android.com/guide/topics/connectivity/bluetooth-le.html

[5] Darwish A, Hassanien AE. Wearable and Implantable Wireless Sensor Network Solutions for Healthcare Monitoring. *Sensors*. 2011;11(6):5561–5595.

[6] Schonle P, Bunjaku P, Fateh S, Huang Q. Modular multi-sensor platform for portable and wireless medical instrumentation. In: 2014 IEEE Biomedical Circuits and Systems Conference (BioCAS). Lausanne, Switzerland, 22–24 October 2014. pp. 165–168. doi: 10.1109/BioCAS.2014.6981671

[7] Brunelli D, Farella E, Rocchi L, Dozza M, Chiari L, Benini L. Bio-feedback system for rehabilitation based on a wireless body area network. In: Fourth Annual IEEE International Conference on Pervasive Computing and Communications Workshops (PERCOMW'06). 13–17 March 2006, Pisa, Italy. doi: 10.1109/PERCOMW.2006.27

[8] Hegde N, Fulk GD, Sazonov ES. Development of the RT-GAIT, a Real-Time feedback device to improve Gait of individuals with stroke. In: 2015 37th Annual International Conference of the IEEE Engineering in Medicine and Biology Society (EMBC), 25–29 August. 2015, Milan, Italy. pp. 5724–5727. doi: 10.1109/EMBC.2015.7319692

[9] Johansson A, Shen W, Xu Y. An ANT Based Wireless Body Sensor Biofeedback Network for Medical E-Health Care. In: 2011 7th International Conference on Wireless Communications, Networking and Mobile Computing (WiCOM), 10 October 2011, Wuhan, China. pp. 1–5. doi: 10.1109/wicom.2011.6040656

[10] Fulk GD, Sazonov E. Using sensors to measure activity in people with stroke. *Top Stroke Rehabil.* 2011;18(6):746–757.

[11] Najafi B, Vaziri A, Boloori A-R. Ambulatory system for measuring and monitoring physical activity and risk of falling and for automatic fall detection [Internet]. US9005141 B1, 2015 [cited 7 December 2015]. Available from http://www.google.com/patents/US9005141

[12] Hao Y, Foster R. Wireless body sensor networks for health-monitoring applications. *Physiol Meas.* 2008;29(11):R27–R56. doi: 10.1088/0967-3334/29/11/R01

[13] Baldus H, Klabunde K, Müsch G. Reliable Set-Up of Medical Body-Sensor Networks. In: Karl H, Wolisz A, Willig A, editors. Wireless Sensor Networks [Internet]. Springer Berlin Heidelberg; 2004 [cited 17 April 2015]. pp. 353–363. (Lecture Notes in Computer Science). Available from http://link.springer.com/chapter/10.1007/978-3-540-24606-0_24

[14] Gomez C, Oller J, Paradells J. Overview and Evaluation of Bluetooth Low Energy: An Emerging Low-Power Wireless Technology. *Sensors.* 2012;12(9):11734–11753.

[15] Mikhaylov K, Plevritakis N, Tervonen J. Performance Analysis and Comparison of Bluetooth Low Energy with IEEE 802.15.4 and SimpliciTI. *J Sens Actuator Netw.* 2013;2(3):589–613.

[16] Mikhaylov K. Simulation of network-level performance for Bluetooth Low Energy. In: 2014 IEEE 25th Annual International Symposium on Personal, Indoor, and Mobile Radio Communication (PIMRC), 2–5 September 2014, Washington, DC, USA. pp. 1259–1263. doi: 10.1109/PIMRC.2014.7136361

[17] Mikhaylov K. Accelerated Connection Establishment (ACE) mechanism for Bluetooth Low Energy. In: 2014 IEEE 25th Annual International Symposium on Personal, Indoor, and Mobile Radio Communication (PIMRC), 2–5 September 2014, Washington, DC, USA. pp. 1264–1268. doi: 10.1109/PIMRC.2014.7136362

[18] Balogh A, Imre S, Lendvai K, Szabó S. Service Mediation in multihop Bluetooth Low Energy networks based on NDN approach. In: 2015 23rd International Conference on Software, Telecommunications and Computer Networks (SoftCOM), 16–18 September 2015, Split, Croatia. pp. 285–289. doi: 10.1109/SOFTCOM.2015.7314123

[19] Context. Android Developers [Internet]. [cited 31 January 2017]. Available from https://developer.android.com/reference/android/content/Context.html#sendBroadcast(android.content.Intent,%20java.lang.String)

[20] TI BLE Stack [Internet]. Bluetooth low energy software stack and tools. Available from http://www.ti.com/tool/ble-stack

[21] Nagaraj Hegde, Edward Sazonov. SmartStep 2.0 – A completely wireless, versatile insole monitoring system. In: 2015 IEEE International Conference

on Bioinformatics and Biomedicine (BIBM) [Internet]. 2015. pp. 746–749. Available from https://www.researchgate.net/publication/284229018_Smart-Step_2.0__A_completely_wireless_versatile_insole_monitoring_system

[22] Git at Google [Internet]. [cited 10 December 2015]. Available from https://android.googlesource.com/platform/external/bluetooth/bluedroid/+/8dc42c603f5571047878640705eee18fa5396f11%5E%21/

[23] Max packet size allowed for BLE communication - Bluetooth® low energy Forum - Bluetooth® low energy - TI E2E Community [Internet]. [cited 1 February 2017]. Available from http://e2e.ti.com/support/wireless_connectivity/bluetooth_low_energy/f/538/p/108807/384844

[24] Specification. Adopted Documents. Bluetooth Technology Special Interest Group [Internet]. [cited 16 December 2015]. Available from https://www.bluetooth.org/en-us/specification/adopted-specifications

[25] Heydon, Bluetooth Low Energy: The Developer's Handbook [Internet]. [cited 28 March 2017]. Available from https://www.pearsonhighered.com/program/Heydon-Bluetooth-Low-Energy-The-Developer-s-Handbook/PGM127761.html

[26] CYW4339 Single-Chip 5G WiFi IEEE 802.11ac MAC/Baseband/Radio with Integrated Bluetooth 4.1 and FM Receiver | Cypress Semiconductor [Internet]. [cited 28 Mar 2017]. Available from http://www.cypress.com/documentation/datasheets/cyw4339-single-chip-5g-wifi-ieee-80211ac-macbasebandradio-integrated

[27] Trepn Profiler [Internet]. Qualcomm Innovation Center, Inc.; 2015 [cited 11 July 2015]. Available from https://play.google.com/store/apps/details?id=com.quicinc.trepn

[28] Davidson KT Carles Cufí, Akiba, Robert. Getting Started with Bluetooth Low Energy [Internet]. [cited 6 July 2017]. Available from http://shop.oreilly.com/product/0636920033011.do

[29] Measuring Bluetooth® Low Energy Power Consumption [Internet]. Available from http://www.ti.com/lit/an/swra347a/swra347a.pdf

[30] CC2640 SimpleLink ultra-low power wireless MCU for Bluetooth low energy | TI.com [Internet]. [cited 1 February 2017]. Available from http://www.ti.com/product/CC2640/description&lpos=Middle_Container&lid=Alternative_Devices

[31] Bluetooth 5. Bluetooth Technology Website [Internet]. [cited 1 February 2017]. Available from https://www.bluetooth.com/specifications/bluetooth-core-specification/bluetooth5

Chapter 5

Fog computing middleware for distributed cooperative data analytics

Maria Valero[1], *Jose Clemente*[1], *and WenZhan Song*[1]

The Internet of things (IoT) has experienced an exponential growth in the past few years with billions of devices connected to the Internet. At the same time, the storage and the computation capabilities of these devices approximately double every 18 months to support intensive data analytics, while the communication bandwidth does not grow at the same speed and is limited by the spectrum availability. In applications over distributed environments (such as sensor networks), bandwidth limitations make it infeasible to send all data to a central place (e.g., cloud) for post-processing. Furthermore, latency becomes an issue when IoT systems need to transfer data in real time. To overcome the bottlenecks of bandwidth and latency deficiencies and take full advantage of powerful computation of current sensor devices, the fog computing (also called fogging or edge computing) paradigm was introduced. This paradigm brings data processing, networking, storage and analytics closer to the devices and applications.

In this chapter, we introduce you to a novel fog computing middleware to support scalable and flexible distributed cooperative data analytics (DCDA) at the edge of a sensor network. The proposed middleware allows processing, computing, analysing and sharing information at the edge for high-level situation awareness. The proposed fog computing middleware provides services that allow the analysis of the most time-sensitive data at the network edge, minimize latency, conserve network bandwidth and allow collecting and processing data across a wide geographic area under different environmental conditions. This chapter is structured as follows: first, we introduce key challenges and solutions for performing distributed data analytics over sensor networks. We then discuss the architecture and model formulation of the middleware for in-network analytics and cooperation between nodes. Next, we introduce you to some candidate distributed applications for fog computing middleware and present two study cases for seismic imaging using sensor networks. We conclude this chapter with some results and opportunities of the middleware for scalable fog computing systems.

[1]Center for Cyber-Physical Systems, College of Engineering, University of Georgia, Athens, GA 30602, USA

5.1 Introduction

The demand for Internet throughput can only increase as the number of Internet-enabled devices continues to grow and this trend, according to [1] and other similar studies, will continue. Currently, many of these new devices are acting as dumb portals to a scalable cloud back-end; but as the number of devices increases and the bandwidth available for each device diminishes, we may reach a point where connecting to the cloud is no longer feasible (Figure 5.1). One way this could occur is if the data to be processed exceeds available bandwidth; for example, an autonomous car may generate up to 1 GB of data per second [2] but currently, long-term evolution (LTE) can only handle a max upload speed of 5 MB/s. The cloud connections are also not feasible for applications with stringent latency requirements, like the smart grid and many other industrial control systems [3]. The only options to resolve these problems are to increase the bandwidth to the cloud or to improve bandwidth utilization. Local computation resources, if available, can process the data streams closer to the physical phenomena leading to a reduction in the required bandwidth [4]. Edge computing, cloudlets, fog computing and cyber-foraging are some of the terms that are used for mechanisms built upon this notion of dual-purpose sensing and computation nodes at the edge in the city, closer to the physical phenomenon being observed or analysed.

Fog computing presents paradigm-shifting research challenges because the system is built from devices, which are distributed across large spatial regions, and may suffer from fluctuating connectivity among the modules. In this chapter, we introduce the concept of decentralized cooperative data analytics middleware as an alternative to process information over sensor networks in the era of IoT. This middleware aims to answer common questions of nowadays fog computing: (1) What are the ways to design distributed applications such that they can perform resiliently and adapt in response to resource fluctuations? (2) Given a set of fog computing resources and a set of analytical tasks, which operate on spatially correlated data, how should we assign the tasks to the resource to minimize performance uncertainty? and (3) How can we make the applications running on this decentralized architecture resilient in response to failures?

The proposed DCDA middleware manages three different levels of data processing and analytics for the cooperation among edge devices connected to the network. These levels are illustrated in Figure 5.2. On the low level, DCDA middleware

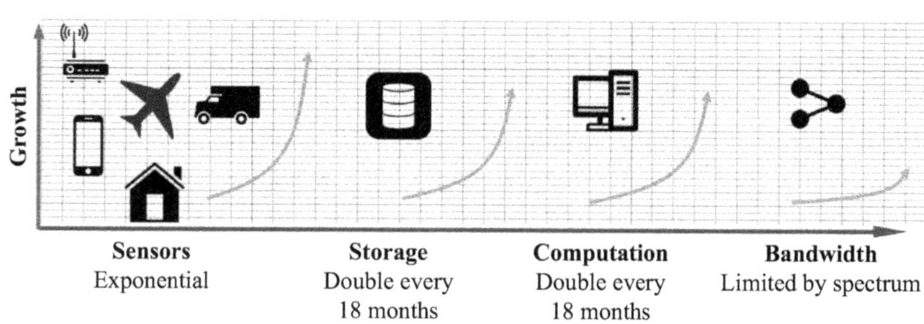

Figure 5.1 IoT components growth comparison

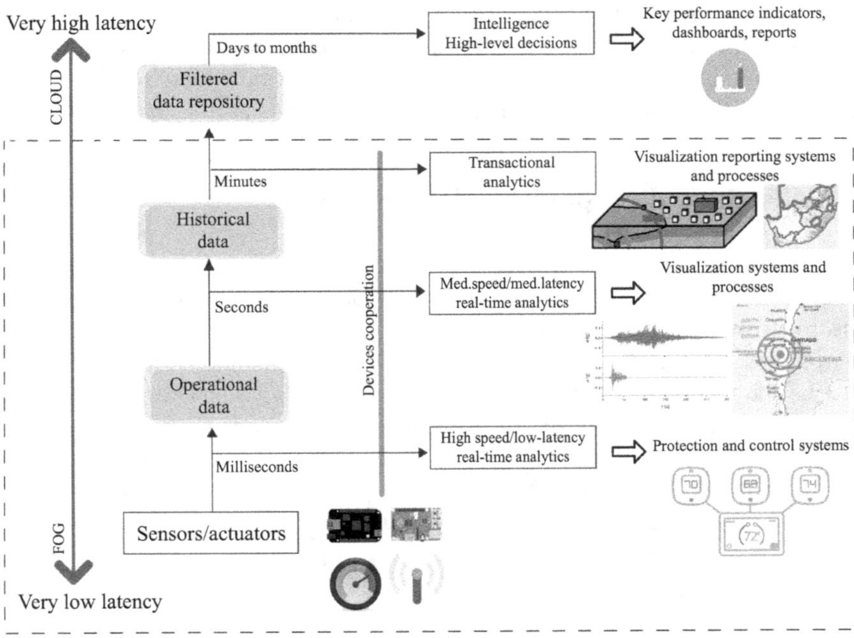

Figure 5.2 Levels of data processing and analytics inside DCDA middleware

intends to generate and analyse real-time data for protection and control systems, for example, alerts and notifications that require immediate actions in the network. On the intermediate level, DCDA handles operational data, as well as historical data to generate real-time analytics for visualization systems and processes. An example is the real-time monitoring of data on smart grids, connected vehicles, seismic imaging, etc. Finally, on a high level, the middleware aims to generate transactional analytics for the decision-making process, for example, real-time 2D and 3D visualization, events detection in the network and compromised functionality.

DCDA middleware enables a mesh network composed of many devices to share resources and capabilities. Every edge device, here called node, is unique and independent of other nodes. Every node handles a set of functionalities to address different problems and its own database stores compressed data. Nodes may cooperate with others to solve a global optimization problem when they are sharing a task (*task-sharing mode*), or provide determined services to other nodes when they are working and cooperating between them (*cooperation mode*). For instance, two nodes might work together to solve a linear least-squares system of equations by sharing and analysing information among them. On the other hand, one node (A) might provide assistance to another node (B) that needs to use a specific function that is available in node A. These mechanisms provide a comprehensive fog middleware for distributed applications that focus on data analytics.

In the literature, there exist research studies that focus on distributed computation for solving problems at the edge of the network. We have pioneered the development of such computing methodology for seismic applications [5–10].

These works present a generalized middleware framework that allows applications to execute different kinds of algorithms, communicate partial results with neighbour nodes, share resources among them and solve a wide variety of problems by guaranteeing the quality of service [11–13] and analytics at the edge [14,15].

5.2 Distributed cooperative data analytics

Cooperation between edge devices has become a topic of significant interest in recent years [15,16]. Challenges of distributed computation and decision-making in sensor networks are not trivial because of potentially harsh, uncertain and dynamic environments, along with energy and bandwidth constraints. In this section, we describe the main challenges of developing a distributed cooperative middleware and how we propose data analytics at the edge of a network.

5.2.1 Challenges and solutions

In the design of a middleware for distributed cooperative solutions in a network, multiple challenges have to be considered. First, in cooperative signal and information processing, we have to consider the trade-off between performance and resource utilization. Sharing more data usually provides high performance in distributed algorithms but also requires more communication resources (energy and bandwidth). Second, the information shared between one node and another needs to be combined with local information in a certain way. This fusion of information may range from simple rules to model-based techniques. Finally, another important issue in cooperative analytics is visualization. Users should be able to see in real time what is happening at the edge of the network and how partial results are contributing to final results.

The DCDA middleware proposed in this chapter attempts to overcome the aforementioned challenges by using a mechanism for exchanging processed information instead of raw data, allowing different types of algorithms that combine information from different nodes and provide a simple interface to interactively query the nodes. Also, the proposed DCDA provides an architecture in which every node has local databases for raw and processed data allowing fast querying and information visualization in real time. We also design an intuitive graphic interface that helps in monitoring the edge network.

5.3 Fog computing middleware architecture

The design of the fog computing middleware architecture is based on the accomplishment of the following requirements: (1) a modelling and design environment for developing and evaluating multimodal applications that can adapt to the resource variations; (2) decentralized dynamic discovery of resources and optimized placement of multimodal components in fog computing networks at runtime; (3) allowing decentralized cooperative computing algorithms that can achieve global optimization

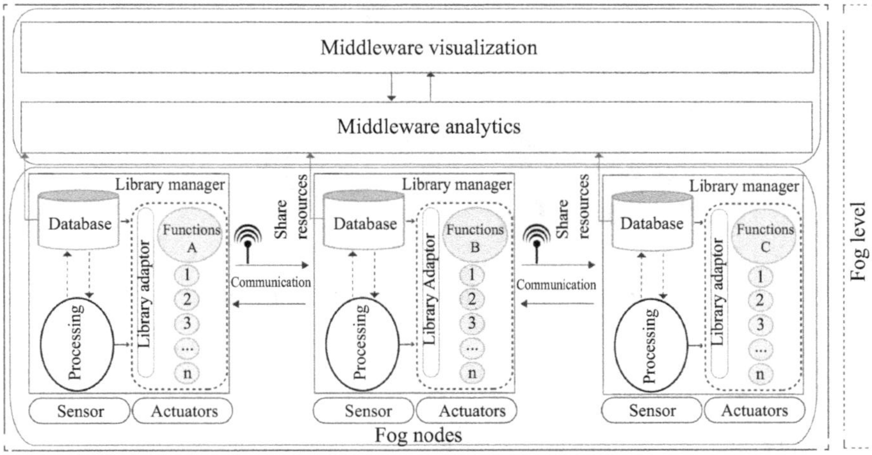

Figure 5.3 Fog middleware architecture

while self-adapting to the fluctuating network and computing resources and (4) resilience to tolerate faults.

The general architecture of the proposed middleware is shown in Figure 5.3. Every node has a relatively small but powerful computational unit (e.g., Beaglebone Black, Raspberry Pi). Nodes are able to store raw and processed information in the database that is designed for compressed data. Inside each processing unit, nodes are able to manage input data, generate real-time information and analyse and develop analytics. Every node is also capable to communicate and cooperate with neighbours using wireless, XBee or Bluetooth. Additionally, DCDA middleware allows users to observe in real time the State of the Network by a simple visualization and analytics interface.

5.3.1 Database

We use a SQL-based database at every node of DCDA middleware. However, any other database can be easily adapted to our middleware. We have one database for storing stream data (transactional database) and another for processed data (analytic database) as it is illustrated in Figure 5.4. The transactional database stores the node measurements in real time. Tables in the transactional database contain a field called *traceBuf* that saves a binary array with a header, a buffer structure and the data itself. The binary data are compressed using *zlib* library achieving a reduction in the size of 80% compared to original raw data. The analytic database is configurable; more types of data can be added to allow specific analytics that the node wants to perform over them. For example, we can create a table to store the location and/or magnitude of an earthquake that has been detected by a mesh of sensors deployed in a wide geographical area. Furthermore, the DCDA middleware database is flexible to store any kind of analysed data that comes from different types of analytic algorithms.

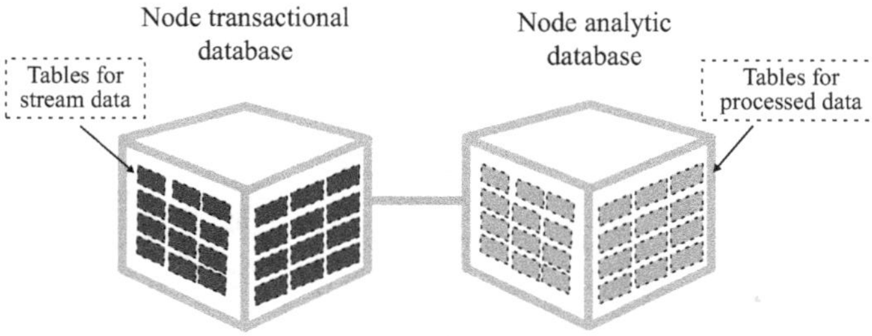

Figure 5.4 Fog node database structure

5.3.2 Library manager and adaptor

The library manager within a fog node is responsible for ensuring the correct algorithm selection in the *Cooperation mode*. For example, if DCDA is configured to execute an algorithm for solving a distributed optimization problem, the library manager orders the processing unit to solve the problem and communicate the results with other neighbours at each iteration. The library adaptor is mainly used in the *Task-sharing mode* to select which library and functions will be used to cooperate with other nodes.

5.3.3 Processing unit

The processing unit is responsible for executing a determined algorithm within the fog node. This unit is also in charge of communicating with neighbours to share information at each iteration or communicate the final results.

5.3.4 Middleware visualizer

We integrate a visualization tool to DCDA middleware in order to monitor results in real time. This tool is based on a Java Servlet framework that is capable of (1) connecting via IP addresses to different edge devices to visualize node states, characteristics and data; (2) reading directly from the device database to perform off-line data analysis; (3) visualizing partial and final results of algorithms that are performed on different nodes in real time and (4) visualizing data analytics over the network. This tool is expandable to incorporate more features using a simple language such as Java.

5.4 Fog computing middleware formulation and modes of operation

DCDA middleware model is based on a mesh network (F) composed of fog nodes (Υ). Υ_j represents a specific fog node j in F, where $j = 1, .., m$ and m is the total number of nodes.

Definition 1 We define Υ_j as a finite sequence of terms such as

$$\Upsilon_j = \left(I_j, S_j, U_j, \tau_j, \Lambda_j, \Gamma_{j[p]}\right),$$

where I_j is the unique id number of Υ_j in F, and S_j is the state of the node Υ_j which indicates Υ_j is joined in the distributed computation. Notice that S_j is useful to determine if the node is up or down in certain stage of the computation. Furthermore, U_j is the degree or number of neighbours of Υ_j, τ_j denotes the timestamp (specific hour, minute, second and millisecond in which the datum is saved or recorded), Λ_j represents hardware information of Υ_j, and finally, $\Gamma_{j[p]}$ represents a list of p libraries available inside Υ_j.

Definition 2 The hardware information Λ_j of each fog node Υ_j is defined as a finite number of elements that describe the current status of processor, memory and battery, such as

$$\Lambda_j = (P_j, M_j, B_j, \rho_j, \alpha_j, C_j),$$

where P_j is the total capacity of the processor in Υ_j in Hz, M_j denotes the total memory in Υ_j by time τ_j in Gb, B_j is a Boolean variable to indicate if the battery is enough according to a knowing threshold and ρ_j is the available percentage of the processor in Υ_j by time τ_j. α_j is the percentage of memory available in Υ_j by time τ_j, and finally, C_j denotes the communication type (for instance, Wi-Fi, Bluetooth, XBee, etc.)

Definition 3 We define Γ_j as the specific library function available in a node Υ_j,

$$\Gamma_j = (\psi_j, \kappa_j, \mu_j, \omega_j),$$

where ψ_j is the *id* of the library, κ_j and μ_j are Boolean parameters that indicate the library is in the node Υ_j and it is available for computations, respectively, and ω_j denotes the type of library (mathematical operation, filter, image processing, etc.).

Nodes in the proposed middleware can work together on the same task (*cooperation mode*) or can work among themselves to provide services to each other (*task-sharing mode*).

5.4.1 Cooperation mode

In *cooperation mode*, a set of m nodes need to solve a specific problem by sharing information among them. For instance, consider this general problem: every node j privately holds an objective $F_j : \mathbb{R}^n \to \mathbb{R}$, which describes the data acquisition process at node j. The goal is to find a global consensus solution $x \in \mathbb{R}^n$ to minimize the optimization problem $\min_{x \in \mathbb{R}^n} \left\{ F(x) := \sum_{j=1}^{m} F_j(x) \right\}$. This general problem is applied for distributed travel-time tomography [17].

Under these circumstances, our middleware allows the nodes to construct a mechanism to combine the information of the neighbours with the local one. This mechanism consists of a 'handshaking' in which a node shares its information Υ_j for constructing weighted matrices to assign the contribution percentage of a node in the computation. Note that here the information of S_j and U_j is useful to know whether a node is contributing to the computation and its degree. After the 'handshaking' process is completed, the nodes start to communicate using a communication type ($\Lambda_{j_{(C_j)}}$ of node Υ_j) to obtain the final result.

5.4.2 Task-sharing mode

In *task-sharing mode*, one node Υ_j may need to compute some analytics, in a determined timestamp, using libraries and algorithms available in other nodes Υ_{i1}, Υ_{i2}, ..., where $i1, i2, \ldots \in N_j$ and N_j is the neighbourhood of Υ_j. Notice that $|N_j|$ is the number of neighbours of node Υ_j and here denoted by U_j. The node Υ_j decides which neighbour node it will help to compute the algorithm according to (5.1)

$$E_j = \max_{i=1:U_j} \left(\Gamma_{i[x]_{(\kappa_i)}} \cdot \Gamma_{i[x]_{(\mu_i)}} \right) \cdot \left(\frac{1}{l_{ij}} + \left(\Lambda_{i_{(P_i)}} \cdot \Lambda_{i_{(\rho_i)}} \right) + \left(\Lambda_{i_{(M_i)}} \cdot \Lambda_{i_{(\alpha_i)}} \right) \right) \cdot \Lambda_{i_{(B_i)}}, \quad (5.1)$$

where x is the specific library to execute and l_{ij} is the latency between node i and node j. Notice that l_{ij} is calculated using the timestamps between nodes i and j. Also, the node Υ_i with the maximum E_j is the neighbour node with more resources available to meet Υ_j requirements. Υ_j transfers the compressed and preprocessed data to Υ_i and obtains a result as answer.

5.5 Case studies in subsurface imaging

The seismic imaging process involves high data rates (up to three channels per sensor, sampled at 50–500 Hz) and larger scale network (e.g., hundreds or even thousands) if essential details of geophysical dynamics are required. It is virtually impossible to collect all raw seismic data to a central place (even with compression) due to the severe energy and bandwidth constraints, and the disruptions caused by the harsh environment factors (e.g., rugged terrain, harsh weather and occasional ash eruptions). Instead of delivering raw data for post-processing, sensors have to be designed to perform adaptive, in-network processing of high-resolution seismic data to compute the subsurface imaging model for direct visualization.

We present two case studies where we successfully used DCDA middleware to retrieve subsurface images using different distributed algorithms. The first case involves the location (2D and 3D) of earthquakes; and the second case uses the ambient noise for generating a 2D velocity map of the subsurface.

5.5.1 Case 1: travel-time location

Tomography can be defined as the science of computing reconstructions in 2D and 3D from projections [6]. Then, travel-time tomography uses information of the time of travel of the signals generated by an earthquake to reconstruct an image of the

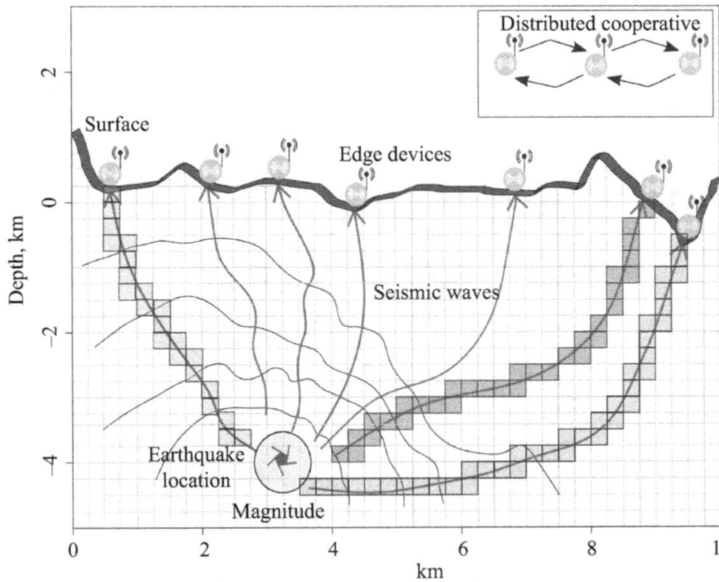

Figure 5.5 Travel-time tomography

subsurface [18]. Travel-time tomography involves many steps as illustrated in Figure 5.5. Those include (1) seismic sensors or edge devices (top circles) that measure the vibration after the occurrence of an earthquake and calculate the arrival time of the p-wave [19]. This process is called arrival time *picking*; (2) next, the earthquake location is estimated; (3) the magnitude of the earthquake is then calculated by combined information of all edge devices; and (4) lastly, the rays are traced from earthquake to nodes and the tomography inversion is performed.

Generally speaking, earthquake locations can be determined by inverting the arrival times of P- and S-waves measured at the stations of a seismic network [20]. The accuracy of seismic event locations is controlled by several factors, including the network geometry, available phases, arrival-time accuracy [21] and knowledge of the subsurface structure [22]. In order to improve the velocity accuracy, station and source terms can be accounted [23], and hypocentres and velocity structure can be jointly inverted [24].

We have used the details of the work presented in [6] for distributed travel-time location. In this chapter, we use distributed algorithms for calculating the arrival time picking as described in [19], and well-known algorithms for earthquake location (Greiger's method) [25] and magnitude calculation method described in [26].

5.5.2 Case 2: ambient noise tomography

Another type of seismic tomography is called ambient noise tomography (ANT). ANT uses vibrations of the ambient noise to estimate the local phase speed of the waves and reconstruct an image of the subsurface. ANT involves the steps shown in Figure 5.6. Those include (1) seismic sensors or edge devices (top circles)

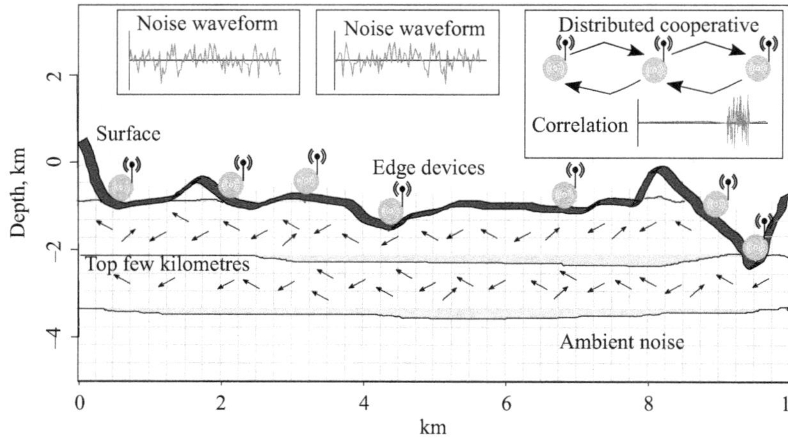

Figure 5.6 Ambient noise tomography

measure the vibration of the ambient noise; (2) edge devices calculate the cross-correlation of the signal waves with their neighbours and perform a frequency-time analysis to obtain travel-time measurements of the ambient noise signal and (3) using a method known as Eikonal tomography [27], a speed map is built.

In ANT, the data are recovered from ambient seismic noise, which implies no need for active energy sources like earthquakes. Ambient noise imaging can be applied to regions with nonexistent seismicity, and it produces reliable measurements at frequencies that are particularly difficult using earthquakes or explosions due to scattering and attenuation. This advantage represents an attractive cheap scenario since producing active energy sources (explosions) in nonseismic areas is very costly. Moreover, utilizing the persistent nature of the background noise, several studies [28] have demonstrated the possibility to determine temporal structure variation and to monitor volcanic activities using noise cross-correlation. The complete formulation of the method can be found in [27] and the distributed implementation is described in [29].

5.5.3 Decentralized algorithms

The main idea of decentralized cooperative computing is leveraging the computational capacity of each node in the network and performing computing and analytics close to where data are being produced. Neighbouring nodes in the network cooperate with each other by sharing partial analytics based on their local data. The key technical challenge thus lies in minimizing the communication complexity among neighbouring nodes in order to obtain low-latency real-time analytics.

We have developed a series of innovative decentralized algorithms and embedded them on our DCDA middleware to solve a global problem in *cooperative* and *task-sharing* modes. These algorithms have been applied to travel-time location and tomography, for example, in Zhao *et al.* [17,30,31] and to ambient noise tomography, for example, in Valero *et al.* [29]. The aforementioned algorithms exhibit (1) low communication complexity, (2) balanced computation and communication, (3) multihop communication free, (4) simple implementation, (5) no need of central coordinators and (6) fault-tolerance and preservation of privacy. In addition, our proposed decentralized mechanisms are scalable, transferable and capable of solving a large class of Big Data computing problems.

5.6 Middleware evaluation

5.6.1 Equipment

We applied our proposed solution to seismic and ambient noise tomography case studies and evaluated using a fog computing testbed and real seismic devices deployed in the field. We developed a fog computing testbed consisting of 64 BeagleBone black (BBB) as shown in Figure 5.7(a). These computing units are interconnected via wireless network forming a cluster. Each BBB operates a Linux-based OS and has 512 MB DDR3 RAM, 16 GB flash storage and 1 GHz ARM Cortex A8 processor. They are credit card sized, low-power computing units that cost under USD 70. For field testing, we have used 25 R1+ devices which are small, light-weight seismograph nodes as shown in Figure 5.7(b). R1+ is designed for data acquisition in the field and is equipped with an internal, industry-standard Linux processor. R1+ also has an external antenna for XBee communication that allows a communication range up to a few km with a high-gain antenna. It also can communicate via wireless protocols like transmission control protocol (TCP) and user datagram protocol (UDP). If the data transmission volume and power consumption are almost the same between XBee and Wireless, then analytics can be performed independently from the medium.

(a) (b)

Figure 5.7 (a) Testbed consisting of 64 BeagleBone black forming a cluster and (b) R1+ seismograph nodes

5.6.2 Datasets

For travel-time location, we use real data acquired by R1+ devices. We deployed these devices in a wide range area inside the University of Georgia campus. We artificially generated earthquake signals using a hammer and a metal pad over the field. The data recorded by the instruments were processed in real time by taking advantage of our proposed DCDA middleware. For ambient noise tomography, we used ambient noise data collected from 500 stations of the EarthScope Transportable Array (USArray)[1] database deployed over our BBB cluster. The data were compiled between September and November of 2007. We use distributed cooperative computation for generating ambient noise velocity maps inside our middleware.

5.6.3 Analytics processing

A detail description of how to execute the analytics with the two study cases used to evaluate the middleware are presented here.

5.6.3.1 Compute travel-time tomography in the proposed middleware

First, a network of sensor nodes are deployed on the field. Sensors start to read the information from the medium. In DCDA, nodes work in a *cooperative mode* to solve the problem of calculating arrival picking and event location and magnitude estimation.

Arrival picking

There are two types of body waves that propagate through the earth. One is Primary waves (P-waves) which is the fastest seismic waves, and it is essential to calculate the location and magnitude of a specific earthquake. P-wave arrival time is the feature that is used to perform these analytics. Examples of pickings of P-wave arrival times are represented with vertical lines in Figure 5.7. P-wave arrival time on sensors is different, due to different wave propagation delays. The time difference between sensors is several seconds in a deployment up to ten kilometers. Manual picking requires data preprocessing, and it is time-consuming in a large sensor deployment. To reduce the communication cost, to avoid raw data transmission and to meet the requirements of a real-time system, each unit performs an algorithm to automatically detect the events and calculates its travel time.

The used algorithm in this study case is called STA/LTA method [21]. The STA (short-time average) is sensitive to rapid fluctuations in the amplitude of the time series, whereas the LTA (long-time average) provides information about the background noise [32]. The arrivals are picked on the maximum value of the

[1]http://www.earthscope.org/science/observatories/usarray

derivative function of the STA/LTA ratio curve. The STA and LTA at the *i*th time sample are commonly defined as

$$\mathrm{STA}_i = \frac{1}{ns} \sum_{j=i-ns}^{i} \mathrm{CF}_j, \tag{5.2}$$

$$\mathrm{LTA}_i = \frac{1}{nl} \sum_{j=i-nl}^{i} \mathrm{CF}_j, \tag{5.3}$$

where i,j are the sample number, *ns* and *nl* represent the short and long window lengths, respectively. CF stands for characterization function which characterizes waveform properties that each node extracts from the signal [33].

Event Location and Magnitude

The second real-time data analytics that is calculated is the localization and magnitude of the earthquake. It is estimated using the arrival time from different nodes. Based on the system architecture, when an arrival picking is calculated at node, it is sent to the sink node. Sink node receives and processes in real time these arrival times to estimate the earthquake location. It is important to mention that the location of an earthquake by a sink node is only possible if the nodes that transmit arrival time are relatively close. Several factors as signal quality and sensor noise can affect event detection causing false alarms. However, these are discarded by earthquake location because a minimum of four arrival times with differences of a few milliseconds are required for the estimation of an earthquake location and its magnitude.

5.6.3.2 Compute ambient noise tomography in the proposed middleware

In this study case, a network of sensor nodes is also deployed on the field. Nodes work in *Cooperative mode* to generate the subsurface image based on ambient noise, but also in *Task-sharing mode* to share the libraries that are used to correlate data and generate interpolation.

Ambient noise velocity map generation

The ambient noise raw data gathered from each individual sensor need to be prepared to get a suitable individual waveform for future cross-correlation. Nodes communicate and cooperate with each other to calculate the cross-correlation between each the signal of pair of nodes s_1 and s_2

$$C_T(\tau, s_1, s_2) = \frac{1}{T} \int_0^T u(t, s_1) u(t+\tau, s_2) dt, \tag{5.4}$$

where T is the total time recorded (usually a window is used), u is the wave signal and τ is the displacement. If one node doesn't have the correlation library, it can ask the library of a neighbour node to share the task. When the ambient noise has been correlated for the nodes for a long period, nodes apply another method called Eikonal Tomography Function [27] to generate the velocity map. In the same way, if there is a node that does not have the library to apply the eikonal method, this node can share the task with other nodes. The complete explanation of ambient noise tomography over sensor networks can be found in [34].

5.6.4 Results visualization on DCDA middleware

Figures 5.8–5.10 show the results of running distributed algorithms for travel-time tomography and ambient noise tomography. Figure 5.8 shows raw data recorded by a node (top) and the arrival time picking calculated within the same node (bottom). Figure 5.9 shows 2D and 3D locations and magnitude of different detected

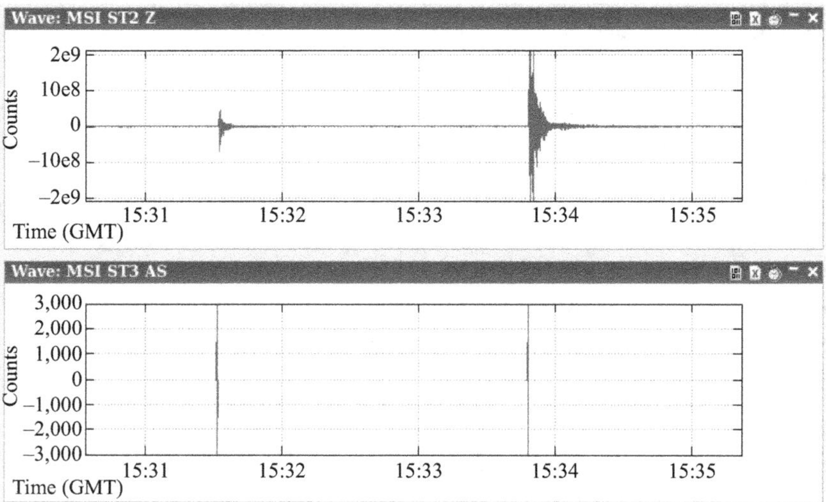

Figure 5.8 Raw data and picking time visualization

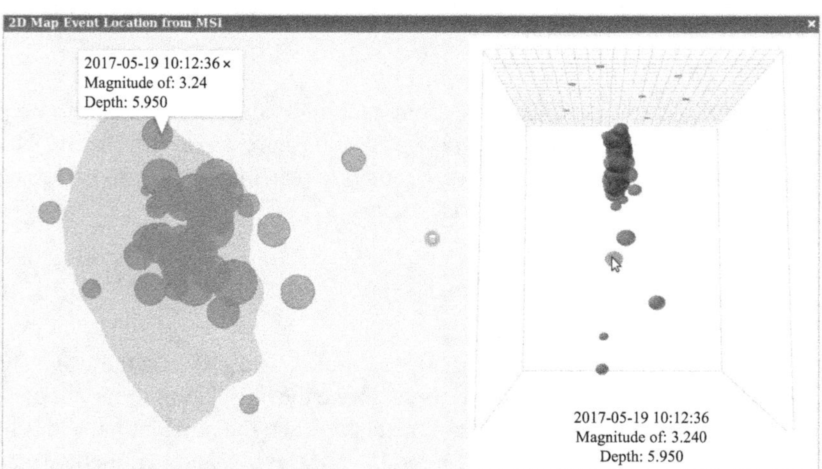

Figure 5.9 2D and 3D locations and magnitude of different detected earthquakes (circles) using DCDA middleware

Figure 5.10 Ambient noise velocity map constructed using our DCDA middleware

earthquakes (circles) using DCDA middleware. Figure 5.10 illustrates the velocity map of ambient noise using USArray data in DCDA middleware.

5.6.5 Middleware versus central approaches time evaluation

We measured the speedup between computing travel-time and ambient noise tomography in our middleware and the centralized system like Cloud, where all IoT devices send the data to a central place for post-processing. Figure 5.11 illustrates the improvements in terms of computation time achieved using the DCDA middleware. Note that for some processes such as arrival time picking and ambient noise imaging, the centralized approach is more than twice slower compared to our middleware. This is due to the amount of data to transfer for calculating the time of arrival (time picking) and the ambient noise image. For processes with less amount of data to transfer, the speedup of our middleware is not too high. This scenario confirms our premise that for real-time applications, computing in the edge of the network is faster and reduces latency. Furthermore, with DCDA middleware we only transfer processed data which implies an efficient use of available bandwidth in the network.

5.6.6 Scalability, robustness and flexibility

The proposed DCDA middleware can be augmented to handle increasing workloads. We can easily add more nodes to the network and integrate them into the cooperation mode or task-sharing mode. Most importantly, we can increase the number of algorithms performed within our middleware. We reduce the bandwidth bottleneck by ensuring the processed data flow only between nodes. This also guarantees the system scalability.

One of the important characteristics of DCDA middleware is its fault-tolerance, and here we validate it by simulating node and link failure. We evaluated these properties of DCDA middleware in Figure 5.12 using the magnitude calculation

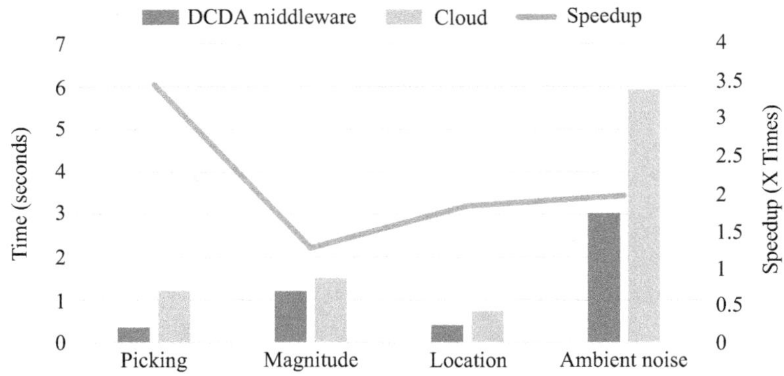

Figure 5.11 Time evaluation and speedup comparison between the proposed DCDA middleware and centralized approach

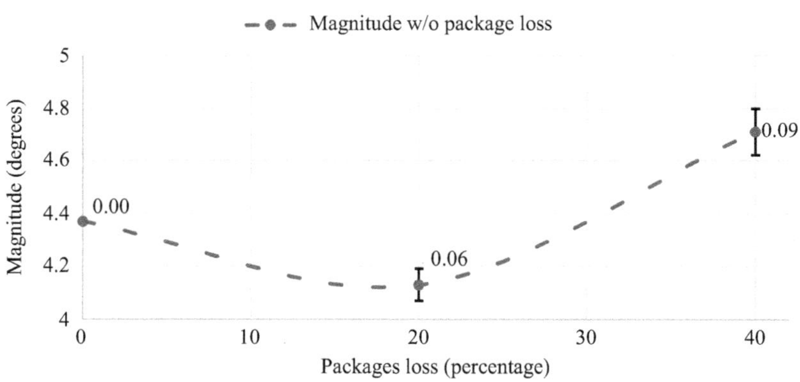

Figure 5.12 Performance of DCDA middleware under different cases of node/link failure

algorithm. We run each experiment with three different cases: Case (1) no failure; Case (2) 20% of the nodes fail for 10% of the time and Case (3) 40% of the nodes fail for 10% of the time. From Figure 5.12, we observe that there is no significant effect on the error regarding a scenario with no package loss. However, some algorithms require all nodes to be running and hence its speed depends on the slowest node. This property is extremely important when choosing algorithms for edge computing.

Finally, DCDA middleware is flexible to handle different kinds of distributed algorithms. However, the performance is directly dependent on the algorithm design.

5.6.7 Energy consumption

We conducted tests to measure the amount of energy consumed by fog nodes when computing information using the middleware and when sending raw data to be processed at the centralized approach. Figure 5.13 shows the battery usage for different nodes and its average. Even though with the middleware the nodes need to compute in situ and communicate with neighbours, the process consumes less power than to transmit raw data to the cloud because the communication cost is higher than the computation cost.

5.6.8 Communication cost

We measured the communication cost based on the number of packages transmitted during computation. Figure 5.14(a) shows that at the centralized approach (Cloud in this case) the communication cost is high because all nodes send raw data for further processing. In contrast, Figure 5.14(b) illustrates the communication cost using the proposed DCDA middleware. Notice that the cost in DCDA middleware is low since nodes communicate only processed data, and picks occur only in the nodes that receive these data for processing final data analytics.

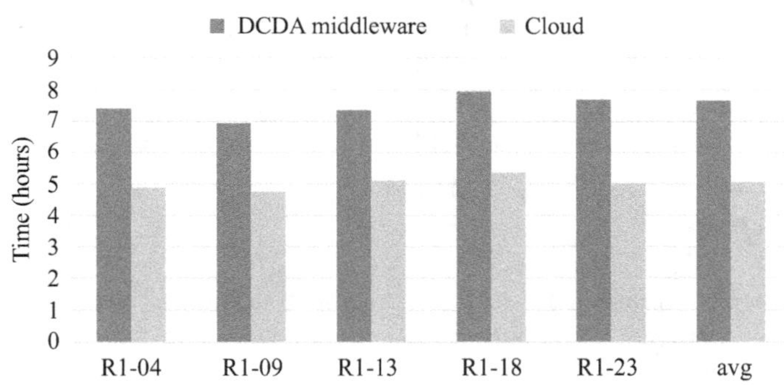

Figure 5.13 Battery usage in hours using DCDA middleware versus centralized approach

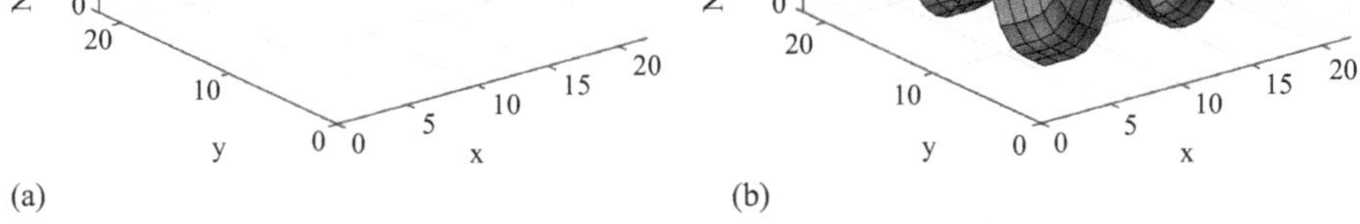

Figure 5.14 (a) Communication cost sending data to a central place (e.g., Cloud) and (b) communication cost in the proposed DCDA middleware. Axes x and y represent nodes numbers

5.7 Opportunities

There is a wide range of opportunities for research and applications by using the proposed middleware. One of the most outstanding is the elastic scalability and adaptability of the system. Consider the variability in the potential use cases of fog, the proposed middleware architecture can enable elastic scaling of modest deployments up to large mission-critical deployments. This scalability is essential for a fog computing middleware implementation to adapt to business needs as it relates to workload, system cost and performance. Each fog node is an internally scalable member of a scalable fog network community that can run on its own or as a part of the hierarchical fog fabric. Each fog node's data and functions can be accessed or revoked by any other legitimate fog node, so that the flexible computing and network resource allocation and the combination can be implemented on top to meet the application QoS (e.g., delay and accuracy) provisions. Individual fog nodes can also scale internally, through the addition of HW or SW resources to the node's basic infrastructure. Fog networks can scale up and scale out through the addition of fog nodes to assist existing heavily loaded nodes, either on the same level of the fog hierarchy or in adjacent levels. The fabric can be scaled up or down in a demand-driven elastic environment. Storage, network connectivity and analytics services should be able to scale with the fog infrastructure.

5.8 Conclusion

In this chapter, we presented a novel fog computing architecture, which enables decentralized cooperative computing in sensor networks. In fog computing paradigm, computing and data analytics are pushed into the edge of a network [35]. A mesh network is formed and it consists of many edge devices. These devices share resources and capabilities, and they might cooperate to solve a global problem (usually requires full knowledge of the whole network). A natural question is how to design a way of cooperation between edge devices. Fortunately, our proposed decentralized cooperative computing algorithms can address this issue and fit into the developed fog computing middleware [35]. The proposed middleware (with our decentralized algorithms embedded) has been tested in seismic and ambient noise imaging applications, and the evaluation results demonstrate that it is a promising approach for real-time situation awareness under bandwidth and energy limitations, particularly in harsh, uncertain and dynamic environments.

References

[1] Lucero, S. (2016). IoT platforms: enabling the internet of things. An IHS Technology Whitepaper.

[2] Mearian, L. (2013). Self-driving cars could create 1gb of data a second. *Computerworld*. July 23. Available from https://www.computerworld.com/article/2484219/self-driving-cars-could-create-1gb-of-data-a-second.html

[3] Weiner, M., Jorgovanovic, M., Sahai, A., and Nikolie, B. (2014). Design of a low-latency, high-reliability wireless communication system for control applications. In *2014 IEEE International Conference on Communications (ICC)*, IEEE, pp. 3829–3835.

[4] Schmidt, D. C., White, J., and Gill, C. D. (2014). Elastic infrastructure to support computing clouds for large-scale cyber-physical systems. In *2014 IEEE 17th International Symposium on Object/Component/Service-Oriented Real-Time Distributed Computing*, pp. 56–63.

[5] Kamath, G., Shi, L., Chow, E., and Song, W.-Z. (2015a). Distributed tomography with adaptive mesh refinement in sensor networks. *International Journal of Sensor Network*, 23(1):40–52.

[6] Kamath, G., Shi, L., Song, W.-Z., and Lees, J. M. (2016b). Distributed travel-time seismic tomography in large-scale sensor networks. *Journal of Parallel and Distributed Computing*, 89(1):50–64.

[7] Kamath, G., Song, W.-Z., Ramanan, P., Shi, L., and Yang, J. (2015b). Dristi: distributed real-time in-situ seismic tomographic imaging. In *14th International Conference on Ubiquitous Computing and Communications (IUCC)*, Liverpool, UK.

[8] Shi, L., Song, W.-Z., Xu, M., Xiao, Q., Lees, J. M., and Xing, G. (2013). Imaging volcano seismic tomography in sensor networks. In *The 10th Annual IEEE Communications Society Conference on Sensor and Ad Hoc Communications and Networks (IEEE SECON)*, pp. 304–306.

[9] Zhao, L., Song, W.-Z., Shi, L., and Ye, X. (2015a). Decentralized seismic tomography computing in cyber-physical sensor systems. *Cyber-Physical Systems*, 1(2–4):91–112.

[10] Kamath, G., Agnihotri, P., Valero, M., Sarker, K., and Song, W.-Z. (2016a). Pushing analytics to the edge. In *2016 IEEE Global Communications Conference: Selected Areas in Communications: Internet of Things (Globecom2016 SAC IoT)*, Washington, DC, USA.

[11] Saurez, E., Hong, K., Lillethun, D., Ramachandran, U., and Ottenwälder, B. (2016). Incremental deployment and migration of geo-distributed situation awareness applications in the fog. In *Proceedings of the 10th ACM International Conference on Distributed and Event-based Systems*, ACM, pp. 258–269.

[12] Brogi, A. and Forti, S. (2017). QoS-aware deployment of IoT applications through the fog. *IEEE Internet of Things Journal*, 4(5):1185–1192.

[13] Brogi, A., Forti, S., and Ibrahim, A. (2017). How to best deploy your fog applications, probably. In *International Conference on Edge and Fog Computing*, pp. 105–114.

[14] Orsini, G., Bade, D., and Lamersdorf, W. (2016). Cloudaware: a context-adaptive middleware for mobile edge and cloud computing applications. In *IEEE International Workshops on Foundations and Applications of Self* Systems*, IEEE, pp. 216–221.

[15] Bonomi, F., Milito, R., Natarajan, P., and Zhu, J. (2014). Fog computing: a platform for internet of things and analytics. In *Big Data and Internet of*

Things: A Roadmap for Smart Environments, Nik Bessis & Ciprian Dobre, Switzerland, pp. 169–186.

[16] Bessis, N., Xhafa, F., Varvarigou, D., Hill, R., and Li, M. (Eds.) (2013). *Internet of Things and Inter-cooperative Computational Technologies for Collective Intelligence*, Volume 460. Berlin/Heidelberg: Springer-Verlag.

[17] Zhao, L., Song, W.-Z., Ye, X., and Gu, Y. (2017). Asynchronous broadcast-based decentralized learning in sensor networks. *International Journal of Parallel, Emergent and Distributed Systems*, 33(6):589–607.

[18] Kamath, G., Shi, L., and Song, W.-Z. (2013). Component-average based distributed seismic tomography in sensor networks. In *The 9th IEEE International Conference on Distributed Computing in Sensor Systems (IEEE DCOSS)*, pp. 85–95.

[19] Liu, G., Tan, R., Zhou, R., Xing, G., Song, W.-Z., and Lees, J. M. (2013). Volcanic earthquake timing using wireless sensor networks. In *The 12th ACM/IEEE International Conference on Information Processing in Sensor Networks (IPSN13)*, pp. 91–102.

[20] Thurber, C. H. and Rabinowitz, N. (Eds.) (2013). *Advances in Seismic Event Location*, Volume 18. Dordrecht, the Netherlands: Springer Science & Business Media.

[21] Akram, J. and Eaton, D. W. (2016). A review and appraisal of arrival-time picking methods for downhole microseismic data. *Geophysics*, 81(2): KS71–KS91.

[22] Pavlis, G. L. (1986). Appraising earthquake hypocenter location errors: a complete, practical approach for single-event locations. *Bulletin of the Seismological Society of America*, 76(6):1699–1717.

[23] Spence, W. (1980). Relative epicenter determination using p-wave arrival-time differences. *Bulletin of the Seismological Society of America*, 70(1): 171–183.

[24] Kissling, E., Ellsworth, W., Eberhart-Phillips, D., and Kradolfer, U. (1994). Initial reference models in local earthquake tomography. *Journal of Geophysical Research: Solid Earth*, 99(B10):19635–19646.

[25] Geiger, L. (1912). Probability method for the determination of earthquake epicenters from the arrival time only. *Saint Louis University Bulletin*, 8(1): 56–71.

[26] Ristau, J. (2009). Comparison of magnitude estimates for New Zealand earthquakes: moment magnitude, local magnitude, and teleseismic body-wave magnitude. *Bulletin of the Seismological Society of America*, 99(3): 1841–1852.

[27] Lin, F.-C., Ritzwoller, M. H., and Snieder, R. (2009). Eikonal tomography: surface wave tomography by phase front tracking across a regional broad-band seismic array. *Geophysical Journal International*, 177(3):1091–1110.

[28] Brenguier, F., Campillo, M., Hadziioannou, C., Shapiro, N., Nadeau, R. M., and Larose, E. (2008). Postseismic relaxation along the Aan Andreas fault at Parkfield from continuous seismological observations. *Science*, 321(5895): 1478–1481.

[29] Valero, M., Kamath, G., Clemente, J., Lin, F.-C., Xie, Y., and Song, W.-Z. (2017). Real-time ambient noise subsurface imaging in distributed sensor networks. In *The 3rd IEEE International Conference on Smart Computing (SMARTCOMP 2017)*, pp. 1–8.

[30] Zhao, L. and Song, W. (2016). Decentralized consensus in distributed networks. *International Journal of Parallel, Emergent and Distributed Systems*, 33(6):550–569.

[31] Zhao, L., Song, W. Z., and Ye, X. (2015b). Fast decentralized gradient descent method and applications to in-situ seismic tomography. In *2015 IEEE International Conference on Big Data (Big Data)*, pp. 908–917.

[32] Li, F. and Song, W. (2017). Automatic arrival identification system for real-time microseismic event location. In *SEG Technical Program Expanded Abstracts*. Society of Exploration Geophysicists (SEG); Houston, TX, USA, 2017, pp. 2934–2939.

[33] Wong, J., Han, L., Bancroft, J., and Stewart, R. (2009). Automatic time-picking of first arrivals on noisy microseismic data. *CSEG*, 1(1–2):1–4.

[34] Valero, M., Li, F., Wang, S., Lin, F.-C., and Song, W. (2018). Real-time cooperative analytics for ambient noise tomography in sensor networks. *IEEE Transactions on Signal and Information Processing over Networks*, pp. 1–4.

[35] Clemente, J., Valero, M., Mohammadpour, J., Li, X., and Song, W. (2017). Fog computing middleware for distributed cooperative data analytics. In *World Fog Congress 2017 (WFC 2017)*, pp. 1–6.

Chapter 6

IoT-enabled water monitoring and control for smart city

Joyanta Kumar Roy[1,2]

In a world when resources of living are scarce, urban people are suffering from a crisis of potable drinking water. The massive increase and concentration of human population in urban areas are liable for consumption of the vast majority of resources. Therefore it is highly pertinent to make cities greener and more sustainable. Advanced systems developed to improve and automate processes within a city will play a leading role in smart cities. From the intelligent design of buildings, which capture and harvest rainwater for future use, to smart control systems, which can monitor infrastructures autonomously, the possible improvements enabled by sensing technologies are of utmost importance. Internet of things (IoT)-enabled sensing and control poses numerous challenges, which are of a technological and social nature. The IoT solutions of smart water can help water utilities to coordinate analysis of supply and demand to provide smooth control actions across large numbers of distributed smart assets to monitor and deliver optimized performance to entire water production and distribution network. This ensures minimum water loss, assured water quality and quantity to every consumer. It is an excellent challenge in technology, and IoT-enabled smart assets will play a vital role towards the solution. The plugin Internet interface modules are developed to convert existing sensor and actuator system for secure and speedier communication of data to the central decision system either situated in the plant or in a cloud. All smart devices for water monitoring systems are M2M enabled and can share information directly or through the central system. The smart water solution technology is applicable for improvement of overall performance, quality and reliability of water and wastewater treatment and distribution system. The topic includes remote-based pipe monitoring and leak detection system to minimize nonrevenue water and to restore water supply in 24×7 manners with active operational management in damaged assets.

[1]Automation Consultant, GKW Consult GmbH, Kolkata, India
[2]Department of Applied Physics, University of Calcutta, Kolkata, India

6.1 The concept of smart city

The cities are called the engines of growth of the nation. India is a country of a very high population. The record of census 2011 said that 31% population lives in urban areas of India and contributes 63% GDP of the country and expected GDP growth to 75% by 2030. It indicates that the comprehensive development of physical, Institutional, social and economic infrastructure is required to improve the quality of life. The Smart City Mission should provide necessary core infrastructure to give a decent quality of life to its citizens with a clean and sustainable environment through the application of 'Smart' solutions. The core infrastructure elements in a Smart City are the combination of adequate water supply, assured electricity, efficient urban mobility and public transport. The robust IT connectivity, digitalization, sanitation and solid waste management with affordable housing, especially for the poor are necessary for a smart city. The better kind of governance especially e-Governance and citizen participation, sustainable environment, safety and security of citizens especially for women, children and the elderly are the essential requirements.

The smart solutions of all related smart city infrastructure is an expandable illustrative list and depicted in the given Figure 6.1.

The involvement of information technology plays a significant role in the smart solution. The application of Internet is everywhere for each type of services, and all are connected to each other. The E-Governance and citizen services are IT-based for efficient use. Large numbers of IoT-enabled high definition electronic

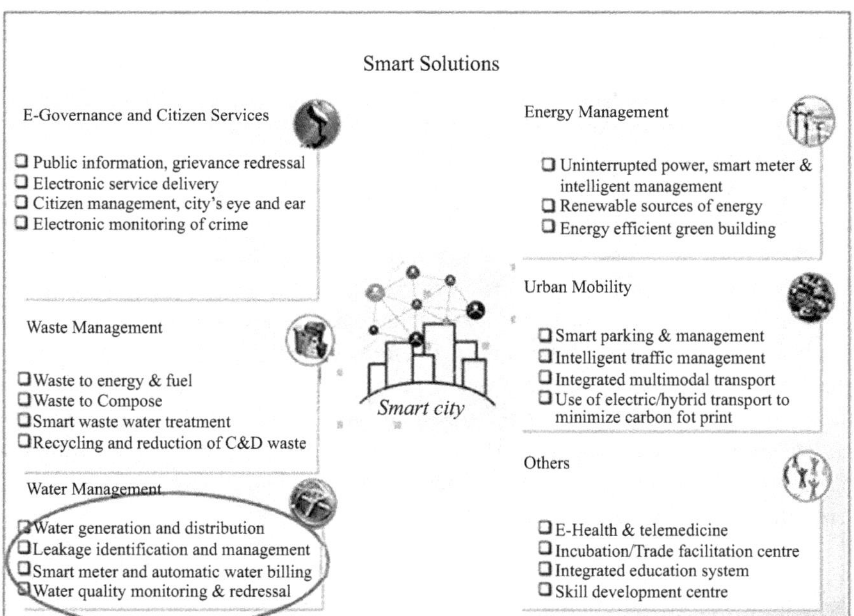

Figure 6.1 The solutions towards smart city challenges

cameras are deployed in every location of public places in a smart city to monitor and manage crime. The cloud-based smart E-Governance software with Artificial Intelligence is custom built as per requirement for efficient E-Governance and Citizen Services. The quantum of waste depends directly on the population and is very difficult to manage when a population is very high. Technology is required to convert solid waste to compose fertilizer or high-tech plant is needed to burn out the waste and extract energy and fuel may give an alternate source of electricity as backup power. The Waste Water of all types is collected through underground drain network and treated for recycling and reduction of C&D waste through IT and sensor-enabled fully automated Smart Waste Water treatment plant. It will ensure safe public hygiene in a smart city. The IoT-enabled Smart bins will replace ordinary dumping bin for solid waste, which are capable of sensing bio-degradable or nondegradable waste and can communicate with the waste management team for efficient waste mobility. The electricity is the prime mover of the Smart city; all infrastructures are highly dependent on uninterrupted and quality power. The backup power generation from renewable sources is vital to maintain the sustainability of all smart city infrastructures which require electricity. To avoid the enormous electrical load of all infrastructures low-power electrical system are encouraged in the Smart city, for example, use of LED in lighting, energy efficient inverter-based air-conditioner, green building, low-power building automation and Variable Frequency Drive (VFD) operated pumping machinery or prime movers, etc. All buildings in the smart city will have the rooftop solar panel installation, which will store electricity during daytime, will give backup power during peak time generally at summer night and all electrical installation in building complex form smart microgrid and will be connected to the main smart grid through the smart meter which is Internet-enabled. The excess energy generated by individual building or a complex of buildings will become prosumer, instead of a consumer. The electricity bill of the consumer will be adjusted automatically by IoT-enabled net metering. Thus the revenue collection for energy utilization will be fully automatic. The urban mobility is also an important issue and challenge. With the increase of population of city dwellers, the number of the vehicle in a road is also increasing in a lousy fashion. To reduce car accident several IoT approaches will be taken, such as smart sensor-based traffic management and automatic car-parking management so that the driver can visualize the nearest available car parking space through his smart mobile App.

It is expected that the smart city must have a pollution free environment with a minimum carbon footprint. The fuel-driven vehicles are an excellent source of air pollution and increase the carbon footprint. So the electric or hybrid cars are preferred in the urban mobility. A hybrid vehicle uses two or more types of power. The fuel is used to drive the internal combustion engine and an electric generator that powers an electric motor. During start-up, more power required when IC engine will work and during running electric motors provides the power to the engine drive resulting in fuel efficiency and less carbon footprint. Most of the smart city vehicle will be Internet-enabled with GPRS system. These cars will have the facility to communicate with the other vehicle, and IoT-enabled driver assistance system will reduce car accident significantly. There are other infrastructures like Telemedicine and e-Health portal.

Smart hospitals and many other foundations will play for the comfort of the citizen. The focus of this chapter is water and its management using the IoT.

6.2 The water management in a smart city

Water is the synonym of life. On an average, the body of an adult human-being contains 60% water. Most of the water in the human body is contained within our cells. In fact, our billions of cells must have water to live upon. The most of the diseases are water-borne. The waterborne illness is caused mostly by infections that are predominantly transmitted through contact or through consumption of disease-infected water. Many of the Protozoa infections, parasitic infections, bacterial infections and viral infections are waterborne. The overall bacteriological contamination plays a major role in a nation's health. Besides the bacteriological contamination, our natural resources of water are continuously contaminated by industrial waste. One of the Millennium Development Goals is to improved access to safe water which is the engine for socio-economic growth. Therefore in this context, the importance of information on water resources systems is very valued and cannot be underestimated given the environmental status of our world as well as the socio-economic development process of nations. Access to clean water for drinking and sanitary purposes is a precondition for human health and well-being. Unpolluted water is also essential for the ecosystems. The water is of three types. The Blue water is the fresh water of river, pond or water bodies in the surface of the earth or aquifer beneath the ground. The Green water is the soil moisture used by the plant. The Grey water stands for the waste water.

6.2.1 The water contamination

The surface water polluted by industrial waste, cattle waste and contaminated with protozoal, parasitic, bacterial and viral infections causes many fatal waterborne diseases that threaten life. It causes diarrhoea, cholera, typhoid and not suitable for human drinking. The water pollution can take in many forms and have different effects:

- Faecal contamination: The familiar source is sewage. It makes water unsafe for human consumption and recreational activities, such as swimming, boating or fishing. Many organic pollutants, including sewage effluent and farm and food processing wastes, consume colossal oxygen, suffocating the aquatic life. Ultimately these contamination affect groundwater resources used for drinking water purposes.
- Nutrients: Nitrates and phosphates from farm fertilizers to household detergents can over-fertilize the water causing the growth of large mats of algae, some of which can be toxic. When the algae die, they sink to the bottom, decompose, consume oxygen and damage ecosystems.
- Pesticides and veterinary medicines from farmland and some industrial chemicals can threaten wildlife and human health. Some of these damage the hormonal systems of fish, causing 'feminization' (endocrine disruption).

- Metals, such as zinc, lead, chromium, mercury and cadmium are incredibly toxic. Copper complexes are less toxic, and cobalt and ferrous complexes are only weak toxicants. Concentrations of cyanides in waters intended for human use, including complex forms, are strictly limited because of their high toxicity.
- The arsenic contamination of groundwater usually occur in two ways: Anthropogenic activities and aquifer naturally contain groundwater contamination and cross the permissible limit of WHO (0.01 mg/l); excess arsenic has been detected in tube well or hand-pumped drinking water supply. In many countries, about 200 million people are under risk of arsenic poisoning. The arsenic found widely in the earth's crust. It is mostly in organic, inorganic and in the gas form. The organic arsenic is abundantly found in seafood. It's not harmful, but the inorganic arsenic found in groundwater is detrimental to health because it gets stored in the body, thus adversely affecting multiple organ systems and form many diseases including cancer that leads to death. Consuming the arsenic contaminated water amounts to slow poisoning with often a fatal end.
- Organic micropollutants, such as pharmaceuticals, hormones and chemical substances used in products and households can also threaten health.
- Chlorinated hydrocarbons exist in the natural systems, several of which are highly toxic to humans. These molecules persist in the environment for a longer time and threaten to contaminate aquatic and soil systems.

6.2.2 The conventional water treatment plant

So the natural water needs treatment to remove contamination within the acceptable human limit as per WHO or ISO standard. The primary objectives of traditional drinking water treatment are removal of microbial pathogens (coliforms), particles (colour and turbidity) and total dissolved solids. There are several treatment processes to remove a contaminant from water. The surface water has no arsenic or has arsenic but within the acceptable limit, therefore surface water after the treatment process can be used in the drinking water supply scheme.

Figure 6.2 is the process flow diagram of a conventional water treatment plant. The treatment plant consists of several plant components at different houses. The traditional water treatment plant is consisting of (a) raw water pump house, (b) filter house, (c) chemical house, (d) clearwater pump house and (e) sludge pump house. The other process infrastructures are the open-air type such as the stilling chamber, flash mixture, clarifloculator, backwash reservoir and centrifuge, as shown in Figure 6.2. The raw water pump house is usually situated close to the river. The raw water pump abducted water from the river and fed to the stilling chamber either through the open channel or pipeline. The massive particles in the water settle down in the stealing chamber, and the water flows under gravity through an open channel towards clarifloculator. In this channel, the measured amount of coagulant mostly alum solution/polyelectrolyte solution or lime solution are added through electronic dosing system. The raw water with coagulant flows to the flash mixer for appropriate mixing. The clarifloculator is a big circular water tank having the inner and outer chamber. The raw water with coagulant enters to the inner annular tank from the

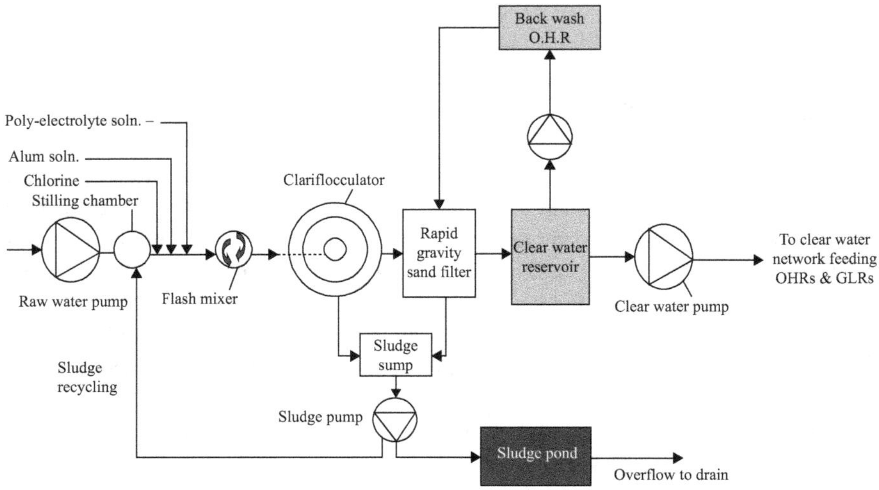

Figure 6.2 Conventional water treatment plant with rapid gravity sand filter

bottom. Motor-driven mixing paddles are rotating inside this inner tank. The water flows from inner tank to the outer tank. In the outer tank flocculation reaction takes place, and most of the dissolved solid comes out from the water and settles down to the bottom of the outer tank. The turbidity of the raw water highly reduced after flocculation. The clarified water from the outer tank comes out from the peripheral launders and flows to rapid gravity sand filter and finally the filtered water is stored in the clear water reservoir for distribution. The clear water pump house has some pumps – some are working and some kept as standby. These pumps pump out water from the clear water reservoir and feed to water distribution mains. The filter beds need backwash on a regular interval. The loss of head (LOH) and flow through the filter – these two parameters decide the filter bed health and gives an indication of filter bed efficiency and provide the alarm to start the backwash mechanism.

The flocculated and filtered sludge are collected to the sludge tank from the clariflocculator and filter beds. The sludge water and solid are separated using centrifuge system. The solid sludge is deposited to the sludge pond and sludge water is fed back to stilling chamber for recycling. The clear water pump house has some high-speed pumping machinery. A few are working and others kept in standby mode. The inlet of these pumps is connected to a common header for suction of water from clear water reservoir and push water to clear water delivery mains for distribution to the clear water network. The disinfection process has several methods, such as

1. Destruction or removal of vegetative pathogens (microbes, spores, cysts, viruses, etc.) by autoclaving or membrane filtration;
2. By physical methods, for example, by the application of radiation, that is, UV, X-ray or Gamma-ray and
3. By chemical method using chlorinated component, that is, chlorine, chloramines and chlorine dioxides.

The conventional treatment process uses the chlorine gas for disinfection. The chlorination is made at two stages. At a prechlorination stage, the chlorine gas is mixed with water, then added to the raw water. At this stage of disinfection, most of the bacteriological contamination gets destroyed. At the postchlorination stage, chlorine added to clear water reservoir with a limit of 0.6–1 mg/l for further protection from the vegetative pathogen.

The production and distribution of potable drinking water for the large community in the smart city is a significant challenge to the water sector industries. The underground-water is the natural source of water but due to excessive pumping, the arsenic contamination is taking place owing to the extreme reduction of partial pressure in the underground water level. Moreover, the surface water is not able to penetrate through the land-bed to reach up to the underground water level to balance the deficiency of water quantity. The country abounds with rivers enriching the large population by its natural river water resource. River water is much safer than the underground water in today's world scenario. Contamination of arsenic or other heavy metal has a little chance in this water source. Regardless of these facts, the other chemical contamination has also very minimum chance to occur. So, it has been decided by the planner of Smart City that a first step to implementing the water treatment plant supported by Industry 4.0 standard distributed SCADA control system shall be equipped with IoT-enabled smart sensors and actuators situated at the bank of the river.

6.3 Industry 4.0

Industry 4.0 means the current trend of automation and data exchange in manufacturing technologies. It uses cyber-physical systems, the cloud and cognitive computing.

Industrialization started with steam and gave rise to the first machines that mechanized some of the work that our ancestors had done manually. At that time water pumps were operated with steam.

After that, we had the electricity, assembly lines and the birth of mass production (Figure 6.3). The pumping machinery was powered and controlled by electricity. The third generation of industry came in with the advent of computers and the beginning of automation when robots and machines along with SCADA system began replacing workforce on those assembly lines. Now, this is a new world of Industry 4.0, in which Internet and cloud-based communication and automation have come together in a novel way, with devices connected remotely to computer systems, which equipped with machine learning algorithms can control the distributed pumping machinery from remote with minimum human support.

Industry 4.0 has highly intelligent connected systems that create an entirely digital value chain. It is mainly software controlled hardware-based cyber-physical systems. The cyber-physical system integrates communications, IT, data and physical elements. These devices convert the traditional plants into smart factories. The primary objective is that the machines talk to other devices through Internet and

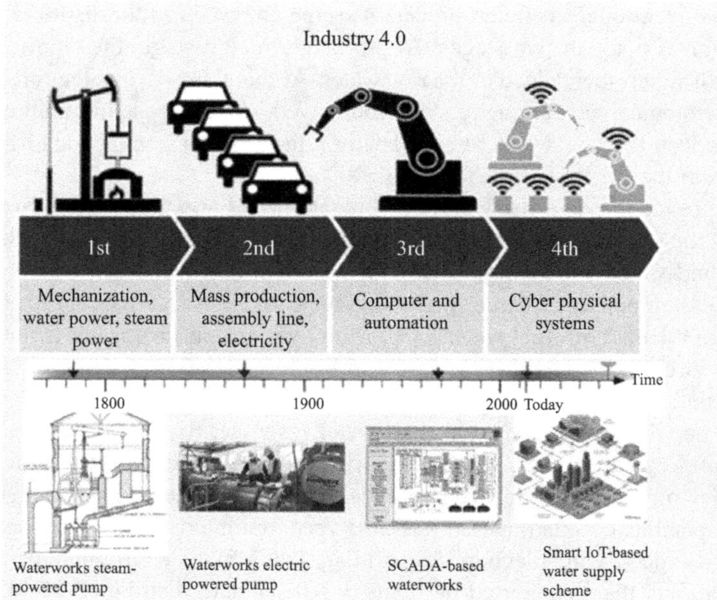

Figure 6.3 History of industrialization

products and the information is processed and distributed in real time resulting in profound changes to the entire industrial ecosystem.

New technologies are revolutionizing traditional water supply scheme. Recently, 'smart methods' based on information and communication technologies are employed to deliver the integrated supply-demand side management with better efficiency. The similar development in the area of water distribution network comparable to Smart Grids in power distribution network is less publicized. Incorporating these new and smart methodologies into water distribution networks contribute a sustainable, low-carbon society, thus helping achieve progress towards the European 2020 targets on climate and energy. This picture provides an example and an overview of a typical Smart water network.

The primary source of information towards the smart water grid is real-time and near-real-time meter/sensor data. The report includes pressures, flows and water levels at selected points throughout the water supply/distribution network. It is based principally on instantaneous flow, pressure and level at working physical structures such as intakes, water treatment plant, service reservoirs, booster pump stations, etc. controlled and monitored through the supervisory control and data acquisition system (SCADA), which consists of high-end programmable logic controller and related software. The information generated is augmented by flow and pressure measurements from meters at the entry to district metered areas (DMAs) along with selected pressure monitoring points within the DMA. The data typically transmitted every 30–60 minutes over a cellular communication network.

Therefore the intelligent water distribution network as described comprises controlling and monitoring facilities of valves/pumps (actuators). The sensors (measurement values) data are physical parameters, that is, the flow of water, reservoirs water level, consumers demand and information flow of the actuator commands and sensors. Such an infrastructure is often referred to as a cyber-physical-system. For smart water management Industry 4.0 technology is the combination of Internet-enabled smart sensors and actuators, most rugged IP-based Internet-enabled automation controller and software (PLC/SCADA) system. These elements or things related to smart water management referred as the IoT. The number of IoT devices in bundle may be called a system, which is Internet-enabled, and then it is called the Internet of system. This technology is ubiquitous and distributed type based on machine to the machine of the machine to man communication.

6.4 Internet of things

The IoT is the integration of sensors, devices, etc. on the Internet. The term 'thing' may be said to anything, like people, objects or information in the domain of IoT. It forms a things-oriented and an Internet-oriented perspective. The IoT defined as a world where things can automatically communicate to computers and each other providing services with a semantic-oriented perspective [1,2]. The IoT is 'a worldwide network of interconnected objects/things' with uniquely addressable, based on as follows:

- Dynamic and self-adapting: IoT devices and systems may have dynamically adapted to the changing contexts and may take action based on their operating condition;
- Self-configuring: IoT devices should have self-configuring capability allowing large number of IoT devices through standard interoperable communication protocols;
- The ubiquitous nature of connectivity: such as ubiquitous wireless/ broadband /4G/ 5G;
- Pervasive computing: that involves wireless communication and networking technologies, mobile devices, embedded systems, wearable computers, RFID tags middleware and software agents;
- Unique identity: each IoT device has a unique identity or a unique identifier, for example, the global IP-based identification of everything (IP V6) and
- Integrated into information network: the ability of each thing to send and receive data across the Internet or the private network. IoT devices can be dynamically discovered in the network by other devices in the network.

6.4.1 Physical design of IoT

Figure 6.4 shows the typical generalized block diagram of an IoT device. The IoT device is design based on System on Chip (SOC) or modular. The system built in high-end microcontroller system having multichannel A/D and D/A interface for external sensors and actuators. For water industry, the sensor may be electrical

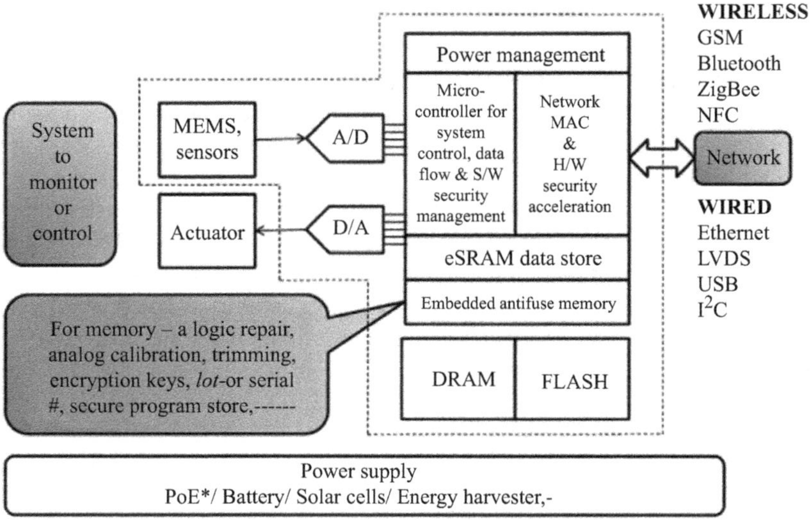

Figure 6.4 Generalized block diagram of IoT device. Courtesy: Texas Instruments

switches, temperature/flow/pressure/level/turbidity/residual chlorine sensors, etc. The actuator may be electrical pumping system, control valve, agitator, mixer, drive motors, and alarms. Of course, the device must connect with sensor and actuator through optically driving interface for voltage or current signal level translation.

The memory may be e-SRAM or DRAM or Flash or maybe the combination of all. There should be power management system with embedded security at a hardware level. The power supply may be power on Ethernet (POE), rechargeable battery and solar cell of energy harvester depending on requirement. The communication is multimodal, may be GSM/ Bluetooth, ZigBee, USB, I^2C, etc. as per requirement and also programmable.

Figure 6.5 is the block diagram of the universal IoT-enabled temperature transmitter for process control application proposed by Texas Instruments. The transmitter accepts RTD, thermocouple and thermistor temperature sensor. The LCD interface is inbuilt. The system has unique power management circuitry and multiple communication interfaces.

6.4.2 Interfacing sensors to IoT

Considering Smart City design, the sensors are essential elements for the highly connected 'IoT' networks. For distributed application where large number of sensors are deployed, the digital interfaces in smart sensors simplify connectivity to the interface of IoT. Hence it enhances the reliability and functionality of systems and also reduces the form factor. The temperature sensors from analog devices, pressure sensors from measurement specialties and humidity sensors from GE sensing are simplifying the interface to the IoT to enhance the reliability and functionality of systems [3].

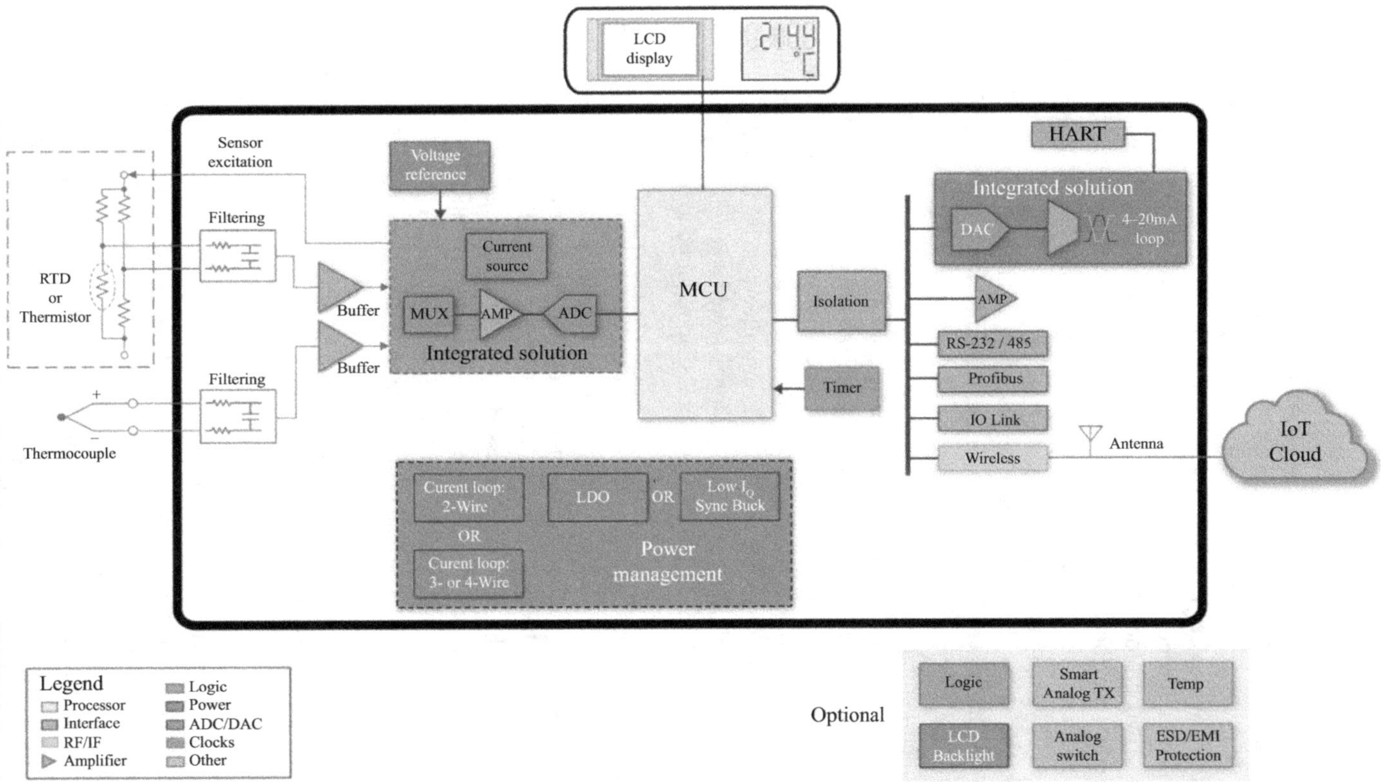

Figure 6.5 Block diagram of temperature transmitter for process control application. Courtesy: Texas Instruments

The IoT starts with measurement, takes data from devices at anywhere in the world, linked up via the Internet to 'big data' information systems through the cloud and provides the foundation for control and optimization.

In some applications, connect in sensors to the IoT are relatively simple but in other cases, needs more consideration in the circuitry. The improvement in CMOS fabrication process and improved packaging allows higher levels of integrations of the sensors and gives regulated digital outputs with protocols acceptable to microcontrollers. The data are then converted to packets for the IoT using the IPv6 protocol for direct connection. In operation, the packetized data flow through the local network that uses an Ethernet gateway or switches to convert header data of the packet into the IPv6 platform for access over the Internet at anywhere anytime.

The data from the microcontroller can handle in many ways, from wired Ethernet to the power-line links over the existing power cabling (Power over Ethernet). The conventional approach is to use a wireless trans-receiver (GPRS/GSM modem) connected to the microcontroller to provide a link back to a central hub and then on to the Internet. The latest emphasis of the IoT is to have more direct access to the sensor node. It permits sensors to be added to a network more efficiently. The number of sensors connected not limited by the address space of an Ethernet hub. The different architecture of the IoT have an impact on the choice of the microcontroller. A simple 8-bit microcontroller can be used directly to the Ethernet hub or switch, but for handling higher-level protocols in IoT application, such as Ethernet or IPv6, 32-bit processor cores are required to manage the protocol stack needed.

However, presently available smart sensors have their own A/D converter optimized to the sensor data resolution by minimizing the power consumption of the node. These can interface to the microcontroller via a serial UART or general-purpose I/O pins, or through an I^2C two-wire interface. It delivers the data directly into the registers in the controller accessed by the protocol stack.

Most of the sensors do not have digital interface inbuilt, they need signal conditioning and digitalization before connecting to the IoT. The signal conditioner for sensor has typically interconnected electronics like preamplifier, scalar, filter and A/D converters. These are now converted into the single chip with voltage current or digital output including the on-chip-embedded microcontroller. The integrated circuit ADT7320 manufactured by Analog Device Inc. [4] is a fast settling high resolution accurate multiplexed temperature sensor having multiplexed sigma-delta ADC inbuilt. Figure 6.6 is the typical interfacing of ADT7320 Digital temperature sensor with microcontroller.

The similar approach can make for digital humidity sensors. HTU21D is a typical digital humidity sensor manufactured by Measurement Specialist. This sensor is low powered, plug-and-play type for humidity and temperature measurement and can be interfaced directly with the microcontroller for real-time application. Humidity sensor calibration is very important. Each sensor needs calibration for temperature and humidity before use. The sensor resolution can be changed from 8–12 bit to 12–16 bit [5] using software command.

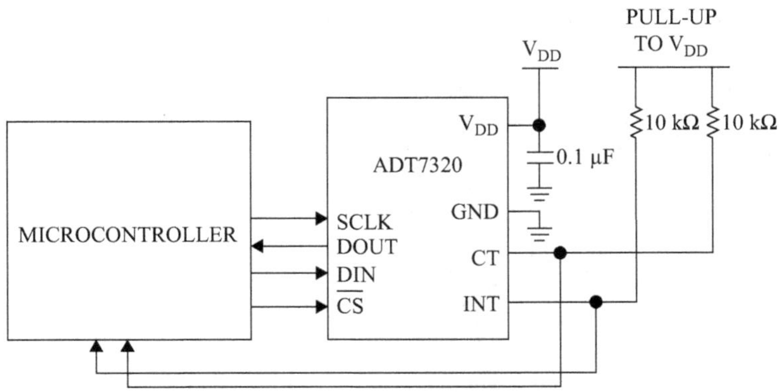

Figure 6.6 The interfacing of ADT7320 digital temperature sensor to a microcontroller in an IoT sensor node [4]

GE Sensing is manufacturing NPA-700 series mem-pressure sensor [6] with the digital interface. It provides digital data directly to the microcontroller. It works with I^2C protocol. The data resolution is very high and is 16 bit.

Many sensors are available with analog output; for them AD7176-2 is suitable. This IC is fast-settling, accurate with high resolution multiplexed $\sum$-Δ ADC for sensors. The analogue input port can be configurable in differential or single mode. The AD7176-2 has three key digital filters to remove line frequency noise. This interfacing IC can be connected directly to a microcontroller, allowing access from anywhere on the Internet. It requires a good knowledge of programming, internal register structure and SPI or I^2C protocol.

6.4.3 Sensors for water quality and hydraulic parameters

In water treatment plant or its distribution network, lots of smart sensors are required to install for monitoring water treatment process, water hydraulics and water quality online in a continuous manner. These transmitters are rugged and designed for harsh environment. They are available with wired or wireless connectivity. They normally operate with two wire communication system with 4–20 mA/HART/MODBUS or ETHERNET communication protocol. The industrial programmable logic controller has used to fetch data from these sensors. For water level measurement in water reservoir noncontact ultrasonic level transmitters are used. The radar type microwave water level transmitter is also used in level measurement with more accuracy. This level transmitter has the local display and has the local controller, which can be used for level control inside the reservoir also. The transmitter operates on the principle of time of flight of sound wave reflected from the surface of the water level. For example, the Prosonic FMU42 sensor manufactured by Endress + Hauser is suitable for the noncontact level measurement in liquid, pastes, coarse and bulk material. The Prosonic FMU42 also used in open channel flow in or at weirs for flow measurement. The two-wire or four-wire compact transmitter can use level measurement in fluid storage tanks, agitator tank, on stockpiles and conveyor belts carrying solid materials. During measurement, the envelope curve is displayed on the on-site display for straightforward diagnosis. Linearization function (up to 32 points) is used for conversion of the measured value into any unit of length, volume or flow rate.

The hydraulic flow measurement may be made in open channel or a closed pipe. For open channel flow measurement system, a combination of a Parshall Flume with an ultrasonic flow meter is used in measuring raw water flows. The Parshall Flume is a specially shaped fixed hydraulic structure that accelerates subcritical open channel flow of water. The flow rate can be determined by level measurement over the flume using the ultrasonic level transmitter. The overall accuracy is typically ±8% to 10% depending on the geometry of the flume, accuracy of the flow meter, installation perfectness, calibration, etc.

The electromagnetic flow meter is using throughout industries for the last 60 years. These flow meters are used for all conducting liquid, such as water, acids, alkalis, slurries, etc. The typical applications are monitoring of liquid flow, filling, dosing and precise measurement during custody transfer. It works in Faraday law of induction and generator principle. When liquid flows through the magnetic field, the charged particle of fluid crosses the magnetic field generated by two electromagnetic coils, a voltage is induced. This voltage is tapped by two electrodes and is directly proportional to the velocity flow, that is, the volume flow of the liquid. The principle of measurement is independent of pressure, density, temperature and viscosity of the liquid. It can measure volume flow of liquid mixed with solid particles, that is, slurries. The electromagnetic flow meters are available from DN2 to DN2200 in size. There are no moving parts hence maintenance free. The biggest challenge of development of IoT-enabled smart sensors for remote monitoring is the supply of power. Due to the nonavailability of electricity, sensors will rely on batteries needing uninterrupted supply of power. The battery has limited amount of energy storage which will get depleted with use. This means that the primary or disposable batteries are not the viable option for the sensor nodes. The energy of the batteries should be replenished before it gets completely depleted. This tells that the rechargeable batteries along with some energy harvesting techniques are the only viable options [7].

Pressure Transmitters are used in hydrostatic level measurement. Differential pressure transmitters are used in flow measurement in Filter beds, loss of head of filter bed, etc. Absolute pressure transmitters are used to measure water pressure in the pipe during the flow of water. Usually, piezoresistive or capacitive sensor elements are used in pressure transmitter depending upon the applications. Location, parameters, units, and sensing systems for and ideal Smart Water Supply scheme are given in Table 6.1.

6.4.4 Embedded controllers for smart water quality monitoring

Different embedded controllers have been used by the researchers for smart water quality monitoring listed in Table 6.2. Each controller has its salient features: mostly they are using external communication modem, for example, XBee, Bluetooth and GSM/GPRS modem for IoT applications [8]. A few examples of use of the embedded controllers for IoT applications are given in Table 6.2.

Table 6.1 Plant and process parameters and its sensing systems in WTP and water distribution circuit

Sl. no.	Location	Parameter	Unit	Sensing system
1		Plant parameter		
	Raw water	(a) Raw water pump status (on-off-standby)	Logical	Electrical contacts with optoelectronic isolators
	Pump house monitoring	(b) Raw water Header water flow	M^3/hour	Electromagnetic flow meter
		(c) Raw water delivery pressure	kg/cm^2	Pressure transmitter
		Process parameter		Smart pH transmitter
		(a) Raw water pH		
		(b) Raw water turbidity	NPU	Insertion type smart
		(c) Run time	Hrs.	Turbidity analyser
				Virtual timer
2	Clariflocculator-status monitoring	(a) Bridge drive motor	Logical	Electrical contacts with optoelectronic isolators
		(b) Mixture motors	Logical	Ultrasonic open channel flow meter
		(c) Flow to clarifllocculator	M^3/hour	
		Process parameter		
3	Filter house monitoring	(a) Filter bed operation (running-shot off–standby)	Logical	Virtual sensor
		(b) Flow through filter	M^3/hour	Open channel flow meter with the capacitive transducer
		(c) Loss of head	M	Differential pressure transmitter
		(d) Flow meter for backwash	M^3/hour	Electromagnetic type of flow meter
		(e) Backwash overhead	M	Ultrasonic noncontact level transmitter
		Reservoir level		
4	Sludge pump house monitoring	(a) Sludge water level	M	Ultrasonic noncontact level transmitter
		(b) Sludge mixing motor status	Logical	Electrical contacts with optoelectronic isolators
		(c) Sludge pump house operation time	Hour	Virtual clock
5	Compressor room monitoring	(a) Compressor and blower status	Logical	Electrical contacts with optoelectronic isolators
6	Chemical House monitoring	(a) Chemical dosing pumps	Logical	Electrical contacts with optoelectronic isolators
		(b) Coagulant dosing status	Logical	Electrical contacts with optoelectronic isolators
		(c) Prechlorination and postchlorination status	Logical	Electrical contacts with optoelectronic isolators
7	Substation monitoring	(a) Incoming feeder voltage (3 phase)	KV	Potential transformer
		(b) Plant feeder voltage (three phase)	V	Potential transformer

(Continues)

Table 6.1 (Continued)

Sl. no.	Location	Parameter	Unit	Sensing system
		(c) Plant feeder current (three phase)	A	Current transformer
		(d) Power factor	Theta	Multiparameter transmitter
		(e) Frequency	Hz	Multiparameter transmitter
		(f) Active power	KW	
		(g) Reactive power	KW	
		(h) Energy consumption	KW hour	
8	Booster pumping station monitoring	(a) Inflow	M^3/hour	Electromagnetic flow transmitter
		(b) Outflow	M^3/hour	Ultrasonic noncontact level transmitter
		(c) Level	Mt	Electrical contacts with optoelectronic isolators, Virtual timer
		(d) Pump status	Logical	Electrical contacts with optoelectronic isolators
		(e) Pump operating hours	Hours	
		(f) Pump status	Logical	
9	Overhead reservoir monitoring	(a) Residual chlorination pump status	Logical	Electrical contacts with optoelectronic isolators
		(b) Residual chlorine value of OHR water	mg/l	Residual chlorine analyser
		(c) Water input to the OHR	M^3/hour	Electromagnetic flow meter
		(d) Water level in OHR	Mt	Ultrasonic noncontact level transmitter

Table 6.2 Use of embedded controller for IoT application [9]

Controller used	References
AtMega	Thinagaran *et al.* (2015); Mitar *et al.* (2016)
PIC	Theofanis *et al.* (2014); Niel *et al.* (2016); Vinod and Sushama (2016)
Raspberry pi + IOT	Vijayakumar and Ramya (2015); Jayti and Jignesh (2016); Sathish *et al.* (2016)
ARM LPC	Francesco *et al.* (2015); Poonam *et al.* (2016)
Arduino	Anthony *et al.* (2014); Christie *et al.* (2014); Pandian and Mala (2015); Gerson *et al.* (2012)
8051	Li *et al.* (2013)
MSP430	Peng *et al.* (2009)
TI CC3200	Geetha and Gouthami [9]

Some controllers are IoT ready as briefly discussed below.

6.4.4.1 Netduino

Netduino is a modified version of Arduino and is an open-source electronics platform, which uses .NET Micro Framework. It features a 32-bit Microcontroller and has a rich development environment. The Netduino Plus has Ethernet cooked ready. There is full TCP/IP stack with application examples to run the controller and Micro-SD card for storing files. It has four serial ports, six PWM channels and a 12-bit ADC inbuilt [10].

6.4.4.2 Raspberry PI 3 B+

Raspberry PI 3 B+ is a credit card size computer having 1.4 GHz, 64bit quad-core processor, dual-band wireless LAN, Bluetooth 4.2/BLE, faster Ethernet and Power over Ethernet support. It supports third-party operating system, that is, UBUNTU MATE, Snappy UbentuCore, Windows 10 IoT Core, OSME Libreclec, Pinet, RISC OS and Weather Station [11].

6.4.4.3 SimpleLink™ MCU

TI SimpleLink™ MCU platform sets a new standard for development with the broadest portfolio of the wired and wireless ARM™ (System on Chip) in a single software development environment. It is delivering flexible hardware software with programming tool option for IoT applications [12].

6.4.4.4 AWS IoT

Amazon is offering AWS IoT which is a cloud service built-in IoT specific service that helps collect data and send data to the cloud. It makes easy to load and analyse that information and provide the ability to manage the devices for the development of IoT product [13].

6.5 Hardware and software security in industrial IoT

In IoT when the total connectivity accomplished, the large data are hopped through different Internet nodes that may be trusted and processed. IoT manufacturer cautions: 'buyer beware' [14] because there are issues that total connectivity opens the opportunity for intentional hacking or malicious data corruption and contamination.

The technologies combat unauthorized access or modification of data using software or hardware security system. The key challenges in IoT security are as follows:

- The IoT devices are small in size. It is difficult to add extra security circuit to integrate within the IoT device;
- Because of the limitation of internal memory, the computational capabilities of complex security algorithms are not fit for IoT device;
- During the manufacturing process of System on Chip for IoT device, the involvement of third-party manufacturing and hardware Trojan may infect the IoT system

Hardware Trojan [15] is the malicious circuit or modification of the hardware of the IC introduced during the design or fabrication process [16]. It results from the malfunctioning of the IC during its runtime. In modern IC design process, most of the stages are not secure well to detect this kind of threats due to its distributed nature. And most of the cases the Trojans are almost impossible to detect in the packaging and testing phases.

6.5.1 Software-based IoT security

The software security system resources kept in the host processor, for example, in IoT host processor usually are the cloud server. The CISCO security agent is one of the examples of the software system. It is endpoint intrusion prevention system [17], rule-based and examines system activity, real-time traffic for determination of system behaviour indicating the attack. But in a System Security Architecture, the software security is considered to be a weak link between a large number of connected systems.

6.5.2 Hardware-based IoT security

The hardware-based IoT security uses a dedicated integrated circuit or a process, which provide the cryptographic function to protect malicious attack. The security IC contains hardware blocks like math accelerator, random number generator, nonvolatile memory, tamper detection circuitry and a physically unclonable function (PUF). An example of such an integrated circuit is DS28E38 manufactured by Maxim that generates PUF to protect against attack [18]. The integrated circuit DS28E38 is an ECDA public key-based secure authenticator that uses ChipDNA® PUF technology. The ChipDNA involves physically unclonable function (PUF) that delivers cost-effective protection from invasive physical attack. It actively monitors Die-shield and reacts to intrusion attempts. The store data are cryptographically protected from discovery. The other features are but not on the list, are unique and unalterable factory programmed 64-bit identification number. This ChipDNA Integrated circuit is tiny in size and has the potential ability in the IoT security application.

6.6 Water monitoring, SCADA and IoT in water sector

6.6.1 Some earlier work on ICT/IoT-based water surveillance

In the year 2014, Temido *et al.* introduced SCADA and smart metering system for water sector companies, which create enormous value addition. The information and communication technology was applied to operation and maintenance of the plant system [19].

The understanding that the contribution of hydraulic pressure for failure analysis in main trunk pipeline for urban sub-network using SCADA system was introduced by Williams *et al.* in the year 2014. The Failure analysis using frequency analysis technique was the central theme of this research work. The data failure recording option is available with the SCADA system used for the forensic investigation [20].

A novel technique for Anomaly Detection System (ADS) using SCADA on the systematic approach of the pattern-based technology developed by Adrian *et al.* in the year 2013. This technique uses graphical interface for model-based SCADA topology and some auto-generated ADS rules. The methodology is user-friendly for anomaly detection in the SCADA network [21].

Gligor *et al.*, in 2012, implemented an orientation-based SCADA system to facilitate the business intelligence suitable to the production sector like water treatment plant, etc. The system developed was economical in all respect. The author has done some analysis regarding various aspects of the SCADA system like safety, implementation, design of the system, etc. For that purpose, the service-oriented architecture (SOA) has been used [22].

The informatics, logistics and governance was introduced successfully by Chowdhury *et al.* (2013) in the Water Treatment Plant using SCADA Architecture. The Artificial Neural Network (ANN) used in water treatment problems to make the system intelligent. They also proposed the diagram of the SCADA architecture for water treatment plant with the idea of water governance. This idea uses the business intelligence for public interest [23].

Muntean *et al.* in 2011 used knowledge discovery in the SCADA Database using the data mining technique. This knowledge discovery used to evaluate performance for better functionality of the whole process plant. The primary purpose of this work was to analyse the classifier that is best suited to the data set provided by SCADA system. Discarding normal values and transmitting warning messages as the anomaly value by the classifier was the theme of the system [24].

Bello *et al.* used fuzzy logic for the development of dynamic modelling and predictive control for coagulation chemical dosing for water treatment plants. The Fuzzy model predictive control (FMPC) strategy was proposed in regulating the dosing treatment. MIMO processes for linearized Takagi–Sugeno (T–S) fuzzy model technique was implemented [25].

Dutta in his article discussed the view over SCADA system for business intelligence in water treatment plants first time in India. Arsenic contamination in drinking water was a significant problem for most of the rural areas in West Bengal. The pure drinking water production plant for rural areas in North 24 Paraganas (at Mongol Pandey Water Treatment Plant) was developed under this mission. The treatment plant has 7.5 MGD capacity, two Booster Stations and fourteen overhead reservoirs distributed over 7.4 million people covering 369 sq. km area. The SCADA system has sufficient amount of plant running data for technical analysis purposes [16].

Wang *et al.*, in the year 2014, analysed water hammer assessment techniques for water distribution systems for the mass community to protect valuable life and wealth that can be destructed by this problem. The business intelligence is one of its sole aspects to calculate the water hazards due to five different reasons discussed in this chapter. The pipe rupture risk factors for maximum water pressure, maximum vacuum, maximum vapour volume and maximum transient force have been analysed by the knowledge of hydraulic transient flow analysis [26].

The application of automation in modified SCADA system empowered the overall plant efficiency and surveillance to the functionality of the plant for the Great lakes-upper Mississippi River to provide service for the public health and environmental aspects [27].

In the year 2013, Vermesan wrote a book '*Internet of Things – Converging Technology for Smart Environment and Integrated Eco-System*', published by River

Publishers Series in communications, IIRC Book open access, 2013, TP- 343. This book discusses vastly on the IoT and its different applications [28].

Vermesan *et al.*, also in the year 2014, wrote another book titled '*Internet of Things – From Research and Innovation to the Market Deployment*', published by River Publishers Series in communications, IIRC Book open access. The idea on the IoT in the market deployment to different plant automation techniques has been discussed in this book [29].

Evans, in the year 2011, wrote a book titled '*The Internet of Things, How the Next Evolution of the Internet Is Changing Everything*', published by Cisco Internet Business Solutions Group (IBSG) [30].

Similarly Karimi, in the year 2012, wrote a book titled '*What the Internet of Things (IoT) Needs to Become a Reality*'. From this book, the basic idea along with the application criteria of the IOT model can be achieved [1].

Bunn, in the year 2007, discussed on 'Closing the Loop in Water Supply Optimization' [31].

Merchant *et al.* [32], in the year 2014, proposed an Analytics-driven water management system for Bangalore city. It was an on-going project for the minimization of the complexity of instrumentation and control mechanism in the water sector. They have developed new software using important customizable performance indicator (KPI) to implement the business intelligence which has great importance in the water sector to handle different problematic situations.

The software developed for this purpose is Smart, reliable and user-friendly for the plant officials. The business level optimization was also a great achievement of this system minimizing different losses like energy loss, produced water loss, etc. which help to build the business intelligence productively. The Minneapolis Province of USA [33] has utilized SCADA technology [34–36] for real-time energy usage connected with other system parameters like storage levels of water and the intake volume. The essence of this system is to have an effective control using minimum staff support.

Nikolić *et al.*, in 2010, implemented a new idea of intelligent decision-making system based on Fuzzy controller [37] for waste-water treatment plant using SCADA system. The authors gave primary emphasis on the development of remote monitoring and control systems for waste-water treatment and its improvement. The development of RTU module for this type of methods is discussed with its intelligent attributes [9].

6.6.2 Smart water supply project: design consideration

Following features from a smart water supply project can be expected:

- The full automation of pumping machinery, valves, motors, actuators, sensors for different stages of water treatment process and distribution, such as abduction, chlorination, filtration, storing and distribution in a smarter way. All components should have communication facilities through the Internet. The status of all assets should be visible. They can communicate with each other and take and execute a command from cyber-physical decision system either available locally or in a remote place.

- The water quality surveillance to be adopted in the system using distributed smart water quality sensors. The data of water quality as per WHO standard should be available 24×7 on real-time basis through ICT anywhere any place.
- The project must ensure minimum water loss during distribution process by implementing IoT-based pressure and flow sensors in all distribution water pipes in metering areas with the use of cognitive water leakage management system.
- Ensure health and life of pumping machinery, filter beds through cognitive intelligent controls and monitoring health with time.
- Minimize energy consumption of treatment and delivery process with the use of a smart electrical system.
- Maintain supply and demand harmony through IoT-based automatic water meter reading of consumers and available treated water data in real time.

For water supply scheme, all sensors and actuators should be smart and IP based. They are available in compact wireless version, or a simple GSM or GPRS modem can convert them to IoT compatible. For local application within the water treatment plant or pump house the sensors may be wired and then connected to the Internet, but for the distribution system, wireless sensors are preferred as sensor network.

6.6.2.1 Proposed smart water treatment plant

The Proposed Smart Water Treatment Plant will be installed beside old WTP at one corner of the city and shall be hydraulically designed for continuous 24×7 operation to produce 40 MLD in 20 hours operation of treated water to specified quality standard (Figure 6.7). The treatment plant scheme shall broadly comprise stream of inlet, Parshall Flume, Flash Mixer, Plate settler and common rapid gravity filter bed and ground level clear water reservoir. The water from ground reservoir will be pumped to elevated tank situated over a hill. From this elevated tank at hill water will be distributed to the distribution system under gravity flow. There are 50 overhead reservoirs, fully automated to distribute house to house. The water from this treatment plant will ensure specified physical, chemical and bacteriological quality. The sludge from plate settler tank flows to the sludge tank and collected and converted to solid sludge at centrifuge system.

The sludge water then will be stored in a dirty water tank and will be used for recycling with the treatment process. There are other two old water treatment plant of different capacity situated at different other corners of the city and will integrate to the proposed single smart water supply scheme with enhancing capacity to fulfil the demand of water to the entire population of the city in a 24×7 manner.

6.6.2.2 Proposed SCADA system

Besides the usual civil and mechanical part of the water treatment plant and distribution, electronic surveillance and control plant and process parameters are very

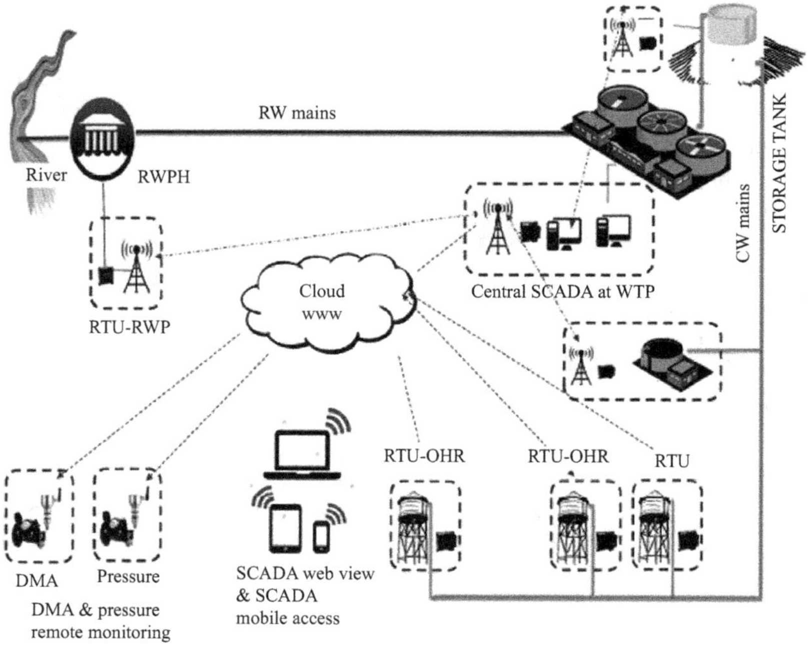

Figure 6.7 The conceptual layout of the smart water supply scheme

important to make the water supply sustainable and steady to produce drinkable water in 24×7 manners. Therefore, distributed supervisory control and data acquisition system have been proposed in this design to cover the following areas:

1. Full control and monitoring of proposed raw water intake using latest PLC-based SCADA system;
2. Monitoring of old raw water intake using latest PLC-based SCADA system;
3. Full automation and monitoring of plant and process of proposed 40 MLD water treatment plant using latest PLC-based SCADA system;
4. Monitoring of plant and process of old 60 MLD water treatment plant using latest PLC-based SCADA system;
5. Monitoring water delivery, water quality and pumping status of 20 MLD old WTP;
6. Monitoring water delivery, water quality and pumping status of 19 MLD old WTP;
7. Full automation and monitoring for 50 overhead reservoirs which are covering 50 distribution zones and
8. DMA assessment using proposed solar/battery-powered GSM/GPRS-enabled electromagnetic bulk flow meters and pressure transmitters, proposed to install at suitable places of 50 distribution zones to prevent leakage and damages and to minimize non-revenue water.

Technology

- A latest cutting-edge technology of 4th generation SCADA
- Fail-proof programmable logic controller (PLC)-based RTU for data.

Acquisition and control

- Hybrid communication technology for SCADA data and control signal

Transmission (OFC LAN, Ad hoc WAN and GSM/GPRS)

- Firewall protected redundant servers and workstation computer system;
- Query-based flexible data handling and compilation system;
- Online IP68 smart transmitters for water quality, flow, pressure and level instrument;
- Web SCADA technology for 24×7 SCADA availability;
- Latest technology of human–machine interface and
- Mobile SCADA technology.

Parameters to monitor

- Substation
  - Incoming/outgoing feeder status
  - Electrical parameters, for example, HT/LT voltage and current, active and reactive power, frequency, QF, KV, KVA and transformer oil temperature
- Raw water intake (new and old WTP)
  - Jack well level, pump ON/OFF fault status, pump working time, pump pressure, raw water flow, raw water delivery pressure, raw water pH and turbidity (only at new WTP)
- New WTP
  - All hydraulic parameters (water level) from inlet well to the sludge tank, all pump and motor working status, working time, all filter bed parameters, clear water delivery flow and pressure to the elevated tank at hill, filter bed status, backwash tank level flow, backwash monitoring, all control valve status, chemical house parameters, clear water tank level, pH, turbidity, residual chlorine, clear water delivery flow and pressure to the distribution... many more.

Old WTP-1 and old WTP-2

- Pump status, CWR water level, delivery pressure and flow, pH, turbidity and residual chlorine of clear water

Monitoring the zonal overhead reservoirs (50 nos.)

- Water inlet flow to OHR
- Water outlet flow to OHR
- Water inlet control valve status
- Water outlet control valve status
- Water level OHR
- Outlet water pressure
- 3 Ph. supply voltages
- UPS output voltage

6.6.2.3 Operational philosophy of monitoring

Monitoring of all parameters of SCADA system related to the Smart Water supply scheme will be available:

1. Local field display accessible to the operator at SCADA central monitoring and control room through large HMI display panel/HMI viewer;
2. At intake, different houses of WTP through regular computer;
3. (SCADA web view)
   Anywhere, any place any time on Internet-enabled desktop PC/Tabs/Smart mobile, etc.;
4. Automatic post chlorination system which doses appropriate chlorine to the clearwater reservoir by considering water volume, chlorine concentration and residual chlorine content in a tank in real time using intelligent and smart control system algorithm through SCADA. Such that there will be the optimal use of chlorine, which ensures the accepted bacteriological quality;
5. Elevated tank at hill level control;
6. Backwash tank level control;
7. Any point of hydraulic overflow generation and automatic shutdown;
8. The alarm for major fault and automatic shutdown of the treatment process;
9. During the shutdown, automation ensures all initial condition of pumps, control valves and actuators such that during restart the WTP runs normally;
10. OHT water level can be set from the remote. A local controller will maintain OHT level and prevent overflowing. No altitude valve is required;
11. Water supply to and from OHT can be controlled from remote.
12. Flow control loop of OHT automation control water flows to the distribution. The set value can be given from the remote. No flow control valve is required;
13. DMA assessment (to minimize non-revenue water);
14. Water authorities and operators face series of challenges daily preserving the supply of water, monitoring the volume conveyed networks, detecting leaks and cost-cutting by reducing the non-revenue-water (NRW);
15. The battery-powered GSM/GPRS-based bulk electromagnetic flow meter along with battery-powered GSM/GPRS-based smart pressure transmitters are proposed and
16. Enhancing district metering areas (DMA) diagnostics and making alerts rapidly to the sudden leaks. Fifty sets of flow and pressure equipment will use for 50 DMA areas.

6.6.2.4 The proposed scheme will help

- *Transferring* water and bulk metering data to central SCADA system;
- Providing comprehensive pressure monitoring information district area diagnostic;
- It allows setting of threshold values, which triggers water leak alert;
- It perfectly adapted to the environmental constraints of the metering chamber;
- It provides long-term results for analysis of water network efficiency and areas with the most significant value of water leakage;

- Record minimum night flow to facilitate leak detection and
- Take into account of key customers for more accurate flow measurement diagnostics.

6.6.2.5 The benefit of the IoT-based SCADA

1. Faithful monitoring and control from the single umbrella of 24×7:
    (i) Smart water supply scheme;
2. Improve the quality of operation and maintenance of plant and machinery of the water supply scheme;
3. Improve the life of plant and machinery through preventive maintenance and scheduling;
4. Can be integrated with other third-party systems, for example, direct metering area (DMA) assessment, pipe leakage, automatic water billing, ERP, etc.;
5. Can control water distribution automatically according to demand and resources and improve distribution rationally;
6. Will help to minimize non-revenue water;
7. Data can be stored for a long time (5–10 years) and can transfer electronically. Secure data backup and restore;
8. Improve the quality of water production through continuous surveillance of plant and process parameters;
9. Minimize chemical inventory through the appropriate dosing;
10. Help in decision making through online analysis and
11. SCADA in real time available on any Internet-enabled computer, laptop, tablet, mobile anywhere, any place and any time.

6.6.2.6 Expected technical features of smart water supply and distribution

- The full automation of pumping machinery, valves, motors, actuators, sensors for different stages of water treatment process and distribution such as abduction, chlorination, filtration, storing and distribution in a smarter way. All components should have communication facilities through the Internet. The status of all assets should be visible. They can communicate with each other and take and execute a command from cyber-physical decision system either available locally or in a remote place.
- The water quality surveillance to be adopted in the system using distributed smart water quality sensors. The data of water quality as per WHO standard should be available 24×7 on real-time basis through ICT anywhere any place.
- The project must ensure minimum water loss during distribution process by implementing IoT-based pressure and flow sensors in all distribution water pipes in metering areas with the use of cognitive water leakage management system.
- Ensure health and life of pumping machinery, filter beds through cognitive intelligent controls and monitoring health with time.
- Minimize energy consumption of treatment and delivery process with the use of a smart electrical system.
- Maintain supply and demand harmony through IoT-based automatic water meter reading of consumers and available treated water data in real time.

6.6.3 The leak detection to minimize nonrevenue water in smart water grid

A typical scenario of water distribution network in a city consists of water pipe network typically categorized as water mains, branch pipelines to overhead reservoirs and pipelines from overhead reservoirs to the consumer taps. It spreads over a few hundred square kilometres. The distribution pattern is just like an electric grid network in a city. Water pipes are usually buried underground, the piping runs kilometres after kilometres. The underground leakage may occur in an old buried pipe due to ageing or sudden water hammer attack. The severe problem started to localize the position of leakage mainly on buried pipelines [38]. The leakage in water pipe which runs over the earth is visible and accessible to detect. Leaks in the system also decrease service water pressure. As a result, more electrical energy is required to send water to the consumer taps. This increase in pressure exacerbates the severity of the leak, which means more water and energy is lost [38].

In earlier days, the water leak survey method most often performed by trained human operators, who find the water leaks by detecting sounds of water leakage by using a type of stick leak detector placed at the ground level [38]. Sometimes this method fails when water leakage is small, and the sound of leakage is too feeble. The investigation time is so ridiculously high that it is not the sustainable method and disrupts the water supply for an extended period. The smart water grid concept is similar to the Smart electric grid. Sensor-based information and communication technology runs parallel with the piped distribution system. IoT-based cyber-physical system or wireless sensor network is playing the vital role of a Cyber-Physical system. The pipelines are not simple pipelines; they are monitored and controlled from the remote. The wireless pressure, flow and vibration sensors installed at specific intervals on the pipes. The pressure, flow and vibration data are collected by the sensor nodes, connected to the Internet and send to the Cyber-Physical system at a cloud (Figure 6.8). A repeater is a device for receiving information from the sensor logger that is installed on a public utility pole or similar facility within the short distance of each sensor logger. The sensor logger is battery powered so that it may be used continuously for a long period without maintenance.

Kei *et al.* (2015) used a large number of the miniature wirelessly connected vibration sensor and data logger in the leak detection of water grid in Japan. These small wireless sensor nodes installed over the entire pipelines of the water distribution circuit with an interval of 200 m.

These WSNs exchanging data with the base station is the wireless communication device for gathering information from the repeater. It uses ISM band licence-free low power radio for communication for approximately 0.5 to 1 km distance in line of sight from each repeater-station. The computers at cloud compute sensor data, communicated via the PTSN to the computing facilities of NEC Japan. The NEC also deployed [39] software called the Water Leak Analysis.

Engine that monitors water leaks is routinely based in combination with the sensor loggers and calculate the certainty of a water leak detection by each sensor logger via a unique index called the Water Leak Score. When a score exceeds a specific value, the engine predicts the possibility of a water leak near to the sensor

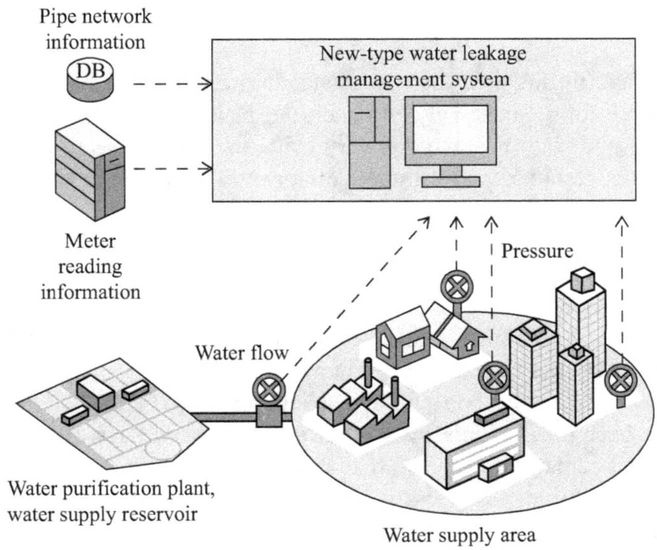

Figure 6.8 Cyber-physical system for water supply scheme

logger and starts a detailed analysis of the data. The complete report consists of a 'correlation analysis' between multiple sensors around the location where a leak is suspected. The water leakage precisely pinpointed with an accuracy of about 1 metre. There has traditionally been an approach to install wireless-vibration-sensor nodes to detect water leakage in their proximities, but it requires the extra work for an operator to open the utility hole at the suspected location and the attachment of water leak detector to identify the precise water leak point.

The disadvantage of this system is that the infrastructure needs dedicated wireless sensor network and dedicated ICT service which is expensive.

'Earnpipe', a novel hybrid water pipeline monitoring (WPM) method that has been proposed by Karraya *et al.* [40], uses Leak Detection Predictive Kalman Filter (LPKF) and modified Time Difference of Arrival (TDOA) based on pressure measurements. The data collected from sensors is filtered, analysed and compressed with Kalman Filter-based algorithm. The sudden escape of water causes a pressure wave along the pipeline. This wave propagates at the speed of sound through the pipeline. Time difference between two WSN is measured, and the application of TDOA algorithm has predicted the leak.

Hydrophone is more sensitive than the vibration sensors and is useful in detecting leaks in larger pipes. The best results are obtained if the sensors installed in 100 m or 150 m after location of the air valves not more than 750 m apart [41].

Virtual district metering area utilizes flow meters to monitor inflow and outflows to different subsections of District metering areas combined with automated meter reading from customer premises. The mass balance calculation can be used to derive the actual water loss (nonrevenue water) in the area. Operators can then focus their leak survey efforts on the areas where high water losses observed.

6.7 Conclusion

It is a lapse that till this time there is no standardized concept of the Smart Water Supply Scheme for a Smart City. As such the planning related to the concerned infrastructure and facilities largely vary with the local scenario in respect of available water resources, topography of location, population density, level of implementable technology, etc. The foregoing text of this chapter attempts to review and discuss the IoT and its role in monitoring and control of water treatment plant and its distribution in a smart city. In combating the challenges inherent in the process, an in-depth discussion on the latest IoT hardware, software and security issues have entered herein. It is believed that the use of IoT in Water Distribution Management with smart concept is an inevitable tool. To strengthen such conviction, relevant case studies on strategic planning of IoT-enabled Water Supply System have been cited. While the subject encompasses issues of great width and complication, it cannot be expected that a single article would show the path with unambiguous foot-prints. Rather it is intended that the treatise will invoke further thought and amplification of conceptual points linked with the subject.

References

[1] K. Karimi, 'What the Internet of things (IoT) needs to become a reality', 2012 Tp. 16, http://cache.freescale.com/files/32bit/doc/white_paper/INTOTHNGSWP.pdf

[2] J. Bradley, C. Reberger, A. Dixit and V. Gupta, 'Internet of everything: a $4.6 trillion public-sector opportunity', 1191 http://globenewswire.com/news-release/2014/04/16/627796/10077088/en/iTalk-to-Launch-Medical-Marijuana-M2M-Enterprise-Solution.html

[3] How to interface sensors to the Internet of things DigiKey, https://www.digikey.in/en/articles/techzone/2013/oct/how-to-interface-sensors-to-the-internet-of-things

[4] http://www.analog.com/en/products/analog-to-digital-converters/integrated-special-purpose-converters/digital-temperature-sensors/adt7320.html#product-overview

[5] https://www.digikey.in/product-detail/en/amphenol-advanced-sensors/NPA-500B-005D/235-1313-1-ND/3757836

[6] https://www.digikey.in/product-detail/en/analog-devices-inc/AD7176-2BRUZ/AD7176-2BRUZ-ND/3777418

[7] R. J. Kumar and S. C. Mukhopadhyay, 'Monitoring water in the treatment and distribution system', *New Generation Sensors and Systems*, S. C. Mukhopadhyay (Ed), Series: Smart Sensor Measurement and Instrumentation, Vol. 16, Springer International Publishing, 2016, pp. 257–288, DOI 10.1007/978-3-319-216713

[8] M. Haefke, S. C. Mukhopadhyay and H. Ewald, 'A Zigbee based smart sensing platform for monitoring environmental parameters', *Proceedings of*

IEEE I2MTC 2011 Conference, IEEE Catalog number CFP11MT-CDR, May 10–12, 2011, Hanghou China, pp. 1549–1556

[9] S. Geetha and S. Gouthami, 'Internet of things enabled real-time water quality monitoring system', *Smart Water*, 2017, 2,1, DOI 10.1186/s40713-017-0005-y

[10] https://www.adafruit.com/product/1104

[11] https://www.raspberrypi.org/products/raspberry-pi-3-model-b-plus/

[12] http://www.ti.com/wireless-connectivity/simplelink-solutions/overview/overview.html?gclid=CjwKCAjwq_vWBRACEiwAEReprHGOIAUalJDqhfIApT9ecOacpYceGg-rjz4caQmN99_LvstqlhxE1RoCX8oQAvD_BwE

[13] https://aws.amazon.com/iot/?sc_channel=PS&sc_campaign=acquisition_IN&sc_publisher=google&sc_medium=iot_nb&sc_content=iot_p&sc_detail=iot&sc_category=iot&sc_segment=154390601933&sc_matchtype=p&sc_country=IN&s_kwcid=AL!4422!3!154390601933!p!!g!!iot&ef_id=Wp43BQAABH6RAxpj:20180424094306:s

[14] B. Baker, 'IoT security: hardware vs software', EDN network, January 17, 2018, https//edn.com/electronics-blogs/bakers-best/4460252/IoT-security—hardware-vs-software

[15] S. Koley and P. Ghosal, 'Addressing hardware security challenges in Internet of things: recent trends and possible solutions', UIC-ATC-ScalCom-CBDCom-IoP, IEEE Computer Society, 2015, DOI 10.1109/UIC-ATC-ScalCom-CBDCom-IoP.2015.105

[16] P. K. Dutta, 'SCADA system for North 24 pgs arsenic areas: rural water supply scheme at Mongal Pandey water treatment plant – first time in India', *Journal of IPHE India*, 2007, 8 (2), 24–31

[17] Wikipedia, 'CISCO Security Agent', https://en.wikipedia.org/wiki/Cisco_Security_Agent

[18] Datasheet, 'DS28E38 Deep Cover® Secure ECDSA Authentication with Chip DNA PUF protection', www.maximintegrated.com

[19] J. Temido, J. Sousa and R. Malheiro, 'SCADA and Smart Metering systems in water companies: a perspective based on the value creation analysis', 12th International Conference on Computing and Control for the Water Industry, CCWI2013, *Procedia Engineering*, 2014, 70, 1629–1638, http://dx.doi.org/10.1016/j.proeng.2014.02.180

[20] G. Williams and G. Kuczera, 'Analyzing SCADA to understand the contribution of hydraulic pressures to trunk-main failure', 16th Water Distribution System Analysis Conference, WDSA2014 – Urban Water Hydro informatics and Strategic Planning, *Procedia Engineering*, 2014, 89, 1452–1459, http://dx.doi.org/10.1016/j.proeng.2014.11.472

[21] D. Adrian Rusua, B. Gengeb, and C. Siaterlis, 'SPEAR: A systematic approach for connection pattern-based anomaly detection in SCADA systems', The 7th International Conference Interdisciplinarity in Engineering, INTER-ENG 2013, 10–11 October 2013, Petru Maior University of Tirgu Mures, Romania, *Procedia Engineering*, 2014, 12, 168–173

[22] A. Gligor and T. Turc, 'Development of a service oriented SCADA system', International Conference Emerging Markets Queries in Finance and Business,

Petru Maior University of Tîrgu-Mures, ROMANIA, October 24th–27th, 2012, *Procedia Economics and Finance*, 2012, 3, 256–261, http://dx.doi.org/10.1016/S2212-5671(12)00149-9

[23] R. K. Chowdhury and W. El-Shorbagy, 'Informatics, logistics and governance in water treatment processes', Department of Civil and Environmental Engineering, United Arab Emirates University, Al Ain, United Arab Emirates, 2013, DOI: 10.5772/54386

[24] M. Muntean, H. Valean, C. Rotar and M. Risteiu, 'Knowledge discovery in a SCADA System,' DATABASE Acta Universitatis Apulensis, 2011, pp. 37–54, ISSN 1582-5329

[25] O. Bello, Y. Hamam and K. Djouani, 'Fuzzy dynamic modelling and predictive control of a coagulation chemical dosing unit for water treatment plants', *Journal of Electrical Systems and Information Technology*, 2014, 1, 129–143, http://dx.doi.org/10.1016/j.jesit.2014.08.001

[26] R. Wang, Z. Wang, X. Wang, H. Yang and J. Sun, 'Water hammer assessment techniques for water distribution systems', 12th International Conference on Computing and Control for the Water Industry, CCWI2013, *Procedia Engineering*, 2014, 70, 1717–1725, doi: 10.1016/j.proeng.2014.02.189

[27] Recommended Standards for Water Works, 2003 Edition, 'A report of water supply committee of the great lakes – Upper Mississippi River Board of State of Provincial Public Health and Environmental Managers', Health Research Inc., Health Education Services Division.

[28] O. Vermesan and P. Friess, 'Internet of things – converging technology for smart environment and integrated eco system', River Publishers Series in Communications, 2013. IERC Book Open Access, 2013, TP-343

[29] O. Vermesan and P. Friess, 'Internet of things – from research and innovation to market deployment', River Publishers Series in Communications, IERC Book Open Access, TP-143, 2014, http://www.internet-of-things-research.eu/pdf/IERC_Cluster_Book_2014_Ch.3_SRIA_WEB.pdf

[30] D. Evans, 'The Internet of things how the next evolution of the Internet is changing everything', 2011, Cisco Internet Business Solutions Group (IBSG), Tp. 11

[31] S. Bunn, 'Closing the loop in water supply optimisation', http://www.schneider-electric.co.uk/documents/support/white-papers/energy-optimisation-software-for-drinking-water.pdf

[32] S. C. Mukhopadhyay and J. A. Jiang, Eds., *Wireless Sensor Networks and Ecological Monitoring*, Vol. 3, Smart Sensors, Measurement and Instrumentation, Springer-Verlag, 2013.

[33] S. C. Mukhopadhyay and A. Mason, Eds., *Smart Sensors for Real-Time Water Quality Monitoring*, Vol. 4, Smart Sensors, Measurement and Instrumentation, Springer-Verlag, 2013.

[34] A. Merchant, M. S. Mohan Kumar, P. N. Ravindra, P. Vyas and U. Manohar, 'Analytics driven water management system for Bangalore city', 12th International Conference on Computing and Control for the Water Industry, CCWI2013, *Procedia Engineering* 70, 2014, 1137–1146, doi: 10.1016/j.proeng.2014.02.126

[35] 'Minneapolis water works uses SCADA system', www.epa.gov/safewater, www.awwa.org, www.health.state.mn.us/divs/eh/water

[36] V. Nikolić, Ž. Ćojbašić, I. Ćirić, E. Petrović, 'Intelligent decision making in waste water treatment plant SCADA system', Facta Universitatis Series: Automatic Control and Robotics, 2010, 9, 1, 69–77

[37] M. S. A. Rahman, S. C. Mukhopadhyay, P. L. Yu, *et al.*, 'Detection of bacterial endotoxin in food: new planar inter digital sensors based approach', *Journal of Food Engineering*, 2013, 114, 346–360, http://dx.doi.org/10.1016/j.jfoodeng.2012.08.026

[38] O. Marlyusheva, 'Smart water grid', Plan-B, Technical Report, Master's Thesis, Colorado State University, Department of Civil and Environmental Engineering, 2014

[39] F. Kei, M. Yuuji, I. Kazuo, I. Yusuke, Y. Ayumi and T. Toshiaki, 'A water leak detection service based on sensors and ICT solutions' *NEC Technical Journal*, 2015, 9 (1), 107–110

[40] F. Karraya, A. Garcia-Ortiz, M. W. Jmala, A. F. M. Obeid and M. Abida, 'EARNPIPE: a test bed for smart water pipeline monitoring using wireless sensor network', 20th International Conference on Knowledge-Based and Intelligent Information and Engineering Systems, *Procedia Computer Science*, 96, 285–294

[41] Public Utilities Board Singapore, 'Managing the water distribution network with smart water grid', *Smart Water*, 2016, 1(4), DOI 10.1186/s40713-016-0004-4

Chapter 7

IoT for smart homes

Anindya Nag[1], Md. Eshrat E. Alahi[1], Nasrin Afsarimanesh[1], Sumedha Prabhu[1], and Subhas Chandra Mukhopadhyay[1]

The usage of smart homes in real-time applications has been one of the state-of-the-art due to the quality of life they provide to the residing life. Internet of things (IoT)-based smart homes are booming in the market where a large number of IoT-connected daily used items are commercially available. These devices are used in the smart homes for ubiquitous monitoring of the different activities of the residing people. This chapter gives an overview of some of the smart devices available for IoT-based smart homes along with some of the research work done on IoT-based smart homes in the laboratory. This also showcases some of the sensors that have been used for biomedical applications and have the potential to be used in smart homes.

7.1. Introduction

With the advancement in science and technology, there have been many ways to improve the quality of life for human beings. The implementation of specific methodologies in terms of electronic assistance had been researched and worked upon to increase the longevity of human life. One of the most popular choices for this goal has been the implementation of smart homes [1], where people with ailment have been kept and taken special care. The importance of smart homes lies in the precise monitoring of the people regarding their day-to-day activities to figure out any abnormality in comparison to their normal life. Among the different types of specialized homes designed for monitoring people in commercial and research-based conditions, smart homes for elderly care has been a standout. In these specialized homes, sensors have been positioned at different locations to identify the activities and movements of the residential elderly people.

One of the biggest advantages of these homes is the automated monitoring system which makes it easier and efficient to keep a track of the minute changes happening in every individual's life. Figure 7.1 shows the schematic diagram of the operating phenomenon of a smart home [2]. It is seen that the sensors located in the

[1]School of Engineering, Macquarie University, Macquarie Park, NSW 2113, Australia

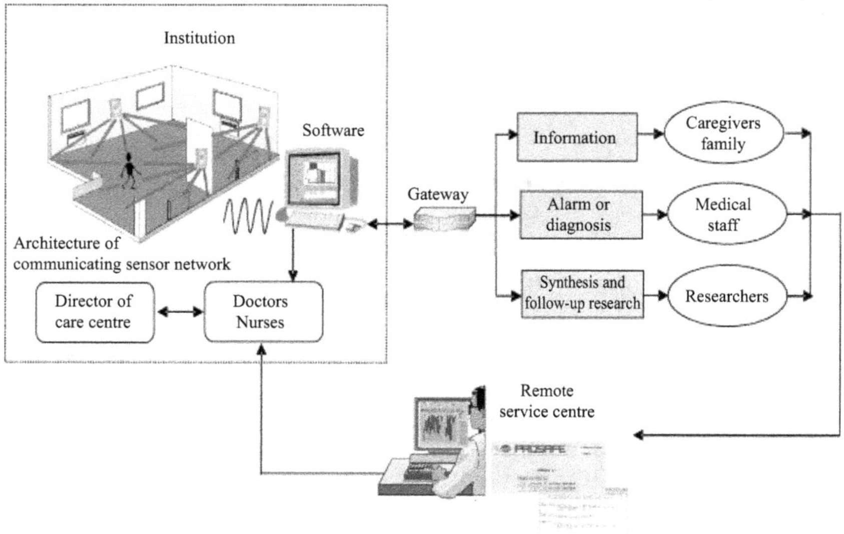

Figure 7.1 Schematic diagram of the working phenomenon of a general smart home [2]

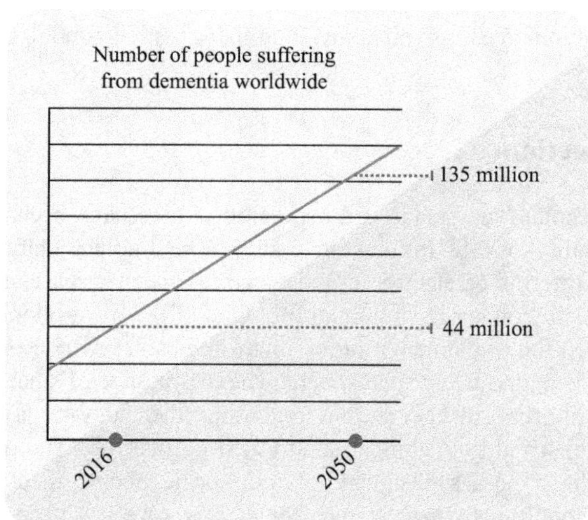

Figure 7.2 Increase in the number of elderly people suffering from dementia between the years 2016 and 2050 [3]

house form a sensor network in order to communicate among each other to send the signals to the monitoring unit via a gateway. The monitoring unit then immediately informs the medical staffs, family members or researchers depending upon the type of assistance required. This approach seems to be an effective way to significantly reduce the probability of casualties during an emergency. Figure 7.2 shows a fact of the increase in the number by a factor of 3 for the elderly people suffering from dementia between the years 2016 and 2050 [3]. As a result of these problems, researchers all around the world have been trying to come up with two solutions. First, smart devices are designed to monitor the specific activities of the elderly people. For example, smart beds, smart ovens and fall detectors are some of the devices that are installed in different locations of the house.

Second, these smart sensing devices employ Internet of things (IoT)-enabled technology to monitor and control the operations inside these homes in an easier and controlled manner. In this chapter, we will talk about some of the significant research work done on IoT-based smart homes. We will also talk about some of the IoT-embedded commercial smart sensing devices that are used in smart homes. Finally, we will also discuss some of the parameters that are being monitored along with their associated sensing systems in these specialized homes.

7.2 IoT-based smart homes

The usage of smart sensing devices in the specialized smart homes designed for elderly people have started since the last two decades [4,5], by the time sensors were available commercially to be used for ubiquitous applications. The sensing devices have been implemented in everyday items, forming a cognitive network for the individuals residing in the house to provide an environment for the people to have a happy and independent life. There are some established research works going on around the world that utilizes IoT-based smart sensing devices in these specialized care centres. Some of them are Gator Tech Smart House, University of Florida [6], Smart Community Alliance, Japan [7], KIDUKU, Japan and Ireland [8], Smart homes, UK [9], etc. The ultimate agenda of these projects is to monitor the normal and healthy lifestyle of elderly people residing in these houses. This section defines IoTs, their utilization in smart homes in integration with some of the commercial smart devices and finally some of the research work going on IoT-based smart homes in laboratory-based environment.

IoT is one of the most common concepts of things that are readable, controllable, addressable and locatable via internet in the twenty-first century. All the surroundings of our daily life can be associated with the Internet due to the growing faster computing and communication abilities. Figure 7.3 shows the schematic diagram of a smart home employed with different IoT connected utilities. IoT-based smart homes consist of several sensors, which are connected wirelessly to develop supporting distributed networks. Each IoT-enabled sensor node in the smart home includes three subsystems: (1) sensor subsystem for environment sensing, such as temperature, humidity and light intensity; (2) processing subsystem, consisting a microcontroller and integrated circuit to process the sensor data for computation and (3) a communication subsystem for exchanging the collected data between different sensors [10–15].

The traditional Wireless Sensor Networks (WSNs) offer specific applications which are mostly designed as a closed system; however, IoT-based applications are independent of specific applications [16] and are more focused to develop a

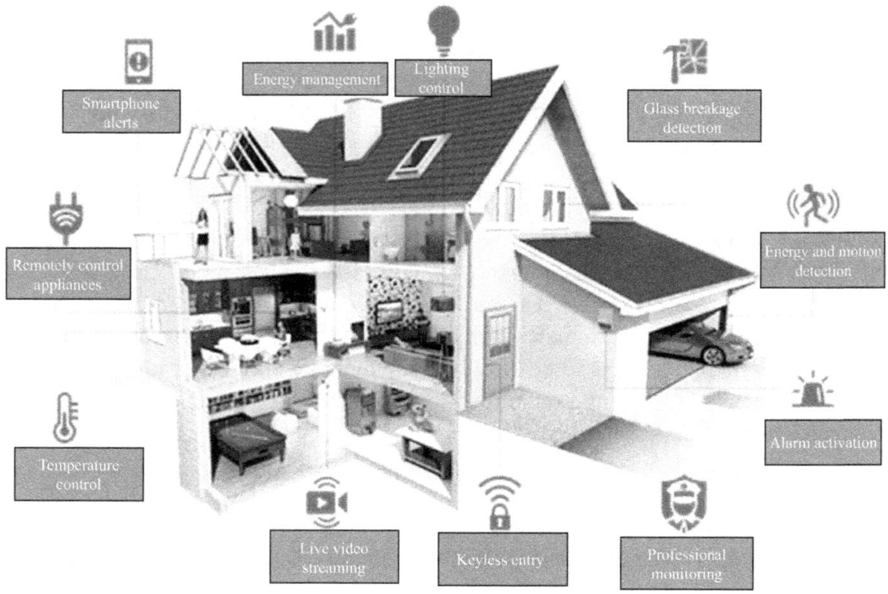

Figure 7.3 Schematic diagram of an IoT-enabled smart home depicting the employment of smart sensing devices in different utilities

large-scale WSNs infrastructure, which could support open standard. In most of the IoT-based smart home systems, the smart actuators and smart sensors are installed within the home environment to control and supervise its operations. It also incorporates smart devices to run the whole home smartly without the human direction. Some of the applications are home appliances, lighting, security cameras and alarm systems. These smart devices are connected to the local server via a wireless medium for data collection and analysis. One of the most significant issues is how securely the smart devices can send data to the appropriate destination. All the smart devices have constraints such as limited resources including power supply, memory, limited data processing capability and range of communication [17,18]. The following sensors are important to develop a smart home:

1. Fire/Smoke detector: Fire or smoke detector is one of the essential sensors to build a smart home and to protect the home from fire. The function of the smoke detector is to detect the first sign of fire or smoke almost as quickly as possible and keep the human lives safe. They always create alarm sounds to alert the inhabitants inside the home. Some of the detectors have notification system which can be sent to all the members of the home. The image of a commercial smoke detector that is used in smart homes is shown in Figure 7.4 [19].
2. Humidity detector: Leak sensor is used in a smart home to detect the water leakage in a supply unit. Moisture detection sensors can alert the people to leaks in home so they can fix the problem immediately avoiding any kind of damage. The sensor can be placed around water heaters, dishwashers, refrigerators,

Figure 7.4 Commercial smoke detectors used in smart homes [19]

Figure 7.5 Commercial humidity detector used in smart homes [20]

sinks, water pumps and anything at risk for water leakage. If the sensor detects unwanted water a notification is sent to the owner of the house, so they can check out the problem quickly and take the necessary actions. Figure 7.5 shows a commercial humidity detector that is used in smart homes [20].

3. Smart thermostat: The smart thermostat provides control over the heating and cooling in smart home – from any location. They are always useful to save money by monitoring the temperature and humidity inside and outside of the home. The temperature of a house changes due to the number of reasons and a smart thermostat can adjust the temperature based on behaviour and room usage. The ideal thermostats adjust the room temperature on a room-by-room basis to maintain the ideal temperature when any bodies are in the room and can change the temperature to an energy saving mode when no one is in the room. Figure 7.6 shows a commercial thermostat that is used in smart homes [21].
4. Motion sensors: A motion sensor detects motion and movement in an area. These sensors can alert immediately if there is any movement within the home, or if the doors or windows have been opened or closed. They can even turn the lights on and off as doors are opened and closed. These sensors work as an extra pair of eyes when nobody stays at home. These sensors are the first line of security for smart home break-ins; some sensors might be useful as they can detect when an intruder breaks a window. They always alert the owner by sending the notifications of potential intruders. They are also useful to save energy and can be connected to lighting or the thermostat to help control the energy usage in a room based on the occupancy of the room. Figure 7.7 shows the motion sensors that are used in smart homes [22].
5. Video cameras: The video cameras allow the owner to visually locate the positions of different people in the house with a smartphone.

Whether the owner is inside at home or at work and someone is at the house, the owner can always know and takes the immediate necessary actions. These devices are always useful to keep an eye on the potential intruders. Figure 7.8 shows a commercial video camera used in smart homes [23].

Figure 7.6 Commercial thermostat used in smart homes [21]

Figure 7.7 Motion sensors that are used in smart homes [22]

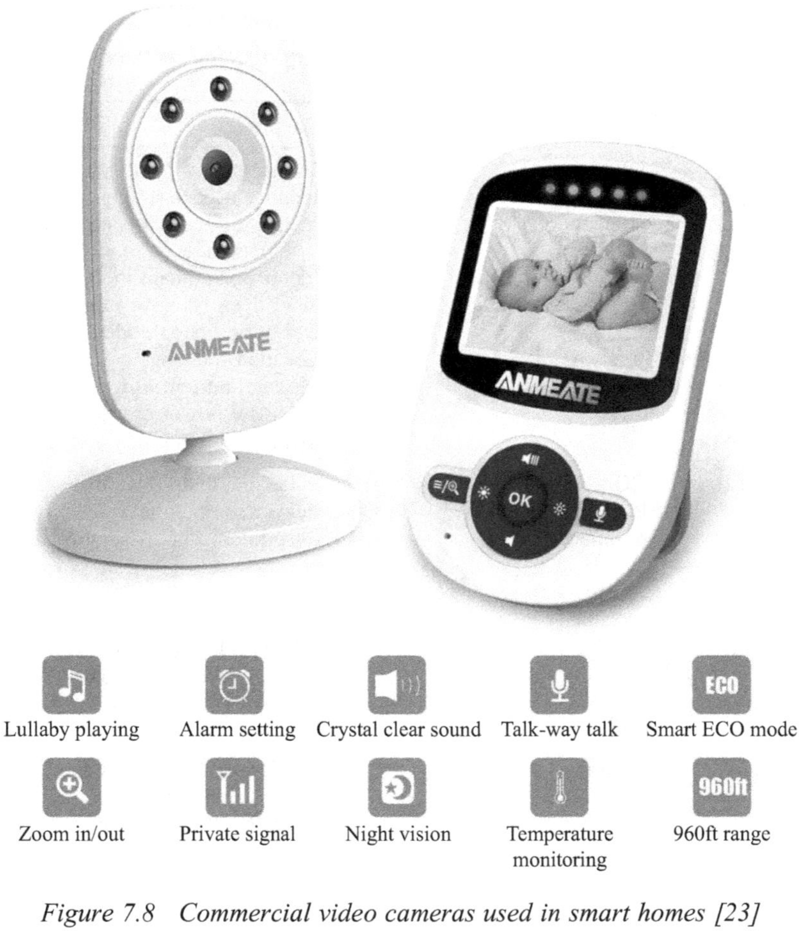

Figure 7.8 Commercial video cameras used in smart homes [23]

From the research point of view, different types of devices are used in the smart homes for monitoring the usage of the devices. Table 7.1 exhibits some of the devices based on the category and purposes they are being used [24]. Even though wireless medium of data transmission is the most favourable one, hybrid media are also used in smart homes with the significance of having a Zeno configuration. Some of the hybrid media are Ethernet, Infrared and Radio-Frequency.

7.2.1 Models for smart homes

Continuous research is being done on smart homes with a range of work done on the behaviour detection of the residing people. The wellness determination [25,26] is one such idea where two functions β_1 and β_2 were introduced to determine the duration of the use of different appliances at a defined time t. The time series modelling was applied to determine the updated time parameters and maximum durations in order to analyse the trend in the usage of household objects for past, current and future conditions. Smart homes over the years have conceptualized different models to understand, analyse and predict the behaviour of the residing people [27]. The prediction of data was also performed using support vector

Table 7.1 Category and purpose of smart home monitoring devices [24]

Category	Name	Purpose
Sensor	Light	Measurement intensity of light
	PIR	Identify user location
	Temperature	Measure room temperature and body temperature
	Pressure	Identify inhabitant location
	Switch sensor	Door open or close status detection
	RFID	Object and people identification
	Ultrasonic	Location tracking
	Current	Measure current usage
	Power	Calculate power usage
	Water	Measure volume of water usage
Physiological device	ECG	Pulse rate and variability
	PPG	Pulse rate and blood velocity
	Spirometer	Respiration rate, peak flow, inhale/exhale ratio
	Galvanic skin response	Sweating
	Colorimeter	Pallor, throat inflammation
	Pulse Oximeter	Measure oxygen saturation of blood
	Sphygmomanometer	Blood-pressure measurement
	Weight	Measure patient weight
	Pulse meter	Monitor heart weight
Multimedia device	Camera	Monitoring and tracking
	Microphone	Voice command
	Speaker or headset	Announce alert and information
	Display (LCD, Plasma panel)	Show visual information

machines (SVM) where the activities of daily living (ADL) were studied using different kinds of wearable devices like kinetic sensors [28]. These sensors determined activities like postural transitions and walk periods and then classify each of the temporal frames into one of the daily activities. A varied range of sensors including motion sensors, activity detection sensors, etc. are fixed at different locations of the house to obtain a large data set, which are then studied with different methods like neural networks [29], predictive algorithm [30], hidden Markov model [31], etc. to determine the anomalous behaviour in the future. One of the interesting approaches towards the IoT-based smart homes is related to the Web-based system which was used by formulating an application framework that supported the concurrent interaction of the residing people [32]. The system consisted of a 6LoWPAN-based WSN inside the smart home that implemented on HTTP caching and push messaging techniques. Different issues like device discovery and service description were addressed using this proposed algorithm. Representational state transfer (REST) algorithm had been employed for the connection of the web with household appliances. The architecture of the web-based system was principally based on three layers: device, control and presentation for the device control and management, processing of the data and representing the available devices dynamically, respectively. Web application description language (WADL) was adapted for the services to the HTTP-based application. The experiments were executed with commercial ZMOTION sensors which were used in the event-based scenarios using push technology. The results were based on the trials with one-hop topology which helped to identify the amount of time required for the motion detector to notify application framework.

Another interesting research work involves the use of vital-radio inside smart homes, which operated as an activity monitoring device via wirelessly sensing of the breathing and heart rate [33]. The operation was based on the change in the signals as a result of the change in position of the chest wall due to inhalation and exhalation. One of the attributes of this system is its wireless sensing capability of physiological parameters of the residing people. Another interesting work involves the evaluation of the elderly people necessities to adapt existing technologies as part of smart home projects [34]. This is a significant work as most of the times the elderly people living in the house wants to have an independent life without any interference and breach of privacy. The work on the management of the devices has also been worked up where a proactive architecture has been proposed [35] that deploys an event–condition–action (ECA) method for the management of the heterogeneous devices in the smart homes. This is a significant work from the point of view of time management as heterogeneous sensing devices are deployed inside a smart home, huge amount of data is generated as a result of which, it is difficult for the monitoring unit to classify the significant ones or the information from similar devices. In addition to the device API and device stub in the system, the modules consisted of SQL statements which communicate with the proactive architecture. The modules generate the SQL statements in accordance with the rules of the system. The IoT-based system consisted of a surveillance device, audio device and an alarm for the execution of the tasks.

7.2.2 Communication protocols in smart homes

The connectivity of the IoT-based smart homes has also been worked upon [36] regarding the different connectivity standards used in smart homes. Some of the common communication protocols used in smart homes to connect the sensors and the gateway are ZigBee, Z-Wave, Bluetooth, Wi-Fi and thread. The most prominent one among this is Wi-Fi, which works on start and mesh topologies and has the highest current consumption (116 mA), giving the highest covered range (150 metres). Some of the problems faced by these communication protocols in the home network are interoperability, self-management, maintainability, signalling, bandwidth and power consumption. For the interoperability challenge, other than Wi-Fi, all the other protocols, especially ZigBee products face this problem the most. This is why, a combined effort solution of ZigBee and thread was released in April 2015 to ensure high interoperability due to the similarity in their basic physical specifications. The inclusion of the Z-wave with ZigBee makes it easier for the sensor nodes operating with this protocol to adjust and adapt to the changes in the environment, while being able to collaborate with other devices efficiently. The maintainability of the network protocol refers to the reliability and dependability of the protocol with the change in environment, activities, nodes and gateways. All of the mentioned protocols have good maintainability to be used in smart homes. The signalling is an important parameter keeping in mind about the losses that happen with the protocols working on long distances. For example, even though Wi-Fi is reliable to be used for medium ranges, where it can have considerable amount of signal losses compared to ZigBee and Z-Wave protocols. One of the main reasons for this is the difference in the range of bandwidth between the protocols as the ZigBee and Z-wave have very low bandwidth. The problems of bandwidth keep on increasing with an increase in the number of IoT-based smart homes creating a huge amount of data in the network server. The lower bandwidth of ZigBee, thread and Z-wave compared to Bluetooth and Wi-Fi does make them a viable open for future applications. The challenge of power consumption in the protocols is somewhat related to the data transfer between the sensor nodes and gateway. Even though the Wi-Fi consumes more power compared to ZigBee, Z-wave and Bluetooth, it does transfer the data at a faster rate, which is sometimes useful in emergency situations. Thread, being a comparatively newer protocol, can be assumed to be in the middle with a moderate speed of data transfer and very low power consumption. But one of the problems that thread faces is the high consumption of power per bit, which makes it a non-viable option for low amount of data transfer.

The usage of IoT in smart homes has been optimized over the years by deploying different communication protocols in these homes. One of the works [37] involves the usage of ZigBee protocol where the XBee modules were connected to the objects as end devices in order to commute their usages via IoT gateway. With the use of ZigBee protocol [38] to send the data to the cloud services, researchers have also worked upon the received data in the cloud using JSON data format to improve the data exchange efficiency. ZigBee has also been used with other technologies like 3G and Ethernet [39] to transfer the data collected by the sensor modules. The work has been done on three layers namely sensing, network and application layer. The hardware consisted of five different parts: ZigBee module, Ethernet module, 3G module, video module and data sampling and control module. The presence of both 3G and Ethernet modules confirmed the transmission of data wirelessly if the latter one fails. Other than the 3G module, researchers have also used Arduino boards and applications to implement IoT-based applications [40]. They operated with two different prototypes namely Bluetooth and Ethernet in indoor and outdoor environments, respectively. The mobile app present in the smartphone was directly linked with the Bluetooth connectivity using serial communication protocol. MAC address of the Bluetooth module was used to pair up with the mobile app in the phone. But two of the biggest challenges they faced with the Arduino board system were the assessment of unique IP address to the heterogeneous devices connect to the Internet in smart homes and the absence of a commercial standard for the integration of the different sensors for the IoT-based devices.

Another concept used in smart homes is the mobile RFID-based IoT systems [41] which is based on the multistandard Near Field Communication (NFC) and Ultra High Frequency (UHF) technologies. The master and slave readers were used in a hierarchical architecture where the slave readers falling within the range of the master reader would respond passively and send the data of the RFID connected objects. The significance of this work lies in the mobile RFID reader, which behaves as an 'Energy Generator' device as it powers up while facing the master and slave readers. One of the research works [42] was done to emphasize on the different services required by the ISO layers to obtain high quality of data from different heterogeneous data sources in the smart homes. The first is the perceptive layer which deals with the controlling of the WSN used for data collection monitoring purposes. RFID, sensors, device control and energy management are some of the attributes at this layer that are analysed, dealt and worked upon. Second is the network layer which deals with the communication of the collected data. Some of the basic examples are Satellite communication, ZigBee, 3G, PLC network, Infrared Ray, Bluetooth and Wi-Fi. The third layer is the software infrastructure layer which is the most significant one as it bridges between the user layer and network layer. The SOA framework and Web Services platform are some of the examples of the type of services used for this layer. The last one is the application layer which is the service provided to the people residing in these homes. Security, medical, data and entertainment are some of the common services offered in these specialized automated homes.

7.2.3 Architecture of smart homes

In terms of architecture, researchers have tried [43] to improve the quality of service by developing a model that categorizes the monitoring system into three different levels. These levels are categorized as the monitoring, control and user-level services. The first category can be defined as the data acquisition level where different kinds of devices are used for ubiquitous monitoring of activities. This is the primary level where the sensor data are collected and calibrated for the second level. The second step is the information processing one where the sensor data are processed with the threshold values set according to user requirements. The last step is the context making where the model generates awareness to the monitoring unit for the required action depending on the information generated by the knowledge engine. Researchers have also worked [44] on the adaptation capability of the IoT-based systems for the activity recognition in the smart homes using a spatio-temporal feature technique which does a semisupervised learning. The model included five major processes: *Initial classification*, *temporal feature-based verification*, *spatial features-based clustering*, *spatio-temporal match evaluation* and *model update*. For the classification part, the data are divided into recognized and unrecognized ones. The recognized data then goes to the second step for verification while the unrecognized ones go for the third step to form clustered instances. The outputs of the second and third steps are then fed into the match evaluation step to join the candidate activities together to form multiple groups of new instances perform a comparative study between each activity of each group for evaluation. The difference of this proposed model from the conventional semi-supervised adaptation model lies in its spatio-temporal features, which gives a higher chance to discover more instances of different activities that are useful for training purposes in order to increase the performance efficiency of the model. The activity recognition was also done using a pattern cluttering technique to a temporal ANN algorithm [45]. The model is divided into two steps, where it starts off with a relevant and efficient unsupervised learning technique named the K-pattern clustering algorithm. The second step is related to the training of the environment for determining and predicting the activities of a person using an artificial neural network based on Allen's temporal relations. Initially, the large amount of data collected from the sensors in the IoT-based smart home goes through an unsupervised learning algorithm. The entire data are computed and grouped into clusters with similar user activity patterns. The algorithm is also able to detect the discontinuous and interleaved activity pattern of the users, resistance towards the noise in the overall data in the set and computing data efficiently by grouping similar activities. The processing of the perceived data was done to detect the temporal relations between them. The second part of the algorithm does the conversion of the perceived data, observation and mining of the most frequency pattern and again grouping of similar patterns. The detection, mining and grouping of the patterns was at a faster rate in comparison to that of the threshold set in the first part of the algorithm. The advantages provided by this approach lies in the decreased amount of time and space required to intercept the activities happening in the smart home environment.

The application layer can also be subdivided into three layers, namely, Kernel layer, Agent layer and Interface layer [46]. The first layer is responsible for the agent management and data transportation from the application layer to the other layers. This layer interacts with the network layer to transfer the collected data for analysis purposes. The interface layer is responsible for providing the different kinds of interfaces in terms of resources to the application layer. The modelling of the resources to the upper layers is done at this level. The agent level is responsible for defining the type of different smart devices used in the smart homes. The most significant level in this architecture is the Kernel level, which manages all the levels in the application layer. It is also responsible for the management, transportation, authentication and authorization of the data and monitoring of its transfer from the application layer to the upper layers. Even though the authorization and authentication are done on the application layer, it is specially controlled by the certification authority (CA), which is centrally located in the smart home for controlling purpose. The CA performs the double-checking process on the call of an agent during an operation. When the CA sends its certification to the agent to validate its decision, the agent sends back its own response to CA through certification. If the responses between the CA and the agent are positive, the CA informs its subordinate agents to work with the other controllers. Research work also includes the use of commercially available ARM microprocessors like SAMSUNG S3C2440A [47] to collect the data from the sensors and process it. These data are then communicated through ZigBee protocol to the monitoring unit. The WSN also prefers using CPLD, which is a complex programmable logic device, but mainly for industrial purposes [48]. Sensors can be controlled using a radio-frequency wireless sensor and actuator network (WSAN) at a frequency of 433 MHz to control, monitor and manage the appliances in the smart homes. The use of CPLD/FPGA is sometimes more preferable than the microprocessors in the smart homes due to their advantages of real-time performance and synchronicity.

Another common phenomenon used in smart homes is the speech recognition of the residing people with specific controllers to differentiate the significant voices from the surrounding noises. One of them [49] used the Griffiths-Jim method in a speech recognition engine where the reception of the speech was done within a specified boundary to immune from the noise and having a high signal-to-noise ratio. Machine learning was done with the engine with limited vocabulary specialized for devices working on Boolean logics like lighting, TV, radio, etc. Another project involved speech recognition using different novel ASR techniques in a multiroom smart home designed for a project named SWEET-HOME [50]. Twenty-one speakers were tested with microphones at different scenarios including activities of daily living. It was found out that the results obtained using techniques like DDA was better than the baseline or other available techniques. The work on the communication protocol has been further enhanced by researchers [51] to showcase the use of 5G for smart homes and smart cities. A four-layered model consisting of the integration of the technologies like 5G, IoT, cloud of things and distributed artificial intelligence was developed for the analysis of handling big data sets obtained from smart homes and smart cities. Another interesting research work employed Bluetooth as the communication protocol in smart homes [52] to transfer the data due to its advantage of reduced energy consumption. The proposed algorithm uses a fuzzy logic mechanism to analyse the sleeping time of each device by determining the ratio of the throughput to the workload. In this way, there is an increase in the lifetime of the devices by 30%. The fuzzy logic approach has been backed up simulation results, which showed there is a considerable reduction in the required power. Another important sector where research work had been done was the development of a protocol for face identification and emotion recognition through image processing techniques. The system comprises two elemental blocks, sensors and decision maker. The first block consisted of the sensors located in the environment and the second element consisted of a single machine which processed the information passed on by the first element. Other than the different kinds of sensors like gas, humidity, temperature, motion, etc. located inside the house, wearable sensors like smartwatches were also used to determine the different physiological parameters of an individual. The processed information was then passed onto the cloud for remote monitoring purposes. For facial recognition, Face tracker was used for mapping process to determine the facial representation using facial points.

In one of the interesting research works, an alternative approach towards the IoT for the replacement of the most commonly used host-centric Internet Protocol with an Information-Centric Networking (ICN) paradigm in a smart home domain [53]. The proposed algorithm does provide a more quick, flexible and efficient technique for data transmission and configuration management. ICN can be used both for providing naming content objects and identification of the functionalities of different IoT connected devices in a logical way, without the need of any of the addresses. The packing of the names with different components is done in a hierarchical manner using this protocol, where it moves the names between two consecutive layers. In some of the systems, the network domain uses a hybrid of ICN and IP protocol instead of singularity to have better communication between the sensors and the gateway for the sensing data. The framework comprised of a configuration manager and authorization manager through which the fixtures are assigned and the credibility to access the fixtures, respectively. The idea of ICN associated with smart home applications does provide a viable option due to the simplified configuration of its associated network and data retrieval capability and, most significantly, it provides security to the network layer to a great extent. The work on the use of ICN on smart homes does provide an interesting insight based on the scalability of the problem, traffic characteristics and handling mobility [54]. It also enhances the storage of the data by giving leverage to the application requirements as much as possible. The privacy and the security of the data are two other issues that get affected on their relation to any communication paradigm. One of the interesting research works depicts the usefulness of video cameras and sensors in terms of the increase in the monitoring capability of the smart homes [55]. The use of Building Control Services (BCS) was done in the smart home to manage the sensors and actuators, having *timewheel* as the central data structure to manage the chain of events. They have employed the Markov Chain Monte Carlo and Hungarian models to analyse the problem faced with the non-overlapping tracking of the residing people in smart homes with a network of cameras.

7.2.4 Security in smart homes

Since, all the IoT-based smart homes involve the uploading of data on the cloud, security is an important factor to ensure the privacy of the personal information of the residing people. Regarding the analysis of the security problems and challenges, another significant work [56] has been done to bridge the gap through a detailed study of the attack surfaces, security issues, forensics and threat models. One of the surveys conducted by Hewlett-Packard in 2014 showed [57] that 80% of IoT-enabled daily used devices in smart homes breaches privacy related to the personal information, while almost an equal amount of communications were not encrypted and had security vulnerabilities. A few of the devices that can be altered are the smart TV, magnetic sensors and the alarm system via different antennas and RF signals. The security issues related to the communication of the data can be addressed by the IoT devices with proper authentication and access control of peers with which the nodes communicate. The surface attacks take place in two networks, public and local with six different communication levels: device-device, device-coordinator, coordinator-gateway and device-controller for the local network and controller-IoT service provider and service-service for the public network. Another work [58] involves the approach towards strong security system by implementing an AllJoyn framework that used an asymmetric Elliptic Curve Cryptography to provide authentication during the system operation. The system ran on a Wi-Fi network where a gateway was used as the central node of the system. Android-based mobile phones were used as end devices which provide a mean for set up, access and control the system. The authentication was done in two rounds to ensure extra stability of the proposed system. In the first round, the user had to load the identity and preshared key of the IoT connected device on the mobile device while in the second round, the identity was uploaded on the home gateway. In the review article by [59] on the IoT applications on smart homes, the problems related to the security was addressed in terms of filtering false network traffic and avoiding unreliable home gateways. One of the ways to circumvent the initial problem is to use a cooperative authentication scheme to prevent insignificant or false data to be distributed in the network [60]. The second problem could be addressed by using a side channel monitoring (SCM) technique [61] to handle the problem at a minimal cost. The technique is the optimized selection of subnet of neighbours for each node along the patch of the router to determine its performance in message transmission in the forward direction. But the source has the reverse direction of message transmission as its primary communication channel.

One of the networking challenges addressed by this group is in terms of the interference of the unreliable nodes during the data communication and filtering of the false data traffic in the community network. Another group has worked [62] on the network-level security of the IoT-enabled devices in smart homes to restrict the illegitimate intruding into residing people's activities. The work was done to augment the device level with the network-level security solutions in order to use the software-defined networking (SDN) technology to dynamically block devices with suspicious behaviour. A three-party architecture was proposed via an open-source SDN platform to determine the efficacy to provide security to the IoT-enable devices. The solutions were based on SDN implementing dynamic security rules that were based on contexts depending on the time or occupancy of people in the house. The protocol was developed on the assumptions that the ISP has the visibility of the devices present in the house and its access was SDN enabled. Floodlight (v0.9) OpenFlow controller was used for operating the ISP network. One of the major ideas proposed for the security and privacy of the smart homes was in terms of a blockchain by eliminating the Proof of Work (POW) and concept of chains [63]. The proposed algorithm was executed by using an online, high-resource device named 'miner', which was responsible for the communication taking place inside and outside the smart home. The concept of POW and bitcoins was eliminated as it faced several challenges like high-resource demand and long latency for the confirmation of transaction, which would result in low scalability and blockage of the whole network due to the broadcasting transactions. The algorithm relied on the hierarchical structure and distributed trust in order to maintain the blockchain and specifying it to the IoT to ensure the security of the home. In order to evaluate the performance of the proposed algorithm, a further simulation was done using three remote sensors and IPv6 communication protocol. The simulation contained handled situations with encryption, hashing and blockchain that was proposed in the algorithm. Although the overlay delay and processing conditions were not considered during the simulation, the transfer of the bytes between the devices, miner and the cloud is three to seven times for the proposed algorithm compared to that of the results obtained from simulation.

The security of the smart homes also dealt with the amount of protection the residing people get from cyber-attacks [64]. This is very helpful from the network point of view as the resource-constraint nature of some of the devices does not permit to provide solutions for any security problems, thus increasing the vulnerability of the information of the users to be hacked or intruded by the third party. Most of the devices were operated in around the GHz range with the use of ZigBee, Wi-Fi (IEEE 802.11), Bluetooth and NFC as the communication protocols. Some of the security problems addressed in this work are the resource and energy constraints, heterogeneous communication protocols resulting in unreliable communication leading to decrease in the computing performance and storage capabilities, execution of the security solutions between the end-devices located inside the smart homes and Internet applications, low reliability of the delivery of the packets, limited availability of energy for communication between different devices and tampering attacks, respectively. Different potential security threats in an IoT-based smart home system in the five different layers in an ISO model showcased the maximum probability of threats and attack framework in the network layer. One of the most significant attacks on the network layer is the Black hole attack on RPL where the attack is initiated with a node and progresses through all the nodes in that pathway, dropping the packets that are transmitted through that pathway. This causes disruption in the flow, thus causing data trafficking in the network. The attack on the application is initiated by the *XMPPloit* command-line exploit tool which enforces the IoT-based smart home devices to remove the encryption on the transmitted data to make it easier for the attacker to modulate them during their transmission. A few of the requirements that can be done to minimize the attacks on the different layers are to possess an intrusion detection system in order to detect unauthorized intrusions and anomalies, employing tamper-resistant devices for the physical layer, employing a device authentication system to classify the authorized devices from the unauthorized ones, have a security-key management system for protecting the sensing devices inside the smart homes that are employed with pre-installed network keys.

The attacks on the security are also based on the location, property, access level and strategy and information damage level [56]. The attack on the location depends on the internal or external position of the attacker in terms of IoT network. When the attacker is external, he gets unauthorized remote access of the network domain from any public network. The attack on device property is based on the low-end and high-end devices used to get access to the IoT network and its associated devices. The low-end devices include the wearable devices used by the people residing inside the homes for ubiquitous monitoring purposes, whereas the end–end devices include the computer, laptop and virtual machines. The access-level attacks are based on active and passive attacks. Active attacks attempt to affect the functionality of the IoT system by disrupting the normal transmission of data. Passive attacks are involved in the unauthorized access to the network domain of a system and perform illegitimate activities to obtain personal information of the residing people. The strategical attacks are defined as the physical and logical attacks which involve the damage and tampering to the IoT devices and affecting the devices in the communication channel, respectively. The vulnerability of IoT-based smart homes for security in terms of privacy was also studied [65] by the researchers in relation to the IoT network traffic created during the data transmission. One of the major concerns remains in revealing the personal information of the residing people happening through metadata and trafficking of the encrypted data. Certain possibilities to deal with the trafficking problem are to separate the traffic into packet streams, labelling the streams by the definite type of device and examining the traffic rates.

7.3 Potential sensors for smart homes

Different kinds of sensors have been used to sense the environment and determine the activities of the people. The fabrication and usage of some of the sensor prototypes have been described in this section. These sensor prototypes have the potential to be used in smart homes for monitoring purposes. A flexible sensor prototype was developed from Multi-Walled Carbon Nanotube (MWCNT) and Polydimethylsiloxane (PDMS) via casting method and used for monitoring physiological parameters of an individual [66]. Figure 7.9 shows the schematic diagram of the fabrication process of the sensor prototypes. Initially, the PDMS layer was cast, desiccated and cured on the PMMA template to form the substrate of the sensor prototype. A nanocomposite layer which formed by the mixing of 4 wt% of MWCNTs into PDMS was then cast on the cured PDMS layer. This layer was then desiccated, cured and laser cut to form the electrodes of the sensor prototypes. The sensor patches were then employed to monitor the movements of the limbs and respiration. The connection of the sensor patches on the individual for limb

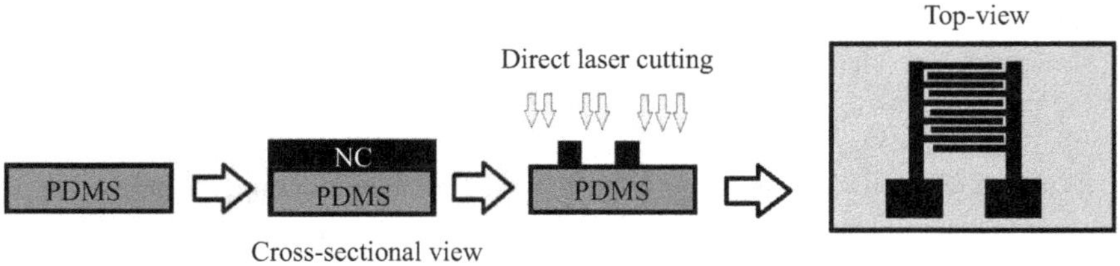

Figure 7.9 Schematic diagram of the fabrication process of CNT-PDMS-based sensor prototypes for physiological parameter monitoring [66]

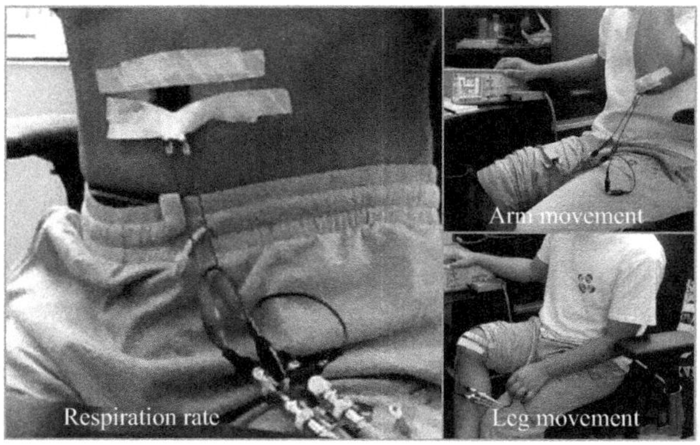

Figure 7.10 Connection of the CNT-PDMS sensor patches to different parts of the body for monitoring purposes [66]

movements and respiration is shown in Figure 7.10. The sensor patches were used as wearable devices where they were attached to the individual with biocompatible tapes. Figure 7.11 shows the respiratory response of the CNT-PDMS sensor patches for two different individuals. It is seen from the response that the fabricated sensor patches are capable of distinguishing the response of the separate individuals. This sensor can turn out to be very useful for smart homes as they can be used in systems for monitoring multiple activities. This would not only help to be used as wearable sensors that can be attached to the clothing of the individuals residing in smart homes but will also help to reduce the overall cost of the sensing system.

Another flexible sensor patch that has the potential to be used in smart homes as a wearable sensor was fabricated by from Graphite and PDMS based on casting on 3D printed moulds. The schematic diagram for the steps of fabrication is shown in Figure 7.12. The 3D moulds were used as templates on which graphite, followed by PDMS was cast to develop the electrodes and substrate, respectively.

The sample was cured to form the prototype which was attached to different parts of the body like neck, elbow, knee and finger to determine their movements based on the stress induced on the sensing area of these patches [67]. Figure 7.13 shows one of the responses where the sensor patch was connected to the neck of an

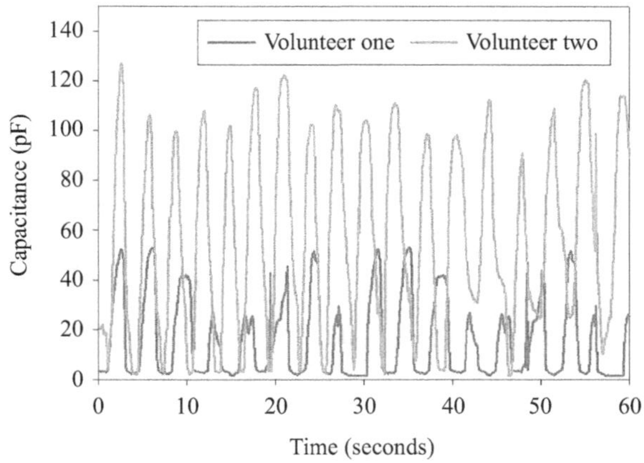

Figure 7.11 Comparative response of the CNT-PDMS-based sensor patches for the respiration of two individuals [66]

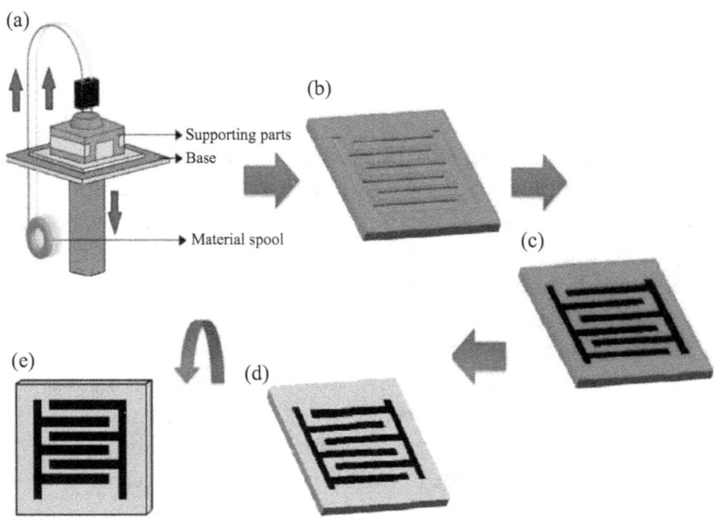

Figure 7.12 Schematic diagram of the fabrication process of the graphite-PDMS sensor patches

individual. Two different situations, namely 'flexed' and 'extended' were considered, which refers the bent and normal/straight positions respectively for the neck. It is seen the sensor patch was capable of monitoring the oscillatory movement of the neck with proper distinction. The oscillatory movements with these sensor patches were done to check the repeatability in their responses.

The graphite-PDMS sensor patches also hold a high potential to be considered commercially for the usage in smart homes due to two factors. First, in the current

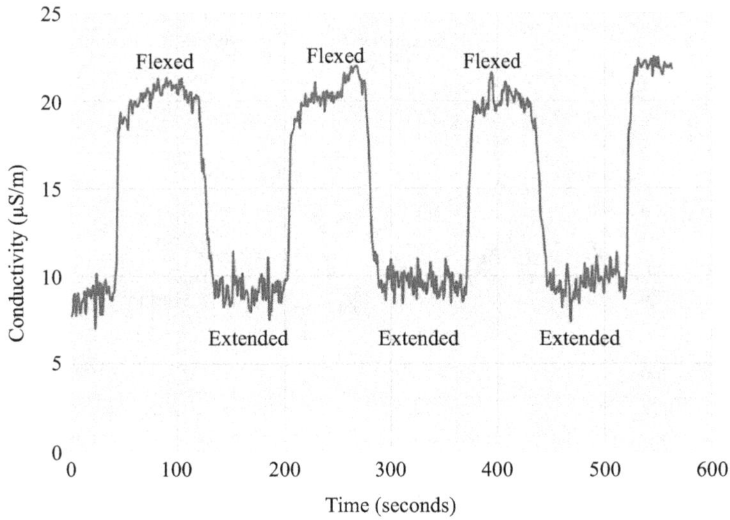

Figure 7.13 Response of the graphite-PDMS sensor patches for monitoring neck movements [67]

era of 3D printing, where most of the electronic devices are being fabricated using this quick and low-cost technique, development of the flexible sensor prototypes for monitoring physiological parameters holds high regard. Second, due to the strain induced nature of this sensor prototype, it can be considered for monitoring patients going through rehabilitation or elderly people in smart homes who are suffering from stroke. Another potential sensing system that can be introduced in smart homes involves the usage of silicon sensors for early detection of osteoporosis. The significance of this work is related to the elderly people who face problems with osteoporosis. In simple words, osteoporosis is a disease that causes bones to become weak and fragile. Worldwide, osteoporosis is a serious problem and it is growing with the ageing population. Early detection of bone loss is necessary to restrict the occurrence of this disease to a large extent. Monitoring the bone turnover markers can be helpful in early detection and monitoring bone problems and to decide the type of therapy. Bone loss can be detected by regular monitoring of serum or urine C-terminal telopeptide of type 1 collagen (CTx-1) [68]. Therefore, fast, portable and low-cost point-of-care (PoC) devices could be helpful for a regular measurement. A highly selective and sensitive portable PoC device was proposed for early detection of bone loss by regular measurement of CTx-I in serum. A MEMS-based interdigital sensor was employed to perform the experiments. The configuration of the sensor was 1-11-50 as shown in Figure 7.14, which means each eleven sensing electrodes is located between two working electrodes and the gap between the consecutive electrodes is 50 μ. Interdigital sensors are designed in a digit-like pattern following the same principle of parallel

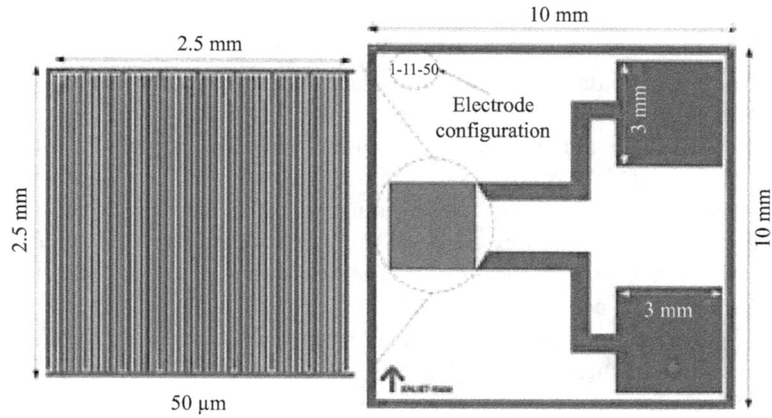

Figure 7.14 Configuration of the interdigital sensor

plate capacitors. There are some advantages in using these sensors such as the uniformity of the electric field, non-destructive, in situ and single-sided access to the material under test (MUT). When a time-dependent AC signal is applied to the sensor, the electric field lines generated which bulge from the positive electrodes to the negative ones and carry some information about the properties of the MUT.

Molecular Imprinted Polymer (MIP) technique was used to prepare the artificial antibodies for CTx-I molecules. The MIP-based artificial antibodies have certain advantages over the natural biological antibodies; they are highly stable in harsh environmental and mechanical conditions and they can be prepared with a very low cost which reduces the total cost of the sensing system. Moreover, no sample preparation is required that is suitable for the development of a rapid system. The artificial antibodies were immobilized on the sensing area of the interdigital sensor using to introduce the selectivity of CTx-I molecules into the sensor. In order to design a portable PoC system for CTx-I monitoring, a microcontroller-based system was designed to quantify the concentration of the target molecule in serum and transfer data to an IoT-based cloud server. The results can be supplied to the medical centre and further analysis and investigation can be done for bone loss detection and treatment. An AD5933 [69] impedance analyser was utilized to measure the impedance, which was produced by varying the concentration of CTx-1. I^2C protocol [70] was employed to collect the impedance information from the MIP-coated sensor. Finally, the measured level of CTx-1 was transferred to an IoT-based cloud server in order to supply the information to a medical person for further analysis. An ADG849 switch was used to calibrate the PoC sensing device before every measurement. The microcontroller generates the impedance information from the impedance analyser and using the standard calibration graph calculates the level of CTx-1. Figure 7.15 shows the first prototype developed as the PoC device for CTx-I detection and quantification. The microcontroller board includes an integrated Wi-Fi that is useful to connect the PoC system to a gateway to transfer data to a remote cloud server. Thingspeak [71] which is a free IoT-based cloud server was utilized to store the data. Thingspeak could be accessible from any location, which helps the medical practitioner to have access to the real-time data for early detection of bone loss. The Arduino Ciao [72] library was used to transfer the CTx-I concentrations to the designated private channel in Thingspeak.

The calibration curve was developed by measuring the CTx-I concentration of five samples, which were prepared by mixing the CTx-I peptide with distilled water. The calibration curve equation was provided to the microcontroller to measure the concentration of CTx-I in unknown samples. Four unknown serum samples collected from sheep blood were tested using the proposed PoC device and the results were validated using a standard ELISA kit. Figure 7.16 shows the correlation analysis between the proposed PoC device and a standard ELISA kit for CTx-I measurement. The developed PoC device employs blood serum as the test sample. However, Blood sampling might not be acceptable and popular for PoC devices. Hence, design and development of a system which uses urine as the test sample can be desirable. A PoC urine testing system could be developed following the principle of the proposed serum PoC device. It can then be installed in for regular and automatic monitoring of bone condition without human interaction. The measured data would be transmitted wirelessly to the health care practitioner for further analysis and actions [73].

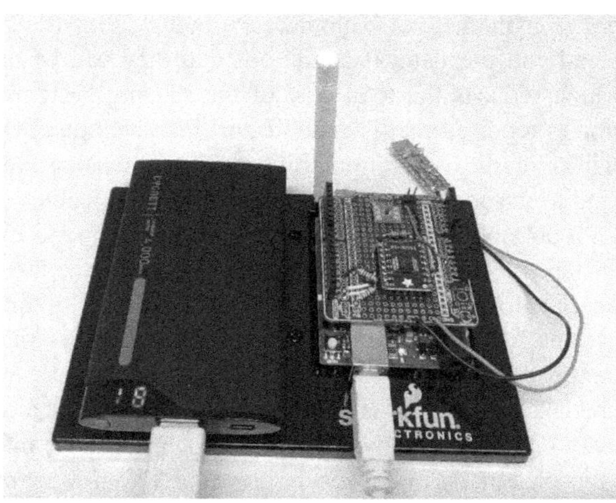

Figure 7.15 The first prototype of the proposed PoC device

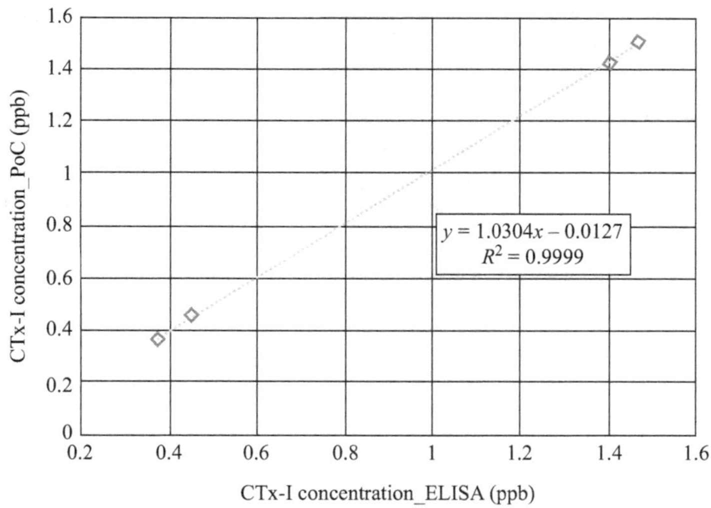

Figure 7.16 Correlation analysis between the proposed PoC device and a standard ELISA kit for CTx-I measurement

7.4 Challenges and future opportunities

Even though a lot of work has been done on the different sectors of IoT-based smart homes, there are still some possibilities to improve the existing approach and algorithms. Some of the possible solutions are discussed here. First, the smart homes operating commercially in today's world are not reachable to the wider range of people but are only accessible to a group of people. One of the major reasons behind this is the cost of living in these homes, which might not be affordable for every person. Second, the information about the advantages of living in these homes should be circulated more among people. The second challenge is the huge amount of data generated in these homes. This requires a lot of memory space in the database, making it difficult to handle it. Algorithms should be developed in such a way that machine learning is done in the most efficient way to determine only the significant data. The third challenge would be the availability of the devices used in the smart homes. Even though some of the commercial devices that are mentioned in this chapter are available to people, they are not being able to comprehend the logic and circuitry of all of them. This makes it difficult for them to use it, in spite of the advantages they provide. The operating principle of the sensing systems should be simple, so that the residing people, especially the elderly ones, are able to comprehend. Another problem with the IoT-based smart homes is the lack of standardization in data collection and processing techniques. Even though so many researchers are working to develop optimized protocols, no standard techniques are yet available to follow. One possible solution for this could be setting a standard by an international organization so that any investor working on developing a smart home would follow those standards. Another problem with these smart homes lies in the high dependability on the Internet connection. If the rate of data transfer goes down, the normal functioning of the monitoring of the different activities goes down. This creates data traffic, thus leading to further problems. Also, the signals from the wireless connections can be interrupted among each other as a result of different heterogeneous sensing devices, leading to false sensed data. One of the possible solutions to this problem could be the division in the frequency bands of the sensing devices to minimize the data interference. If the researchers come up with significant solutions for the above problems, the quality of life can be greatly improved in these IoT-based smart homes.

References

[1] Aldrich FK. Smart homes: past, present and future. Inside the smart home. London: Springer; 2003. p. 17–39.

[2] Chan M, Estève D, Escriba C, Campo E. A review of smart homes: present state and future challenges. *Computer Methods and Programs in Biomedicine*. 2008;91(1):55–81.

[3] Elderly care-smart homes. Available from http://www.jamesdearsley.co.uk/smart-homes-elderly-care/. Last accessed on 1st June 2018.

[4] Chan M, Hariton C, Ringeard P, Campo E, Editors. Smart house automation system for the elderly and the disabled. IEEE International Conference on Systems, Man and Cybernetics, 1995 Intelligent Systems for the 21st Century, Vancouver, BC, Canada. IEEE, 1995.

[5] Tang P, Venables T. 'Smart' homes and telecare for independent living. *Journal of Telemedicine and Telecare*. 2000;6(1):8–14.

[6] University of Florida. Gator Tech Smart House. Available from http://www.icta.ufl.edu/gt.htm. Last accessed on 1st June 2018.

[7] Smart Community Alliance, Japan. Available from https://www.smart-japan.org/english/. Last accessed on 1st June 2018.

[8] KIDUKU, Fijitsu, Japan and Ireland. Available from http://www.fujitsu.com/global/news/pr/archives/month/2013/20130628-01.html.

[9] Smart Homes, U.K. Available from http://www.strath.ac.uk/esru/research/smarthomes/. Last accessed on 1st June 2018.

[10] Ghayvat H, Mukhopadhyay S, Liu J, Babu A, Alahi MEE, Gui X. Internet of things for smart homes and buildings. *Australian Journal of Telecommunications and the Digital Economy*. 2015;3(4):33–47.

[11] He D, Chan S, Guizani M, Yang H, Zhou B. Secure and distributed data discovery and dissemination in wireless sensor networks. *IEEE Transactions on Parallel and Distributed Systems*. 2015;26(4):1129–1139.

[12] Ghormare S, Sahare V, Editors. Implementation of data confidentiality for providing high security in wireless sensor network. 2015 International Conference on Innovations in Information, Embedded and Communication Systems (ICIIECS), Coimbatore, India. IEEE, 2015.

[13] Ghayvat H, Liu J, Mukhopadhyay SC, Gui X. Wellness sensor networks: A proposal and implementation for smart home for assisted living. *IEEE Sensors Journal*. 2015;15(12):7341–7348.

[14] Venkatasubramanian KK, Banerjee A, Gupta SKS. PSKA: Usable and secure key agreement scheme for body area networks. *IEEE Transactions on Information Technology in Biomedicine*. 2010;14(1):60–68.

[15] Poon CC, Zhang Y-T, Bao S-D. A novel biometrics method to secure wireless body area sensor networks for telemedicine and m-health. *IEEE Communications Magazine*. 2006;44(4):73–81.

[16] Li J, Zhang Y, Chen Y-F, Nagaraja K, Li S, Raychaudhuri D, Editors. A mobile phone based WSN infrastructure for IoT over future internet architecture. 2013 IEEE International Conference on Cyber, Physical and Social Computing, Green Computing and Communications (GreenCom), and Internet of Things (iThings/CPSCom), Beijing, China. IEEE, 2013.

[17] Dargie W. Dynamic power management in wireless sensor networks: State-of-the-art. *IEEE Sensors Journal*. 2012;12(5):1518–1528.

[18] Castagnetti A, Pegatoquet A, Le TN, Auguin M. A joint duty-cycle and transmission power management for energy harvesting WSN. *IEEE Transactions on Industrial Informatics*. 2014;10(2):928–936.

[19] Smart smoke detector. Available from https://www.tomsguide.com/us/best-smart-smoke-detectors,review-4472.html. Last accessed on 1st June 2018.

[20] Smart humidity monitor. Available from https://www.amazon.com/AcuRite-Thermometer-Digital-Hygrometer-Humidity/dp/B004K8RF10/ref=br_lf_m_7jtxmhgy5w2c7f3_ttl?_encoding=UTF8&s=home-garden. Last accessed on 1st June 2018.

[21] Smart thermostat. Available from https://www.amazon.de/ecobee-EB-STATE4-01-ecobee4-Thermostat-THERMOSTATS/dp/B06W2LQY6L. Last accessed on 1st June 2018.

[22] Motion Sensors. Available from https://www.officeworks.com.au/shop/officeworks/p/elgato-eve-wireless-motion-sensor-inewindmot. Last accessed on 1st June 2018.

[23] Video cameras. Available from https://www.amazon.com/Monitor-Camera-Temperature-Two-Way-Operating/dp/B06XJQH4CD. Last accessed on 1st June 2018.

[24] Alam MR, Reaz MBI, Ali MAM. A review of smart homes: past, present, and future. *IEEE Transactions on Systems, Man, and Cybernetics, Part C (Applications and Reviews)*. 2012;42(6):1190–1203.

[25] Suryadevara N, Mukhopadhyay SC, Wang R, Rayudu R. Forecasting the behavior of an elderly using wireless sensors data in a smart home. *Engineering Applications of Artificial Intelligence*. 2013;26(10):2641–2652.

[26] Suryadevara NK, Mukhopadhyay SC. Wireless sensor network based home monitoring system for wellness determination of elderly. *IEEE Sensors Journal*. 2012;12(6):1965–1972.

[27] Lotfi A, Langensiepen C, Mahmoud SM, Akhlaghinia MJ. Smart homes for the elderly dementia sufferers: identification and prediction of abnormal

behaviour. *Journal of Ambient Intelligence and Humanized Computing*. 2012;3(3):205–218.

[28] Fleury A, Vacher M, Noury N. SVM-based multimodal classification of activities of daily living in health smart homes: sensors, algorithms, and first experimental results. *IEEE Transactions on Information Technology in Biomedicine*. 2010;14(2):274–283.

[29] Robles RJ, Kim T-h, Cook D, Das S. A review on security in smart home development. *International Journal of Advanced Science and Technology*. 2010;15(1):456–61.

[30] Virone G, Alwan M, Dalal S, Kell SW, Turner B, Stankovic JA, *et al.* Behavioral patterns of older adults in assisted living. *IEEE Transactions on Information Technology in Biomedicine*. 2008;12(3):387–398.

[31] Van Kasteren T, Englebienne G, Kröse BJ. Activity recognition using semi-Markov models on real world smart home datasets. *Journal of Ambient Intelligence and Smart Environments*. 2010;2(3):311–325.

[32] Kamilaris A, Trifa V, Pitsillides A, editors. HomeWeb: An application framework for Web-based smart homes. 2011 18th International Conference on Telecommunications (ICT), Ayia Napa, Cyprus. IEEE, 2011.

[33] Adib F, Mao H, Kabelac Z, Katabi D, Miller RC, Editors. Smart homes that monitor breathing and heart rate. Proceedings of the 33rd annual ACM Conference on Human Factors in Computing Systems, Seoul, Republic of Korea. ACM, 2015.

[34] Portet F, Vacher M, Golanski C, Roux C, Meillon B. Design and evaluation of a smart home voice interface for the elderly: acceptability and objection aspects. *Personal and Ubiquitous Computing*. 2013;17(1):127–144.

[35] Perumal T, Sulaiman MN, Mustapha N, Shahi A, Thinaharan R, editors. Proactive architecture for Internet of Things (IoTs) management in smart homes. 2014 IEEE 3rd Global Conference on Consumer Electronics (GCCE), Tokyo, Japan. IEEE, 2014.

[36] Samuel SSI, Editor. A review of connectivity challenges in IoT-smart home. 2016 3rd MEC International Conference on Big Data and Smart City (ICBDSC), Muscat, Oman. IEEE, 2016.

[37] Kelly SDT, Suryadevara N, Mukhopadhyay SC. Towards the implementation of IoT for environmental condition monitoring in homes. *IEEE Sensors Journal*. 2013;13(10):3846–3853.

[38] Soliman M, Abiodun T, Hamouda T, Zhou J, Lung C-H, editors. Smart home: Integrating internet of things with web services and cloud computing. 2013 IEEE 5th International Conference on Cloud Computing Technology and Science (CloudCom), Bristol, UK. IEEE, 2013.

[39] Bing K, Fu L, Zhuo Y, Yanlei L, Editors. Design of an internet of things-based smart home system. 2011 2nd International Conference on Intelligent Control and Information Processing (ICICIP), Harbin, China. IEEE, 2011.

[40] Mandula K, Parupalli R, Murty CA, Magesh E, Lunagariya R, editors. Mobile based home automation using Internet of Things (IoT). 2015 International

Conference on Control, Instrumentation, Communication and Computational Technologies (ICCICCT), Kumaracoil, India. IEEE, 2015.

[41] Darianian M, Michael MP, Editors. Smart home mobile RFID-based Internet-of-Things systems and services. 2008 ICACTE'08 International Conference on Advanced Computer Theory and Engineering, Phuket, Thailand. IEEE, 2008.

[42] Li B, Yu J. Research and application on the smart home based on component technologies and Internet of Things. *Procedia Engineering.* 2011;15: 2087–2092.

[43] Kang B, Park S, Lee T, Park S, Editors. IoT-based monitoring system using tri-level context making model for smart home services. 2015 IEEE International Conference on Consumer Electronics (ICCE), Las Vegas, NV, USA. IEEE, 2015.

[44] Wu C-L, Tseng Y-S, Fu L-C, editors. Spatio-temporal feature enhanced semi-supervised adaptation for activity recognition in IoT-based context-aware smart homes. 2013 IEEE International Conference on Cyber, Physical and Social Computing, Green Computing and Communications (GreenCom) and Internet of Things (iThings/CPSCom), Beijing, China. IEEE, 2013.

[45] Bourobou STM, Yoo Y. User activity recognition in smart homes using pattern clustering applied to temporal ANN algorithm. *Sensors.* 2015;15(5): 11953–11971.

[46] Jie Y, Pei JY, Jun L, Yun G, Wei X, editors. Smart home system based on iot technologies. 2013 Fifth International Conference on Computational and Information Sciences (ICCIS). IEEE, 2013.

[47] Gaikwad PP, Gabhane JP, Golait SS, editors. A survey based on Smart Homes system using Internet-of-Things. 2015 International Conference on Computation of Power, Energy Information and Communication (ICCPEIC), Chennai, India. IEEE, 2015.

[48] Chi Q, Yan H, Zhang C, Pang Z, Da Xu L. A reconfigurable smart sensor interface for industrial WSN in IoT environment. *IEEE Transactions on Industrial Informatics*. 2014;10(2):1417–1425.

[49] Moir T, editor. From science fiction to science fact: A Smart-House interface using speech technology and a photo-realistic avatar. 2008 M2VIP 2008 15th International Conference on Mechatronics and Machine Vision in Practice, Auckland, New Zealand. IEEE, 2008.

[50] Lecouteux B, Vacher M, Portet F, editors. Distant speech recognition in a smart home: Comparison of several multisource ASRs in realistic conditions. *Interspeech 2011*. Florence. 2011.

[51] Skouby KE, Lynggaard P, Editors. Smart home and smart city solutions enabled by 5G, IoT, AAI and CoT services. 2014 International Conference on Contemporary Computing and Informatics (IC3I), Mysore, India. IEEE, 2014.

[52] Collotta M, Pau G. Bluetooth for Internet of things: a fuzzy approach to improve power management in smart homes. *Computers & Electrical Engineering*. 2015;44:137–1352.

[53] Amadeo M, Campolo C, Iera A, Molinaro A, Editors. Information Centric Networking in IoT scenarios: The case of a smart home. 2015 IEEE International Conference on;Communications (ICC), London, UK. IEEE, 2015.
[54] Zhang Y, Raychadhuri D, Ravindran R, Wang G. ICN based Architecture for IoT. IRTF contribution, October 2013, pp. 1–15.
[55] Coelho C, Coelho D, Wolf M, editors. An IoT smart home architecture for long-term care of people with special needs. 2015 IEEE 2nd World Forum on Internet of Things (WF-IoT), Milan, Italy. IEEE, 2015.
[56] Hossain MM, Fotouhi M, Hasan R, editors. Towards an analysis of security issues, challenges, and open problems in the internet of things. 2015 IEEE World Congress on Services (SERVICES), New York, NY, USA. IEEE, 2015.
[57] Internet of Things research study. Available from http://www8.hp.com/h20195/V2/GetPDF.aspx/4AA5-4759ENW.pdf. Last accessed on 1st June 2018.
[58] Santoso FK, Vun NC, editors. Securing IoT for smart home system. 2015 IEEE International Symposium on Consumer Electronics (ISCE), Madrid, Spain. IEEE, 2015.
[59] Li X, Lu R, Liang X, Shen X, Chen J, Lin X. Smart community: an internet of things application. *IEEE Communications Magazine*. 2011;49(11):68–75.
[60] Lu R, Lin X, Zhu H, Liang X, Shen X. BECAN: A bandwidth-efficient cooperative authentication scheme for filtering injected false data in wireless sensor networks. *IEEE Transactions on Parallel and Distributed Systems*. 2012;23(1):32–43.
[61] Li X, Lu R, Liang X, Shen X, editors. Side channel monitoring: packet drop attack detection in wireless ad hoc networks. 2011 IEEE International Conference on Communications (ICC), Kyoto, Japan. IEEE, 2011.
[62] Sivaraman V, Gharakheili HH, Vishwanath A, Boreli R, Mehani O, editors. Network-level security and privacy control for smart-home IoT devices. 2015 IEEE 11th International Conference on Wireless and Mobile Computing, Networking and Communications (WiMob). IEEE, 2015.
[63] Dorri A, Kanhere SS, Jurdak R, Gauravaram P, Editors. Blockchain for IoT security and privacy: the case study of a smart home. 2017 IEEE International Conference on Pervasive Computing and Communications Workshops (PerCom Workshops). IEEE, 2017.
[64] Lee C, Zappaterra L, Choi K, Choi H-A, editors. Securing smart home: Technologies, security challenges, and security requirements. 2014 IEEE Conference on Communications and Network Security (CNS), San Francisco, CA, USA. IEEE, 2014.
[65] Apthorpe N, Reisman D, Feamster N. A smart home is no castle: privacy vulnerabilities of encrypted IoT traffic. arXiv preprint arXiv:170506805. 2017, pp. 1–6.
[66] Nag A, Mukhopadhyay SC, Kosel J. Flexible carbon nanotube nanocomposite sensor for multiple physiological parameter monitoring. *Sensors and Actuators A: Physical*. 2016;251:148–155.

[67] Nag A, Afasrimanesh N, Feng S, Mukhopadhyay SC. Strain induced graphite/PDMS sensors for biomedical applications. *Sensors and Actuators A: Physical*. 2018;271(1):257–269.

[68] Afsarimanesh N, Mukhopadhyay SC, Kruger M. Sensing technologies for monitoring of bone-health: A review. *Sensors and Actuators A: Physical*. 2018;274:165–178.

[69] Devices A. AD5933: Impedance Analyzer. Internet: Analog Devices [cited 4th September 2017]. Available from http://www.analog.com/media/en/technical-documentation/data-sheets/AD5933.pdf. Last accessed on 1st June 2018.

[70] Mankar J, Darode C, Trivedi K, Kanoje M, Shahare P. Review of I2C protocol. *International Journal*. 2014;2(1):1–6.

[71] Thingspeak. Thingspeak. 2017 [cited 26th August 2017]. Available from https://thingspeak.com/.

[72] Website A. Ciao. 2017 [cited 26th August 2017]. Available from https://www.arduino.cc/en/Reference/Ciao. Last accessed on 01.06.2018.

[73] Afsarimanesh N, Alahi M, Mukhopadhyay S, Kruger M. Smart Sensing System for Early Detection of Bone Loss: Current Status and Future Possibilities. *Journal of Sensor and Actuator Networks*. 2018;7(1):10.

Chapter 8

Wireless sensor network for landslide early warning and monitoring

Cristian Fosalau[1] *and Cristian Zet*[1]

In this chapter, a wireless sensor network (WSN) application aimed for monitoring and warning about imminent landslides is presented. It consists of multiple complex measurement nodes (MNs) connected in a radio communication network spread over the area to be monitored. Each node is able to measure two types of parameters: relative soil layers displacement and pore water pressure, each of them mounted at different soil depths. The sensors devoted to measure the above parameters are of original conception and feature very good metrological qualities, especially in terms of displacement sensing sensitivity, which is as low as 1 mm per year. The measurement node structure can be tailored according to the underground layer structure and can include several displacement and pore water pressure sensors that are connected together in a wired local network structure with serial communication. The WSN is coordinated by a gateway, which has the role of general administration of the network resources, data collection and storage and information transmission to the cloud, from where a central server accesses and downloads the data. The whole structure of such a network is detailed in this chapter, starting with sensors description, continuing with the network architecture, communication protocols, data structure, alarms generation, data analysis, power management and ending with the user interface and some results obtained by in-field deploying of the network.

8.1 Landslides: phenomenon, triggering mechanisms and monitoring techniques

Landslides are among the most common and disastrous hazards occurring all over the world, consisting in downslope sliding of huge quantities of rock and debris. They often occur under steep slope conditions, but it is possible to find them also in connection to low relief. They can be triggered by natural events or by human activity. The main triggering mechanism is the moistening of the impermeable

[1]'Gheorghe Asachi' Technical University of Iaşi, Faculty of Electrical Engineering, Iaşi, Romania

underground layers during long rains or during snow melting. The impermeable layer keeps the groundmass above and it becomes heavier due to the water accumulation on its grains. Due to a favourable orientation of the layers, a heavier mass on a slippery substrate will start to slide easily. When these triggering conditions reach the threshold, the landslide starts with a slow imperceptible movement, then cracks develop while the resistance of the ground weakens. The movement accelerates while the ground becomes heavier and, in the end, the landslide starts with a large mass of ground glide, engaging everything during its fall.

Monitoring landslide serves for a wide variety of reasons, like finding the magnitude or the style of the slide, for emergency alarming or for assisting the design of remedial measures. The monitoring can consist of ground surveys over long periods of time like automated surveys using GPS and extensometers [1], laser scanning [2], manual or automated underground investigations by inserting inclinometers in boreholes [3,4], accelerometers [5], optical time domain reflectometry [6] or aerial techniques like high-resolution imaging via satellites [7,8] or satellite radar interferometry [9]. Besides these, the soil moistening, the underground water mass [10] or the quantity of precipitation [11] are other parameters to be surveyed.

GPS technique offers, over conventional ground methods, the advantage that the stations may not be visible to each other, resulting in a greater flexibility in selecting the measurement points [1]. Also the measurements can be carried out automatically continuously no matter the weather, while the measurement time is reduced to a few minutes. The information is in 3D coordinates and the resolution is as low as 5 mm. The quality of measured data is affected by the weather condition [11] rainy of foggy days introducing the so-called 'wet delays'. Network solutions using more than three references can increase the measurements' accuracy, but rising the costs. Also, using pseudolites (ground-based station transmitting GPS signals) can significantly increase the accuracy.

Laser scanning is a novel technology based on LIDAR that allows remote measurements of millions of points (Figure 8.1), providing a detailed spatial information. Besides it takes tens of minutes to 'image' the monitored area, this method accuracy is altered by the presence of thick vegetation, the positioning difficulties for the scanner and the position of a stable point where to place the device [12].

Geotechnical sensors are widely used for monitoring landslides providing information about the deformation. They are placed in well-chosen locations in order to ensure that adequate information about the landslide is provided. The main geotechnical sensors are extensometers, inclinometers, piezometers, strain gauges, pressure cells, tilt sensors and crack meters [1]. They offer several advantages: local storing data, automatic logging (cable and radio) and ability to change update rate. Inclinometers and piezometers need borehole to be installed, while crack meters, tilt meters, tilt sensors and extensometers are mounted at the surface. The information they provide is essential for deformation monitoring.

Inclinometers measure the displacement at equally spaced depth by measuring the tilt of an accelerometer hanging in a borehole casing (Figure 8.2). Using in

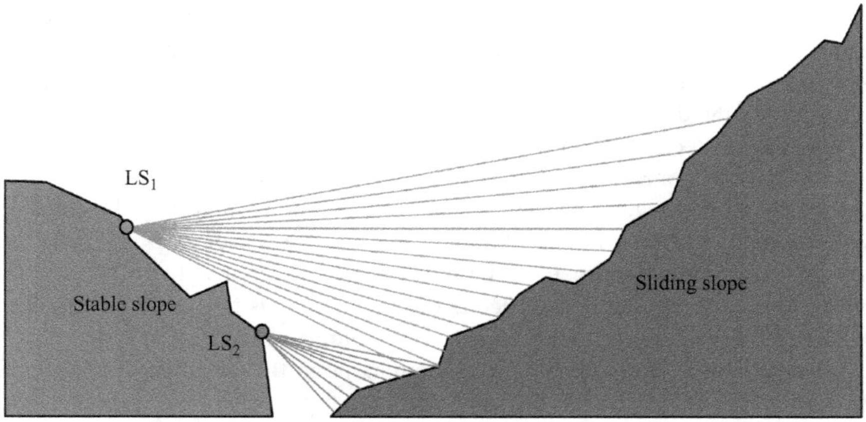

Figure 8.1 Laser scanning technique. Adapted from [12]

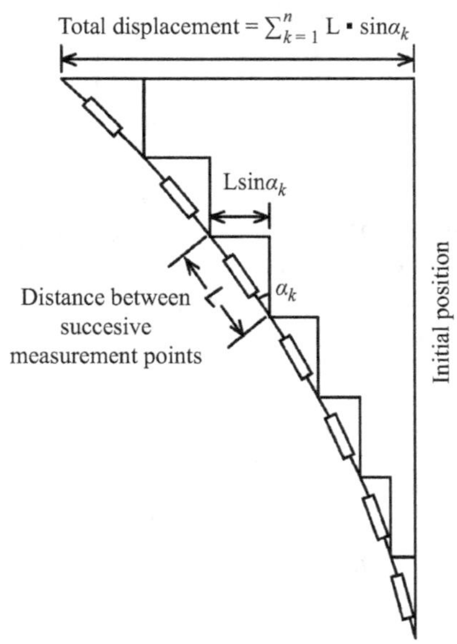

Figure 8.2 Inclinometer working principle. Adapted from [13]

place devices mounted at designated depths, useful information is obtained. The measurement process can be automated and data with accuracies down to 0.02 mm can be logged.

Extensometers are used to monitor unstable parts of land from a stable part [14]. When the ground moves along the crack, the wire is pulled and its displacement is recorded. A major disadvantage is that under large deformation, the cable can break and complete reinstallation is required.

Piezometers, that measure pore pressure, are used for monitoring sites with groundwater threats. Electric piezometers consist of a diaphragm and a porous filter allowing the water to enter a small reservoir. The water pushes the air inside the reservoir and the deflections of the diaphragm are sensed by a vibrating-wire or a strain gauge and converted to a signal depending on the pressure [15].

Optical time domain reflectometry (OTDR) measures the loss that occurs in an optical fibre by detecting the amount of backscattering light [6]. In landslide applications, the fibre is pulled like in conventional extensometers, allowing the measurement of the displacement because of the scattered light in the stressed region. Many sensors can be attached at different intervals to a single optical fibre installed along the direction of the landslide (Figure 8.3).

Synthetic aperture radar (SAR) allows measuring the elevation of the ground using interferometric methods. Such a system uses two antennas placed on the same satellite or aircraft, one emitting a signal and both receiving it, thus creating two images. SAR images contain information on the ground surface in amplitude and phase of the backscattered signal. The amplitude of the image records information about the slope and roughness of the surface, while the phase brings the information about the distance between the satellite and the ground. InSAR uses two SAR images of the same area taken at different time moments (Figure 8.4). If the distance between the ground and satellite changes between the two acquisitions due to surface movement, a phase shift will be detected.

Another solution uses the so-called photogrammetry. The monitored area is photographed from two or more points of known position and camera orientation and the relative position can be measured by triangulation of optical rays. Aerial and terrestrial photogrammetry is extensively used because of their reduced field working time [18]. Additional meteorological sensors are necessary to correlate and to calibrate other sensors like temperature, humidity, dew point, barometric pressure, wind speed or precipitation [1].

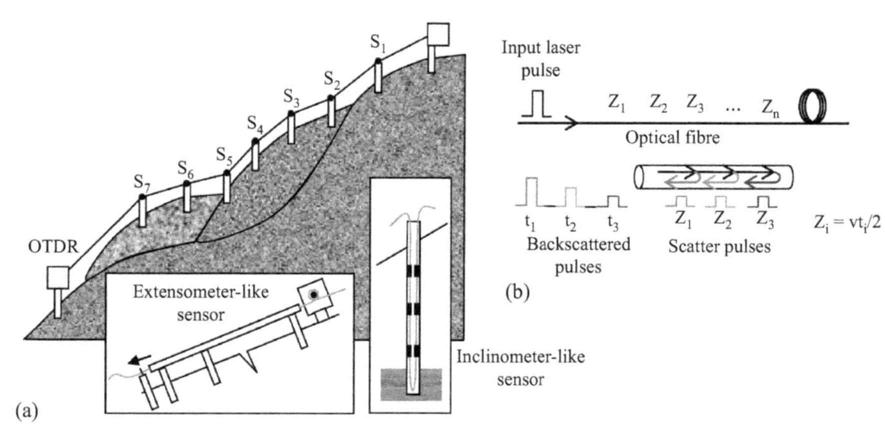

Figure 8.3 OTDR landslide monitoring (Adapted from [6]) and working principle (Adapted from [16])

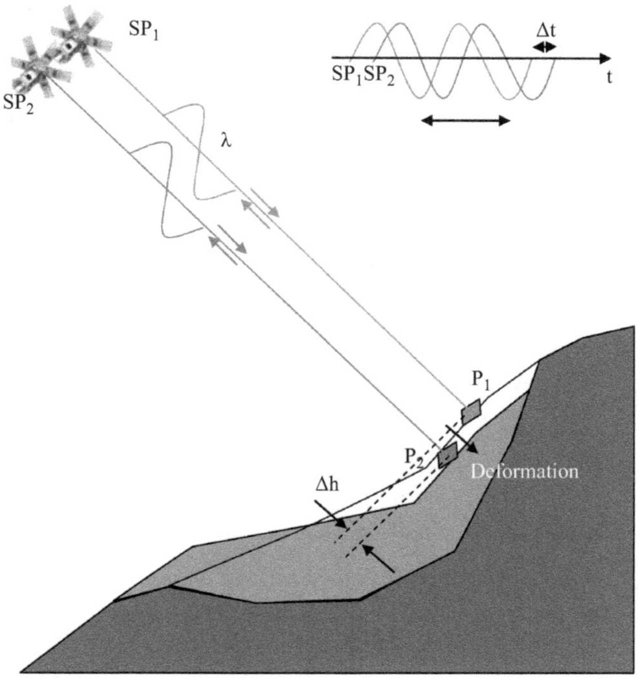

Figure 8.4 InSAR landslide detection. Adapted from [17]

There is no standard setup that will work in any case. The design considerations vary from site to site, but some points have to be considered: the purpose of the monitoring (data recording or public safety), the type of landslide (rapid or slow moving) or the site environment (urban or isolated places).

A complex monitoring network (Figure 8.5) includes various types of sensors (inclinometers, accelerometers, piezometers and rain gauges) with low-range communication capabilities that are deployed all over the monitored area. The information is collected by communication relays (gateways St1 to St5 in Figure 8.5) working as local data collecting units and as intermediate distribution points [19]. They form a so-called 'sensor network'. Their disposal has to be designed according to the obstacles on the site.

Recorded data will be directed to an onsite base station that redirects it towards the data centre. In addition, time domain reflectometers for detecting fractures, laser scanning for surface displacement and aerial SAR for detecting small fractures of the surface can be used for completing the overall picture of the landslides.

In this chapter, a new paradigm of a customized WSN devoted to continuous monitoring of critical parameters related to landslides so that to be able to transmit as early as possible an alarm when such a phenomenon is imminent to start is presented in details. The approach is focused on the hardware components that compose the network, among which a new landslide transducer based on

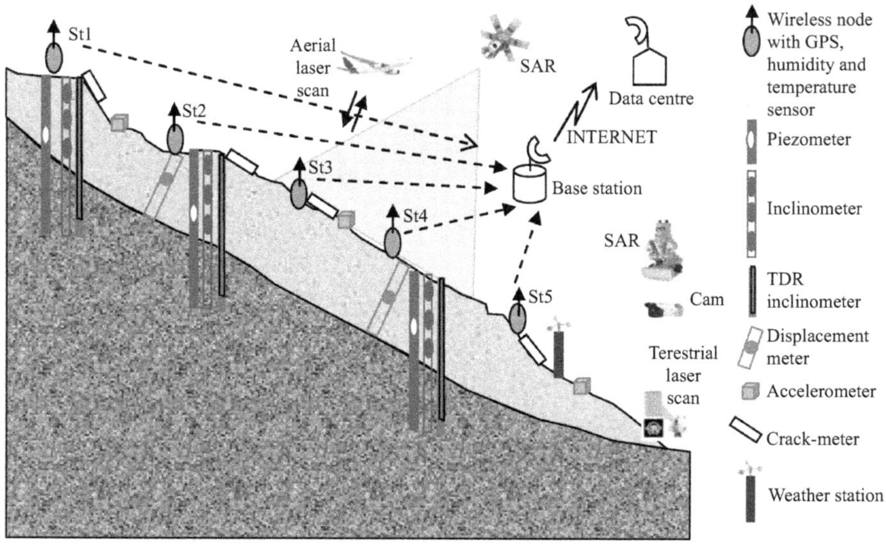

Figure 8.5 Complex spatial network for landslide monitoring. Adapted from [19]

sensing very small deformations is a novelty, as well as on the network architecture and protocols that drive the whole network activity. Some results in terms of data recordings using the WSN for 1 year, along with the graphic user interface dedicated to data analysis and display, are provided in the final section of the chapter.

8.2 WSN setup

In general, a WSN should be built using high-quality components so that the acquired data be precise, reliable, taken in real time and safe, leading thus to a high degree of confidence. A WSN devoted to hazard phenomena monitoring, in particular to landslides, shall be even more reliable, since such an occurrence may bring enormous material damages and even human life loses. The components of the network presented in Sections 8.2.1 and 8.2.2 are trying to respect the above quality and trustworthy principles. In this light, the hardware elements of the customized landslide WSN together with the software implementation of the protocols and procedures at different hierarchical levels are described and discussed.

8.2.1 Network architecture

The deploying particularities of a WSN devoted to continuous monitoring of landslides are first given by the relief on which it is installed, then on the surface to be covered and on the morphology of the investigated soil. Generally, a permanent power supply grid is not available in the area, but the GSM signal exists. Typically, the preferred architecture is the mesh one, as it allows the cooperation between nodes with advantages of re-routing in case of fault with direct benefit over the data confidence, but this solution may be quite costly due to the big number of nodes involved in the WSN in case of a very deformed surface [20]. Such a possible model is depicted in Figure 8.6.

The benefit of this architecture is that it utilizes *dynamic routing protocols* allowing re-routing in case of one or more RN faults and a higher degree of confidence in collected data, but it implies higher costs both for maintenance and for initial investments, for the following reasons:

- Some nodes which might be assigned as *end node* (EN) must be changed into *routing nodes* (RN) (e.g., RN3 and RN4). In this case, they are not allowed to enter *sleep* mode, thus significantly raising the power consumption during their operation. For example, in the present application, an NI WSN-3231 node (from National Instruments) consumes as an EN, 2 W/measurement session, whereas when configured as an RN, its power demand exceeds 12 W/measurement session.
- Because working in router mode much reduces the battery life of the node, alternative solutions must be used for powering the nodes in terms of solar panels or small wind turbines, with a significant increase of the initial costs.

A simplified topology of the landslide WSN is the *multihop routing star architecture*, which is based on fixed routing protocols and is derived from the mesh architecture shown in Figure 8.6 by removing the dashed connections. In this case, RN3 and RN4 become end nodes. For the sake of simplicity, a simplified *multihop routing star* topology has been chosen for this application, which comprises all the base functionalities, namely EN, RN and C. This is depicted in Figure 8.7. Hardware details along with related protocols and procedures are presented in Sections 8.3 and 8.4.

The coordinator C (GW – gateway) performs the whole network management. It is, usually, mounted in a place in the monitored region so that it sees directly as many nodes as possible. This place should also offer good installation conditions for an energy harvester (solar panel or wind turbine). C collects, organizes and redundantly saves data received from the nodes and manages the communication

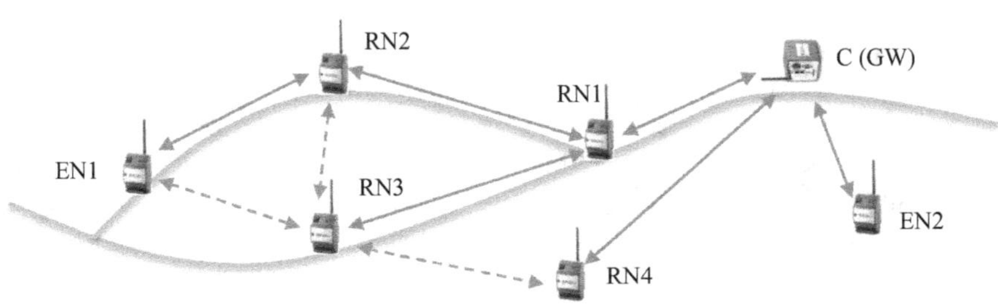

Figure 8.6 A mesh model for a landslide WSN. RN – routing node; EN – end node; C – coordinator (gateway)

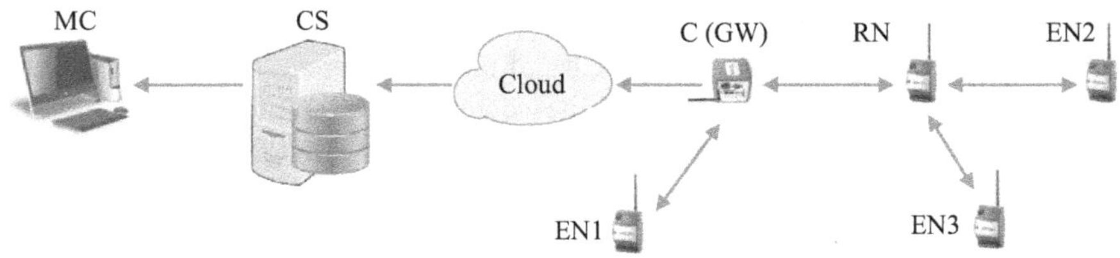

Figure 8.7 Simplified model of a multihop routing star architecture. RN – routing node; EN – end node; C – coordinator (gateway); CS – central server; MC – monitoring centre

with the central server, CS. At this level, it is possible to manually interrogate the network, to make tests and debugging actions at any moment. The nodes carry out the proper data acquisition from sensors. In this case, EN1, EN2 and EN3 are configured as end nodes, devoted only to data acquisition and transmission, whereas RN is configured as router, being charged besides its own data acquisition, with retranslation of information received from neighbouring end nodes or other routing nodes. This way, we can extend this configuration to any complex star or mesh structure. This study does not approach the dynamic routing problem, as the node position is fixed and known. In this case, re-routing is needed for a large surface covered by the network, for which a significant number of nodes is comprised (usually several km^2 and hundreds of nodes). C is equipped with a GPRS modem, which allows sending data to the cloud, from which CS downloads and processes the records and makes them available to the Monitoring Centre, MC, or to the user by means of web pages.

8.2.2 Hierarchical levels

In Figure 8.8, the hierarchical structure of the WSN is sketched. It comprises three levels [21]:

1. *Measurement* level,
2. *Communication* level and
3. *Database and interface* level.

1. The *measurement* level represents the edge of the network. It includes all devices that form the interface between the network and the measurement environment. This is represented by the landslide transducer, along with additional sensors for temperature and pore water pressure that gauge the essential quantities that can influence a real landslide initiation. These devices are organized in measurement nodes (MNs), whose active elements are the measurement points (MPs), that are distributed over the surface of the terrain to be monitored and in soil depth. Inside an MN, MPs communicate with the processing and communication module (PCM) that acts as a local server (LS) mounted on top of MN, at soil surface. Actually, the MPs and the PCM form a local network in which the communication is assured by an RS485 bus. Briefly, the totality of MPs devoted to measure landslide parameters in a certain place, along with their own PCM, comprises an MN.

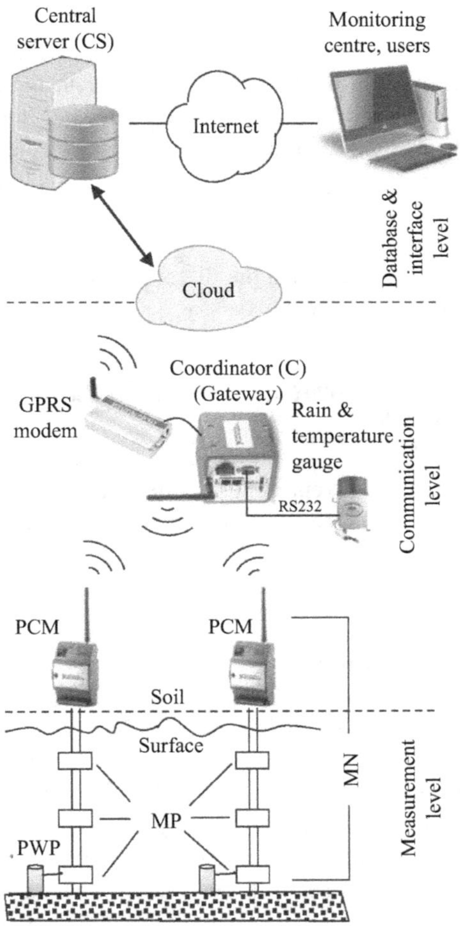

Figure 8.8 Hierarchical structure of the network. MP – measurement point; MN – measurement node; PCM – process and communication module; PWP – pore water pressure sensor. Reproduced from [21]

2. The *communication* level ensures data transmission to the network coordinator (C), and next to the cloud, where they are further accessed by the user through the central server. The communication among MNs, as well as between MNs and C is accomplished by the wireless standard IEEE 802.15.4 in the ISM 2.4 GHz bandwidth. The data uploading to the cloud is performed by C using the general packet radio service (GPRS) by means of a dedicated modem.
3. The *database and interface* level is represented by a central server on which an API collects data from the cloud in the form of a spreadsheet text file. The data are then converted and stored in database formats and processed at request according to the user needs. The user accesses the server by means of HTML

queries and visualises the information through a friendly and comprehensive user interface through a browser. The data uploading to the cloud is carried out synchronously at predetermined intervals of time in regular operation mode (once a day for example) and asynchronously, when the WSN detects significant variations of the measured parameters which might signify a warning or an alarm of a landslide initiation. Warnings are sent when auxiliary parameters like pore water pressure, humidity, amount of precipitation reach significant high values, meaning that a landslide is imminent, but not yet initiated, whilst an alarm is activated when the direct parameters like displacement and orientation exceed a certain level.

8.3 Communication protocols within the landslide monitoring WSN

The *communication protocols* represent the steps that a certain device should follow during its operation for performing a task within the WSN. One may distinguish two types of protocols:

- Node level protocols, which may also be:
  - Protocol for an end node and
  - Protocol for a router node;
- Coordinator level protocols.

  The protocols are implemented with the aid of *procedures* that are specific programs for distinct scenarios. The protocols for the landslide monitoring WSN have been developed on the principle of *state machine*, according to which a node found in a stable state, saying *sleep* or *wait*, will be passed to a new state by an event treated by a procedure, after which the node comes spontaneously to the stable state. In this way, one guarantees an optimal consumption of MN and a higher level of data confidence. Every state is represented by a 'case' structure in which the source code for the task is included.

8.3.1 Protocol implementation for an end node

In Figure 8.9, the state machine diagram for implementing an EN protocol is represented.

As long as the scan time t_s has not elapsed or no procedure is unexpectedly requested by the superior level, the EN remains in the *sleep* state, when the power to the MPs is cut down and its own power consumption is very low (less than 1 mW). The scan time moment $t = t_s$ triggers the measurement session, namely the *Acquisition* procedure, consisting of

- connecting the power to MPs,
  - scanning for all MPs and reading the data (P_d – direct parameters, displacement d and orientation α and P_a – auxiliary parameters, temperature θ, pore water pressure *php* and supply voltage V_s),
  - comparing the values of acquired parameters with maximal limits and

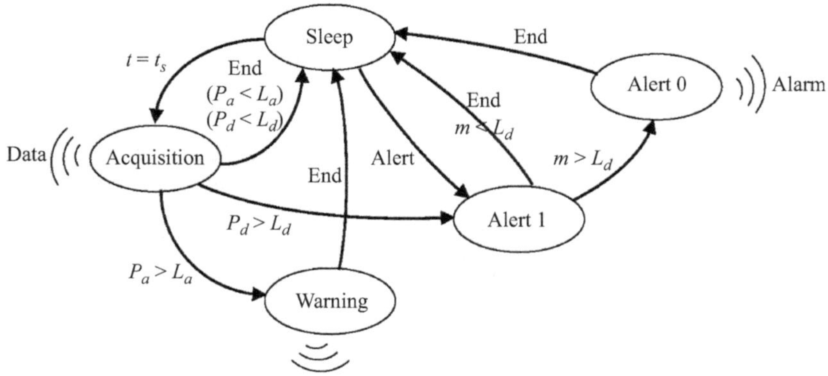

Figure 8.9 State machine diagram for an end node

- sending data to C or to the next RN.

It should be noted here that all the aforementioned parameters are collected and processed at MP level by the μC. Every parameter is averaged over ten samples in order to perform a primary noise filtering. If no parameter exceeds the limits L_a or L_d, after scanning and sending data, the PCM cuts the power to sensors, exits the *Acquisition* procedure and returns to *sleep* state. If all parameters are within normal ranges, the scanning period is established to, let's say, 15 minutes (this value may be changed from the user interface). If any auxiliary parameter (θ or *php*) exceeds the limit L_a, a warning is issued and sent to C and the node goes back to *sleep* state. This is not the case of the voltage supply V_s, whose value is sent directly to C, which compares it with a minimal limit. If V_s is under this limit, C is informed that, in the next future, data may not be available anymore from that node and that the batteries should be changed. If the direct parameter d exceeds the limit L_d, the procedure *Alert 1* is initiated. It means that a landslide is triggered and might become dangerous. Equally, if the measurement has been corrupted, this might be a false alarm. In order to eliminate this possibility, the following actions are scheduled for *Alert 1*:

- reloading the *Acquisition* procedure at a scan time $t_s = 90$ s, ten times; only U_a and U_b are measured this time,
  - computing the values of d and α,
  - averaging the parameters' value over the ten samples and finding the mean m and
- comparing m with the limit L_d.

If the limit L_d is not exceeded, it means that the first measurement has been corrupted and the node goes back to *sleep*. If the limit is exceeded, procedure *Alert 0* is initiated, consisting of sending a message to C by means of the next RN (if any). In this case, C wakes up all the nodes of the network and requests to perform the *Alert 1* procedure. If at least one additional node answers exceeding L_d, C takes the decision that a landslide is imminent in the area and sends the *alarm* message to the central server.

8.3.2 Protocol implementation for a router node

In order to be able to retranslate the information received from a neighbour, a node must be configured as a *router*. In this operating mode, an RN performs the management of information received from its adjacent nodes, as well as the data obtained from its own measurement session. In principle, the protocol for an RN is similar to that of an EN except the routing tasks, which are accomplished after performing its own measurement session. For this, the RN is equipped with additional resources of memory buffer and works in interrupting mode. It means that all requests of routing from the neighbours are queued and treated after doing its own data transmission. Accordingly, the state machine diagram is shown in Figure 8.10.

8.3.3 Protocol implementation for the coordinator

As other protocols, the communication to C is assured by a protocol conceived as a state machine diagram, being implemented in the gateway device NI 9792 using the LabVIEW© RT and LabVIEW© WSN toolkits. The diagram of this protocol is detailed in Figure 8.11, together with the corresponding procedures for each task. C stays always in the *waiting for an event* state. An event triggers execution of one of the following procedures:

1. Data reception;
2. Data transmission;
3. Measurement;
4. User interface and
5. Warning/alarm.

The first procedure consists of receiving data from nodes. C is working in interrupting mode and allows queuing of the node receiving requests. The second one is devoted to transmitting the information to the cloud by the GPRS modem after collecting, processing and storing data in files. The third procedure drives the measurements performed over two important parameters which may affect landslides: outside temperature (θ_{ext}) and quantity of precipitations (Q). For measuring these parameters, a rain gauge device endowed with temperature sensing capabilities is mounted in the vicinity of C to which sends data by serial interface.

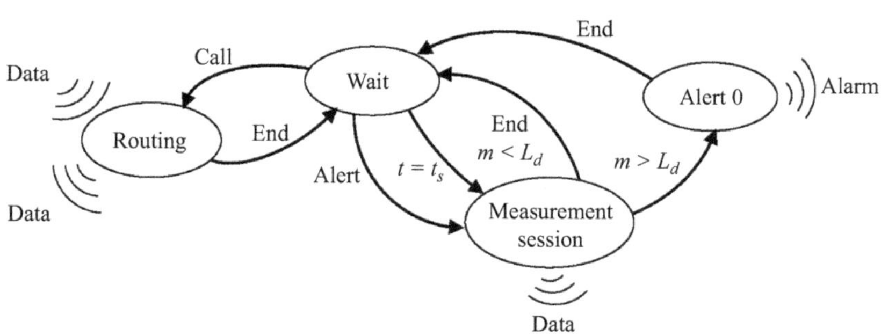

Figure 8.10 State machine diagram for a router node

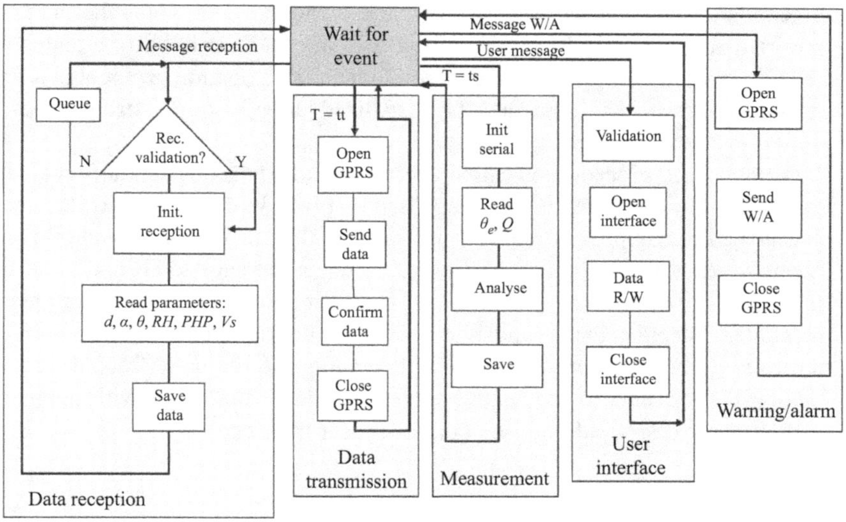

Figure 8.11 Diagram of the protocol at coordinator level

1. *Data reception* procedure is called when an MN is ready to send a message to C. Data reception is performed only after the message validation. This operation is necessary to synchronize reception when more nodes are prepared to send messages or warnings at the same time. If C is busy, the message is sent to a queue. Alarms have priority. As soon as C is ready for reception, it validates the request message and initiates the procedure of data collection, in which it reads the parameters sent by MN: d, α, θ, *PHP* and V_s. In this stage, C analyses the power voltage of all nodes of the network and takes decisions regarding battery maintenance. After receiving all data, they are saved in spreadsheet text files that will be uploaded to the cloud.
2. *Data transmission* procedure is called usually once a day or at intervals of time set from the user interface. At time moment t_t, C initiates the GPRS, uploads the log file to the cloud and waits for confirmation. The file is updated at every MN scan time with new parameters and warnings. After sending data to the cloud, a new file is open. All files are stored into the gateway flash memory.
3. *Measurement* procedure is devoted to read two parameters, such as outside temperature (θ_{ext}) and quantity of precipitations (Q) delivered by a rain gauge mounted in the neighbourhood of the gateway. The information is acquired from the gauge by serial interface. This measurement session is scheduled after receiving data from all nodes. The values are stored in the same log file.
4. *User interface* procedure is called when the user wants, at request, to visualize or load data stored locally in the memory of C at any time between two file uploads. As the gateway is driven by software that runs under a real-time operation system, the user interface procedure is treated separately, without affecting the time-dependent operation of the network. The user connects a

notebook to the gateway by means of the Ethernet port, on which the interface program runs, and sends a user request message. If C is free, it validates the request and sends the requested data through an interrogation protocol that runs on the interface. The data may be visualized directly on the front panel in digital or graphical form.

5. *Warning/alarm* procedure is called every time a node sends an alarm message to the coordinator after performing *Alert 1* procedure and the displacement exceeds a limit. This message has priority over the other messages and it is not queued. When receiving this message, C sends a message to all nodes, asking them to carry out *Alert 1* procedure. If at least one other node sends an alarm message, C decides that a landslide is ready to be triggered and sends an SMS message to the user concomitantly with sending the file under recording containing the last data to the central server. The last data, along with the alarm message, are displayed onto the graphical user interface.

8.4 Hardware components of the WSN

In this section, the hardware components of the landslide WSN are detailed, starting with the landslide transducer, which is a novelty in the field, and ending with the coordinator features.

8.4.1 Measurement node structure

The MN is composed of the highly sensitive landslide transducer (HSLT) that measures the direct landslide parameters such as soil layer displacement and orientation, and a series of specific sensors that measure auxiliary parameters such as soil temperature at different levels and pore water pressure. The structure of an MN is depicted in Figure 8.12. The HSLT consists of a polypropylene pipe on which the MPs are mounted. The pipe is buried vertically into the soil so that the bottom is stuck in the stable and compact layer (usually clay). The depth of burial of the pipe in the soil and the distances h at which the MPs are emplaced are established by prior geomorphological studies and drillings. At the interface between the stable and the gliding layers, a pore water pressure (PWP) sensor is mounted. It provides information regarding the degree of saturation with water of the gliding layer, this being one of the main factors for a landslide triggering [22]. The gliding of soil layers causes the rod deformation. The deformation degree is measured by the MPs and converted in values of landslide displacement (d) and orientation (α). The same MPs keep track of the temperature at their level by acquiring data from appropriate sensors. Thus, the WSN may provide complete 3D information over the landslide parameters both in depth of the soil and over its surface.

8.4.2 Measurement point

Each MP consists of a Wheatstone bridge containing very sensitive devices called highly sensitive strain gauges (HSSG) that are mounted in quadrature along the pipe generators and paired diametrically opposed, as presented in Figure 8.13.

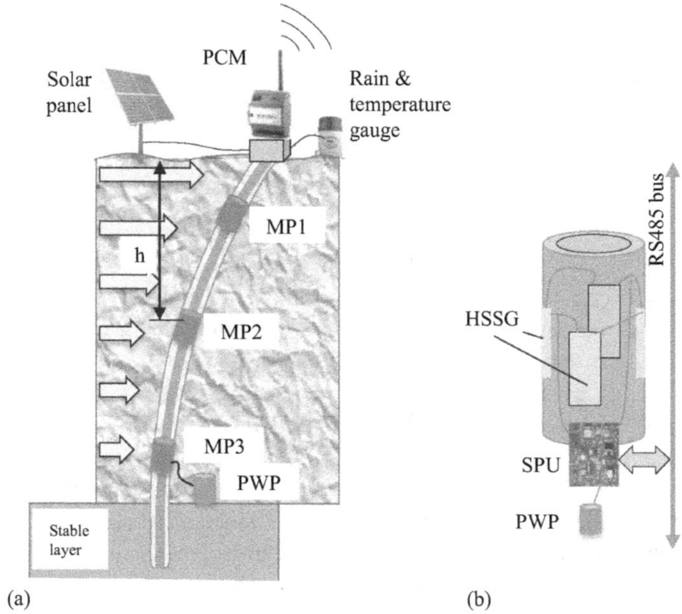

Figure 8.12 (a) Schematic representation of a measurement node and (b) details on a measurement point

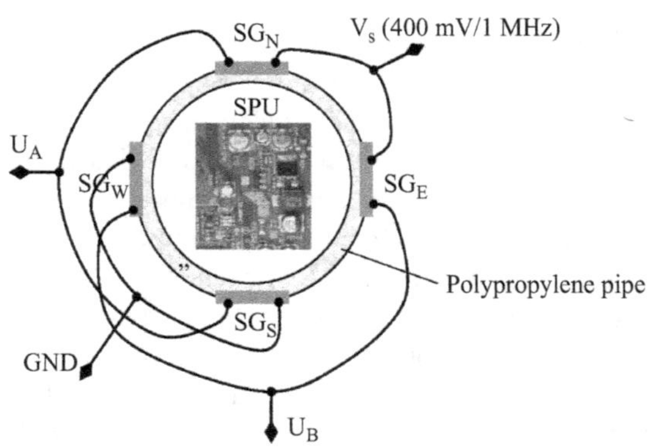

Figure 8.13 Schematic of a strain bridge. $SG_{N, S, E, W}$ – strain gauges mounted along the N, S, E and W directions; SPU – signal processing unit

HSSGs are able to measure with high sensitivity the deformation of the rod subject to the soil layers slides and, hence, the displacement d and the orientation α of the landslide. Every MP is equipped with a signal processing unit (SPU) mounted inside the pipe, to which the ends of the bridge are connected. It contains electronic circuits for processing the analogue signals delivered by the bridge, as well as a microcontroller that acquires the voltages U_a and U_b which are proportional to the deformation, calculates the direct parameters d and α, converts them in digital words and sends them via the RS485 bus to the PCM.

The metrological characteristics of the MP were assessed in laboratory, by calibrating the transducer using an experimental setup [23]. It was proved that the overall error for displacement measurement does not exceed 3%, whereas the orientation may be determined within 12% accuracy.

8.4.3 Highly sensitive strain gauges

Strain gauges are devices widely spread in industry, robotics, civil engineering or mechanics for direct measurement of deformation along one or more axes, but also employed as intermediary sensitive elements for indirect measurement of a large category of quantities such as force, weight, pressure, vibration, torque, small linear and angular displacements, etc. There are two types of commercially available strain gauges: metallic and semiconductor [24,25]. The main parameter characterising a strain gauge is its *gauge factor*, k, which represents the ratio between the relative variation of the output quantity, here a resistance, and the support deformation. Usually, k varies between 2, for a metallic gauge and 200, for a semiconductor one. There are also under research other methods of building strain gauges, based on optical principles, nanoparticles, printing on flexible support, electrospinning and thin film sputtering or nanowires [26–30], each with advantages and drawbacks. In Sections 8.4.3.1 and 8.4.3.2, the construction, operation and characteristics of a new type of highly sensitive strain gauge are presented.

8.4.3.1 Principle of operation

The sensitive part of an HSSG is based on magnetic amorphous microwires (MAMW), employed here as strain sensitive elements.

The MAMWs are metallic alloys in wire form with diameters between several micrometres and more than 100 μm, having as composition Si and B allied with different metals like Fe, Co, Ni, Cr, Mn, Cu and Nb in different proportions. Their manufacturing process consists of jetting the melted alloy, through a quartz nozzle, directly into a rotating cold water layer [31]. In this way, the alloy has no time to crystallize and remains in amorphous state, freezing in residual strains that give rise to a very special shell-core magnetic domains structure [32,33]. Depending on the alloy composition, many interesting effects occur when the MAMWs are subject to external actions such as magnetic fields, as well as to different mechanical stresses, making them appropriate to building a large category of sensors for measuring quantities like magnetic field, current, force, strain, small linear and angular displacements, vibration, tilt, etc. A particular effect arises when the MAMW supports important modifications of permeability and, consequently, of impedance when it is flowed by an alternative current of certain frequency, and it is subject to mechanical stresses such as axial stretching, compression or torsion. This is actually the *stress impedance (SI) effect*, being evidenced more significantly at MAMWs with low magnetostrictivity ($\lambda \approx 10^{-7}$), where also the giant magnetoimpedance effect (GMI) is predominant [34–37]. The SI effect depends on the frequency of the ac current passing through the MAMW. Up to nearly 100 kHz (this limit depends on the geometrical parameters of the wire and on its composition), the impedance variation is mainly due to modification of the inductive part of the sample, according to changes of the magnetization vector direction in the shell, whereas above 100 kHz the skin effect occurs and the impedance variation is mainly produced by the resistive component. The relationship which governs the SI effect is [34]

$$Z = R_{\mathrm{dc}} kr \frac{J_0(kr)}{2J_1(kr)}, \quad k = \frac{1+j}{\delta} \tag{8.1}$$

where R_{dc} is the wire resistance in dc current, r is the wire radius, J_0 and J_1 are Bessel functions, and δ is the skin depth given by [34]

$$\delta = \sqrt{\frac{2\rho}{\omega\mu_\phi(H_x, \xi, \sigma)}} \tag{8.2}$$

In the above equation, ρ is the resistivity of the material, ω is the ac current pulsation, μ_ϕ is the circular permeability in the wire shell depending on the axial magnetic field, H_x, and axial and torsional stresses, ξ and σ, respectively. One may notice that, besides the mechanical stresses, the impedance is also sensitive to magnetic fields, leading to the need of taking special measures of compensation. The compensation with magnetic field, but also with temperature is basically performed electronically, using witness devices.

The two components of Z for a MAMW of length l are given by

$$Z = R_{dc} + j\omega L, \quad L = \frac{l\mu_\phi(H_x, \xi, \sigma)}{8\pi} \tag{8.3}$$

for $f < 100$ kHz and

$$Z = \frac{r}{2\sqrt{2\rho}} R_{dc}(1+j)\sqrt{\omega\mu_\phi(H_x, \xi, \sigma)} \tag{8.4}$$

for $f > 100$ kHz.

One may infer from the above equations that, for low frequencies, the impedance modification is given mainly by variation of the inductance L that depends, in turn, by the magnetic field intensity and axial and torsional stresses, whereas for higher frequencies, both active and reactive components are affected by these parameters.

8.4.3.2 Construction details and characteristics

The construction of an HSSG is given in Figure 8.14 [38].

An HSSG consists of a series of samples of magnetic amorphous microwires (MAMW), of composition $(Co_{94}Fe_6)_{72.5}Si_{12.5}B_{15}$ and a diameter of 120 μm, bonded in parallel along the direction to be gauged on a thin plastic substrate, which is actually a copper-covered foil on which corresponding pads have been etched.

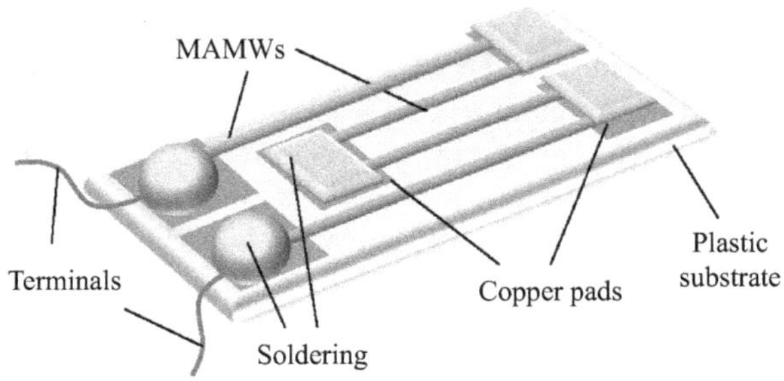

Figure 8.14 Construction of an HSSG

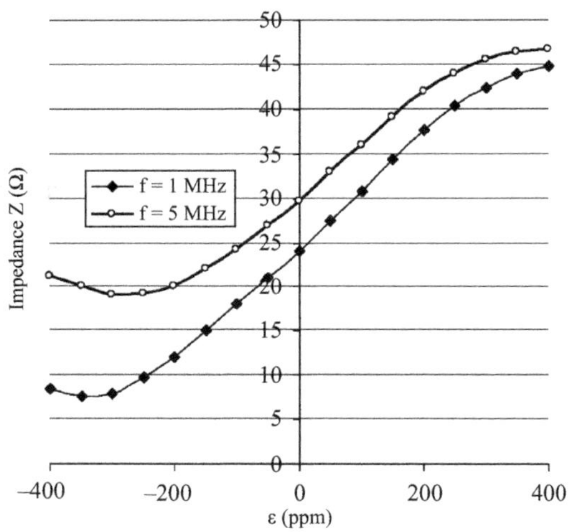

Figure 8.15 Dependence of HSSG impedance on the applied stress, $\varepsilon = \Delta l/l$

The wires are soldered onto the pads under pressure using special soldering paste, being serially connected each other so that their total active length is about 20 mm. The soldering technology of the MAMW to the copper pads must be carefully mastered so that it assures a good electrical contact of the wires to the pad, as well as a firm mechanical resistance to strains. The device is attached directly to the surface whose strain is to be measured, as an ordinary strain gauge.

The main characteristic of an HSSG, representing the dependence of the impedance Z on the applied stress $\varepsilon = \Delta l/l$, traced for two frequencies of 1 MHz and 5 MHz, respectively, is presented in Figure 8.15.

From the above characteristic, one may notice a linear span of about $\pm$ 200 ppm for $f = 1$ MHz, for which the strain gauge reaches a gauge factor of about 2,000, meaning 1,000 times higher than that of a commercial metallic one [39]. This outstanding feature of HSSGs confers to the landslide transducer a very high sensitivity to deformation, being able to sense soil displacements as low as 1 mm or less.

8.4.3.3 Signal processing unit

The signals received from the Wheatstone bridge are processed in the SPU, according to the schematic depicted in Figure 8.16.

The Wheatstone bridge is supplied with a sinusoidal voltage of 400 mV/1 MHz, generated by the oscillator OSC. The ac voltages U_a and U_b collected from the bridge, whose amplitudes depend on the magnitude and direction of the rod deformation, are first amplified with the instrumentation amplifiers IA. Next, they are converted in dc using a peak detector PD and fed to the analogue inputs of the microcontroller μC, which converts them in displacement and orientation values according to the relations [39]:

$$d = \frac{\sqrt{d_N^2(U_A, U_B) + d_E^2(U_A, U_B)}}{2} \tag{8.5}$$

$$\alpha = \tan^{-1}\left(-\frac{(k+2)U_A - V_S}{(k+2)U_B - V_S}\right) \tag{8.6}$$

In the above equations, d_N and d_E are the displacements of the rod along two orthogonal directions on which the strain gauges have been mounted in pairs (e.g., N–S and E–W), and that are dependent on U_A and U_B. Constant k depends on construction parameters of the HSSGs and on their gauge factor. Vs is the amplitude of the supply voltage of the bridge (400 mV). The same μC also provides the value of temperature (θ) by a built-in sensor. At the lowest MP of the rod, the PWP sensor is connected to the μC on one of its analogue inputs. PWP may also be made to work independently of the SPU, being directly linked to RS485 bus. Details on the construction and operation of PWP sensor are given in Section 8.4.4. In the vicinity of every MP, humidity sensors may be mounted and linked to the appropriate μC in

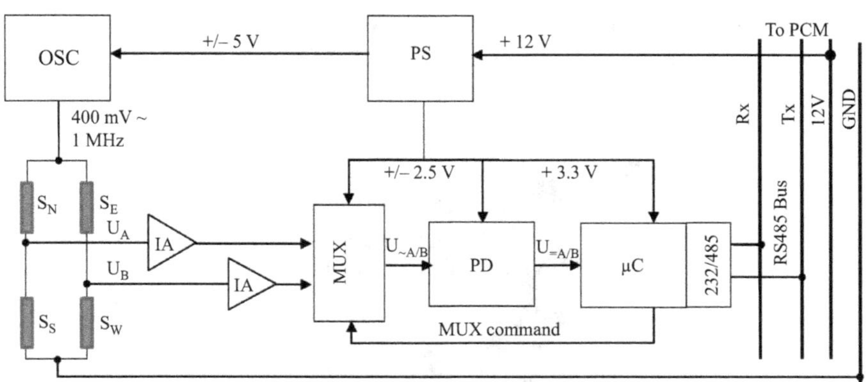

Figure 8.16 Signal processing unit for a measurement point. S_N, S_S, S_E, S_W – strain sensors oriented along the N, S, E and W directions; IA – instrumentation amplifiers; OSC – sinusoidal oscillator; MUX – analogue multiplexer: PD – peak detector; PS – power supply; μC – microcontroller

order to measure the soil humidity at different levels. The data provided by the MP (d, α, θ, pwp, rh and the supply voltage values) are sent to the PCM through the RS485 bus.

8.4.4 Pore water pressure sensor

Often, landslides are triggered by the underground water table accumulated during rainy or snow melting periods. Recorded displacements are then correlated with the water accumulated in the soil [40].

Pore water pressure (PWP), also called *hydrostatic pressure*, is defined as the pressure created by the water that fills the empty spaces between the ground particles (Figure 8.17).

During rain the water tends to infiltrate into the ground. Driven by gravity, the water will go deeper into the ground and will follow small channels or impermeable layers to flow away and cracks to come back at the surface or places to accumulate. When there is no flow, the pore pressure at a depth d below the surface of the water is

$$p_{pw} = \gamma_{\text{water}} \times h \tag{8.7}$$

γ_{water} is a constant equal with the water density multiplied by the gravitational acceleration at that point ($\gamma_{\text{water}} = \rho_{\text{water}} \times g$). The underground level at which the pore pressure is equal with the atmospheric pressure is defined as the water table.

If there is no underground water flow the pressure will have the same level (hydrostatic pressure condition). Thus, the pore pressure at a depth h below the water table is given by (8.7). Below the water table level, the ground is saturated with water. The hydrostatic pressure increases linearly with the depth, as (8.7) shows. If the soil consists of fine grains, the water can rise above the water table because of the capillarity effect. If the ground has a layered structure with alternating permeable and less permeable layers, the pressure variation versus depth may not be linear. If there is a flow due to gravity, it will be a hydraulic

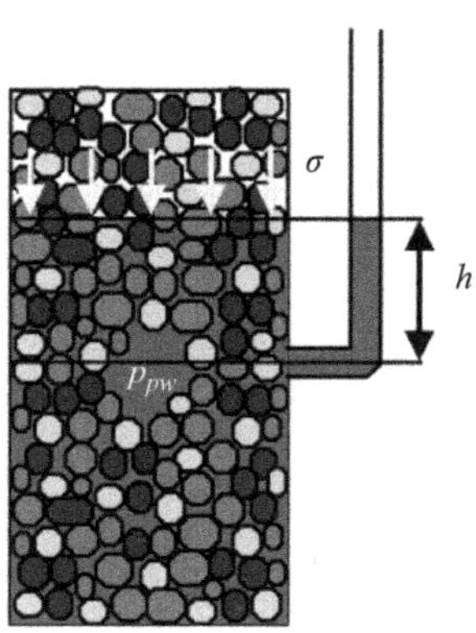

Figure 8.17 Pore water pressure in saturated soil

gradient, so the water level in different boreholes will show a difference. The pressure is higher in the valley than in the top.

PWP is vital for measuring the soil stress. In dry conditions, the stress at a certain level below the surface is given by the mass of soil above (σ_e). If there is water accumulated, then the total stress is generated by the soil plus the water weight:

$$\sigma = \sigma_e + p_{pw} \tag{8.8}$$

Usually, PWP is measured with piezometers [15]. They are complex and costly devices which measure the static pressure using proof enclosures buried into the ground. The pressure transducer is either a vibrating-wire or strain gauge attached to a flexible membrane. These piezometers are connected to ground level electronics or to a portable unit, allowing remote readings.

The PWP sensor presented in this section is an auxiliary device for the landslide transducer which provides valuable information for a very important parameter that may determine a landslide triggering. This approach employs a MEMS device as pressure sensor and a temperature and humidity sensor (T&RH), both connected to a microcontroller-based signal processing unit that acquires and processes information related to PWP measurement and which, in turn, is connected to the RS485 bus of the landslide transducer. As shown in Figure 8.18, it is organized as a sensor node located in a protection tube (5) to be installed in a hole dug into the ground. The tube is terminated at its lower end with an empty container having a filter at its bottom, through which the underground water can enter into the

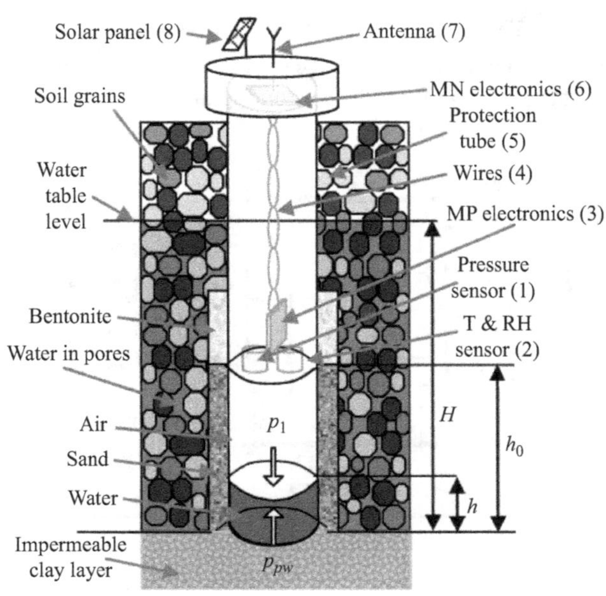

Figure 8.18 Pore water pressure sensor

container. The top of the container is sealed from the rest of the tube and has only two little orifices that allow the sensors (1 and 2) to feel the pressure, respectively, the temperature and the humidity inside the cavity. The main sensor is the pressure sensor, while the T&RH sensor is used for pressure correction with temperature and humidity. Both sensors are connected to the microcontroller that acquires the signals, calculates the PWP and sends its value to PCM via the RS485 bus, as described in Section 8.4.1.

PWP is generated by the height *H* of the water table (Figure 8.18). The water fills the bottom of the cavity and compresses the air inside. At the equilibrium, the air pressure p_1 is equal with the pressure p_{pw} created by the water:

$$p_{pw} = \rho_{\text{water}} \times g \times H \tag{8.9}$$

where ρ_{water} is the water density, g is the local gravitational acceleration and H is the height of the water table. By measuring the p_{pw} the height of the water table H can be calculated.

The pressure sensor has to be a differential one because the measured pressure p_1 depends on the atmospheric pressure, which changes continuously. A solid-state differential integrated pressure sensor (e.g., like MPX5999D) might be chosen [41] (differential, temperature compensated, full-scale 1,000 kPa, sensitivity 4.5 mV/kPa).

The block diagram of the electronic board of the PWP sensor is presented in Figure 8.19. It is built around a low-power microcontroller (like MSP430, STM 32, XLP Pic or ADUCM series), which is a powerful piece of hardware with various resources (ADC, counters, digital IOs, communication interfaces and power down modes). The analogue output of the pressure sensor is acquired by the microcontroller built-in ADC (16 bits), while the T&HR is read via I^{2}C interface. In the end, the pressure value is corrected according to the current temperature and humidity. The dataflow diagram of the sensor operation is presented in Figure 8.20.

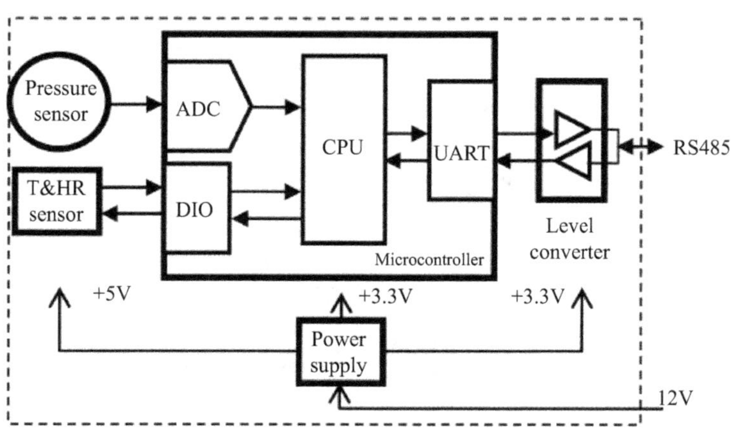

Figure 8.19 PWP signal processing block diagram

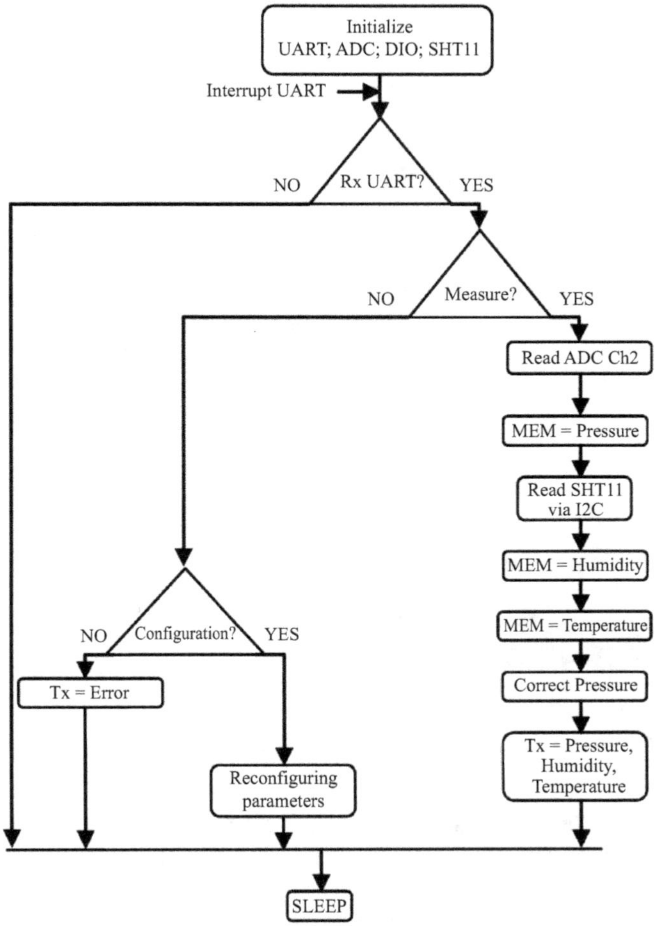

Figure 8.20 PWP software logic diagram

At fixed scan time, both sensors are read, after which the acquired data are locally processed and then transmitted to PCM when asked.

Depending on the ground structure, the landslide transducer may have multiple MPs at different depths, according to the position of the layers. The PWP sensor can work as an independent MN, or can be associated with the MPs of the transducer, thus leading to a complex landslide sensor [42].

The sensor behaviour was tested by sinking it in water. For long-term testing, the sensor output remained constant and was not influenced by the humidity or temperature.

The sensor response presented in Figure 8.21 was compared to a pressure reference sensor in the range 30 kPa to 400 kPa while increasing the pressure (diamonds) and decreasing the pressure (squares). As one may notice, the characteristic shows a

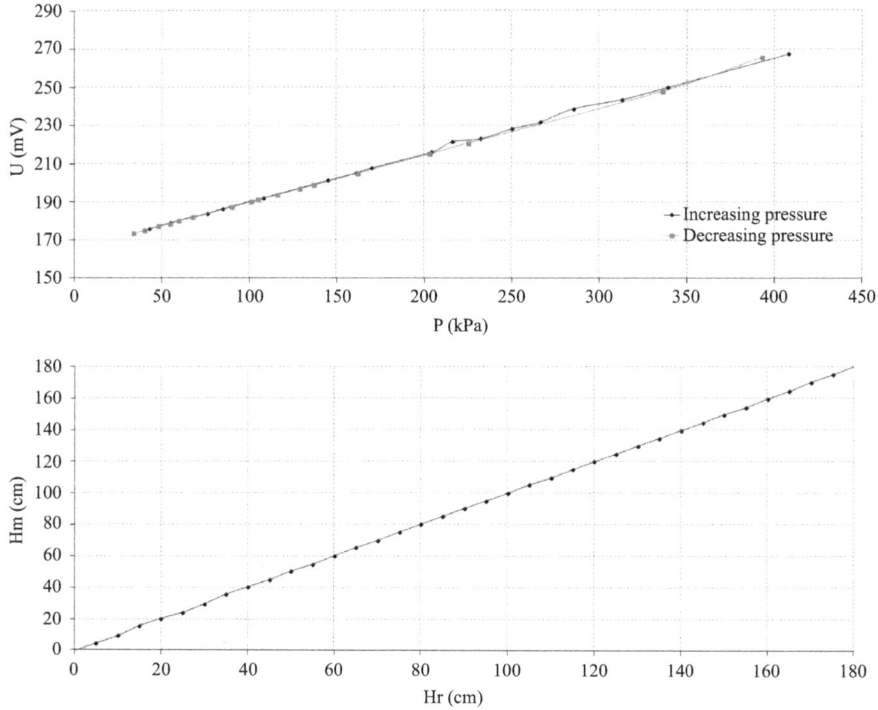

Figure 8.21 Sensor responses (a) in air and (b) in water. Reproduced from [43]

good linearity and no hysteresis. The absolute error has been determined to be less than 1 cm for about 2 m of water column (less than 0.5%).

8.4.5 Processing and communication module

The whole WSN was built using hardware and software equipment from National Instruments company. The hardware modules are from the NI WSN series, while the protocols were developed using the LabVIEW© RT and LabVIEW© WSN toolkits. The NI WSN-3231 module was employed as PCM for collecting and processing the signals received from MPs. The main functions to be fulfilled by PCM are as follows:

- Data aggregation from measurement points;
- Data analysis and processing, warning and alarm issuing;
- Power management for the whole node and
- Data broadcasting to coordinator and routing in case of intermediary node.

An NI WSN-3231 module features the following characteristics:

- Communication with sensors by means of the RS232 and RS485 buses;
- Communication with C or adjacent nodes by the IEEE 802.15.4 protocol, ISM 2.5 GHz bandwidth (200 m in exterior);

- 2 digital I/O and
- Power consumption: less than 1 mW configured as an E with a scanning rate of 1 minute, 200 mW configured as a router node.

In principle, according to the manufacturer data sheet, 4 AA 2,100 mAh alkaline batteries are sufficient for powering only the module configured as an end node for 3 years with a scanning period of 1 minute. However, the whole node (module + transducer + auxiliary sensors) consumes more than 120 mW during a measurement session. Since a measurement session lasts up to 1 second, the battery life may be roughly reduced. It is worth to mention that the PCM, which manages the whole MN activity, sends to sleep all its components between two successive interrogations. In case of a module configured as a router node, even if the node components sleep between measurement sessions, the module only, which has to be awake all the time, consumes 200 mW leading to a battery life less than 6 months. In this case, the power should be assured by an energy harvester such as a solar panel. For a more efficient power management of the WSN, the supplying voltage levels of the nodes are recorded and transmitted to the coordinator every data package.

8.4.6 Coordinator

C manages the whole network of distributed nodes so that the data collection is performed in time, safely and without losses. It plays the role of the network gateway and is responsible for the following functions:

- WSN configuration and supervising;
- Data aggregation from nodes;
- Primary data processing and organization;
- Warning and alarm messages issuing to the user;
- Power management and optimization;
- Local data storage for longer periods of time;
- Data broadcasting to the next hierarchical level (central server);
- Periodic network testing;
- Making data available to the user at request, at any time and
- Performing automated network testing and reporting.

In this application, a NI 9792 module from National Instruments has been employed as coordinator. It displays the following features: processor-type Freescale MPC8347 real time, 533 MHz, 2 GB flash memory, 256 MB DDR2 RAM, integrated web and ftp servers, Real-Time Operating System and radio interface IEEE 802.15.4, 300 m.

Because the consumption of C is high (10 W at maximum load of 36 nodes), it is necessary to assure the energy from a source of at least 20 W. For this, a solar panel of 50 W, a step-down regulator 12 V/1 A – D24V10F12 and a rechargeable lead battery of 12 V/750 W have been employed for this task. For the communication with the cloud, C utilizes a GPRS modem-type HIT65 allowing communication by RS232 serial interface.

8.5 Considerations regarding the power consumption

The power consumption is a very important feature for a sensor network. It becomes critical especially where there is no power network available in the neighbourhood, as for measurements in the field. Some techniques for reducing the power consumption may address: shutting down the components between two measurement sessions, using low-power devices or optimal balance between activities, standby or idle states. Usually, long-term monitoring requires high-energy batteries and optimization of power consumption of the nodes. However, even if an energy harvester is employed to supply the network, a study regarding the power consumption of the hardware components of the WSN in order to find the optimal way for power management strategy may be welcome. In the present approach, the currents to be measured are in tenths of mA range, being either DC or variable. For measuring them, a GMR sensor-type NVE AAV003-10E was employed, having several advantages such as the output is galvanically isolated from the input and it does not introduce any burden voltage to the measured circuit. The input ranges between −80 mA and +80 mA, displaying a sensitivity of about 17 mV/mA, in a frequency band from DC to 100 kHz. The architecture of the current measuring system embedded into an MP is presented in Figure 8.22.

The whole measurement process is automated and driven by a computer through intermediate of a data acquisition board. Using the aforementioned sensor, the currents demanded by an MP of the landslide transducer and by the PCM are recorded and traced during a measurement session.

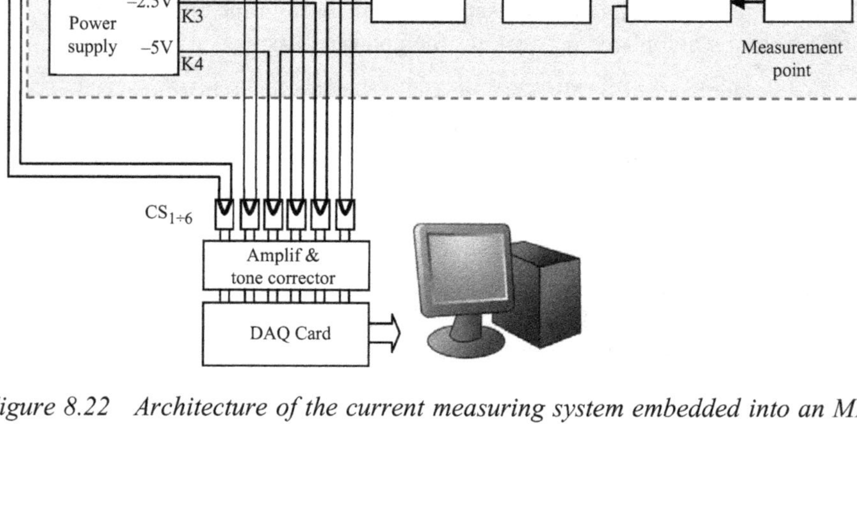

Figure 8.22 Architecture of the current measuring system embedded into an MP

8.5.1 Power consumption at MP level

In order to measure the consumption at the MP level, five power supply currents have been sampled, viz.: ±5 V – the power supply of the RC oscillator, ±2.5 V – the power supply of the multiplexer and peak detector and +3.3 V the microcontroller power supply plus the current flowing into the local power supply source, in total six currents. The six current sensors utilized in the scheme (CS1 to CS6) are connected with the current inputs on the power supply wires of each module of the sensor node. Their outputs are measured with a data acquisition board (DAQ).

The RC oscillator uses an operational amplifier based Wien oscillator with stabilised amplitude. The oscillator drives about 1.1 mA from its power supplies, meaning 0.95 mA × 10V = 9.5 mW power consumption at ±5 V supply, as long as it runs continuously.

The multiplexer and the peak detector are powered with the dual voltage ±2.5 V. A quad operational amplifier is used for the gain stage and for peak detector. It requires about 3.8 mA from the power supplies. This means that this module has a power consumption of about 3.8 mA × 5 V = 19 mW. The multiplexer consumes less than 1 μA and can be neglected.

The microcontroller in the measurement point is an ultra-low-power device-type MSP430AFE253. Its supply current is 450 μA at 3.3 V and 1 MHz clock frequency in active mode. The RS485 transceiver, SN65HVD37, drives, when active, about 30 mA on a load of 100 Ω. When disabled, the supply current falls down to 0.2 μA. The measured current profile for this module is shown in Figure 8.23.

It may be noticed in the figure a constant current consumption of less than 1 mA and six pulses of 32 mA peak. The constant current is taken by the microcontroller (about 600 μA) and the pulses are taken during the data transmission by the RS485 driver. As one can see in the detail (right), the pulse lasts for 2.5 ms as long as the driver is active to send 24 bits of data towards PCM. This means that it takes about 106 mW for 2.5 ms, six times per cycle, being supplied at 5 V. The power consumption of the digital part between two successive transmissions is very small, namely about 2 mW (0.6 mA × 3.3 V).

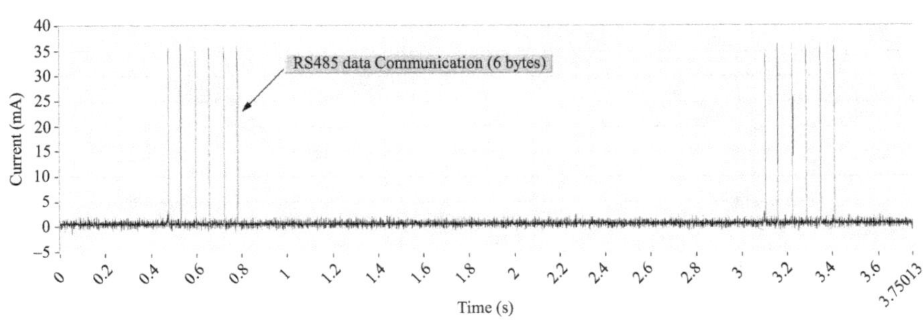

Figure 8.23 Current consumption of the microcontroller block

The MP node takes from the power supply an average energy of

$$W = 0.6\ \text{mA}\times 3.3\ \text{V}\times 66\ \text{ms} + 6\ \text{pulses}\times 32\ \text{mA}\times 3.3\ \text{V}\cdot 1.25\ \text{ms}$$
$$+ (0.95\ \text{mA}\times 10\ \text{V} + 3.8\ \text{mA}\times 5\ \text{V})\times 23\ \text{ms}$$
$$= 1.017\ \text{mWs}$$

During communication with PCM, the consumption of MP goes as high as 550 mW. Figure 8.24 shows the current consumption during a measurement cycle. The following points are marked on the graph: point 1 – the analogue module of the MP is awakened and two measurements are taken for U_a, point 2 – the multiplexer is switched on the second voltage and, again, two measurements for U_b are carried out, point 3 – the analogue part is switched off and point 4 – the microcontroller sends data.

8.5.2 Power consumption at PCM level

The study on the power consumption at PCM level was performed according to the protocol of a measurement session for a node, presented in Section 8.4.1. Briefly, the protocol flows as follows:

- $t = t_S$, power on all MPs;
- wait 25 ms for entering the circuits in nominal operation mode;
- start scanning;
- scanning 3 MPs, 3 quantities of 2 bytes each: send MP address, receive 2 bytes;
- scanning PWP sensor;
- stop scanning;
- power off MN (without PCM);
- radio transmission and
- go to sleep.

The power diagram of the above protocol drawn by measuring the current consumed by the PCM is shown in Figure 8.25. The current was measured using the same system as for MP. Figure 8.25 depicts the power demand of the PCM for a

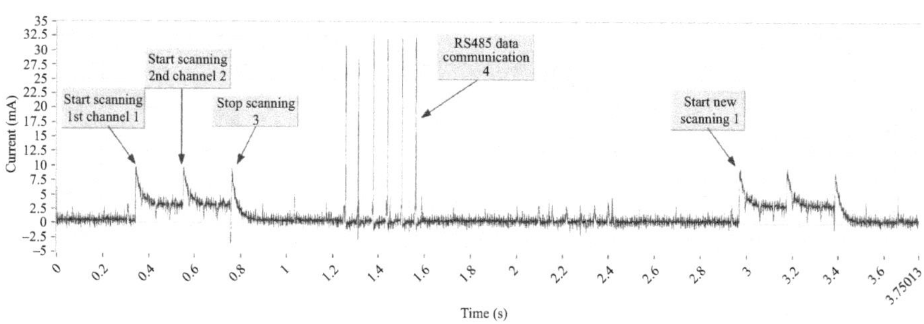

Figure 8.24 Current consumption of MN during a measurement session

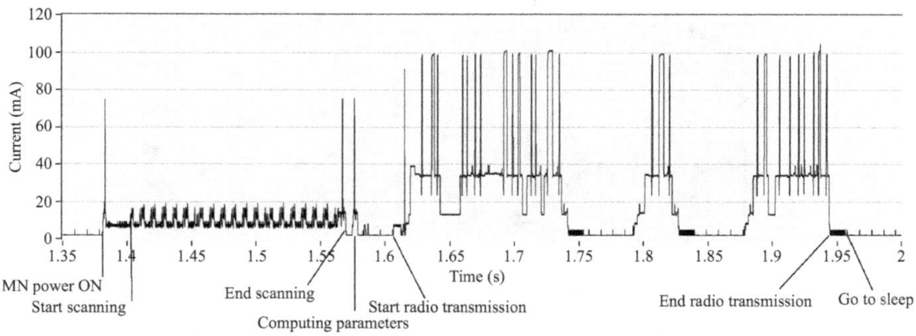

Figure 8.25 Current consumption of PCM during a measurement session

measurement session, meaning acquisition of three parameters, 2 bytes each, from 3 MPs, plus PWP sensor. For every byte, the PCM sends the address of MP through RS485 bus, after which it reads the corresponding byte and computes the direct parameters. The last task is sending data to C. As may be noticed from the graph, sending data by radio needs the most energy of the entire session. Taking into account the supplying voltage of 6 V for the PCM, it results that a measurement session requires at least 1.87 W for a time interval of about 600 ms, meaning a quantity of energy of about 1.15 Ws consumed from the power supply. If the PCM is supplied by a set of 4 AA alkaline batteries of 2,100 mAh each and scan time is 15 minutes, it means that an MN halves the battery life in 235 days, supposing that no alarm happens in that time.

8.6 In-field deployment and data presentation

8.6.1 One year recordings using the WSN

For deploying the WSN into the field, a slope prone to landslides from the village of Mădârjac, Iaşi County, Romania, was chosen. Its picture is shown in Figure 8.26. On an old slide of this area, two drillings were made, F1 and F2, finding a sliding structure of about 7.5–8 m thick to the stable clay layer with the presence of groundwater. A set of transducers was placed in the area in order to monitor the soil activity and to warn or alarm, if necessary, on a landslide triggering. The architecture of this network is similar to that shown and discussed in Section 8.2.1. It includes the coordinator, C (GW), and four measurement nodes of which three are configured as end nodes (MN1, MN3 and MN4) and one as a router node (MN2). Since the WSN was installed in the late autumn 2016, the whole year 2017 was able to be supervised. A continuous monitoring during the year was performed, with particular attention during the months of March to May 2017, when snow melting and accumulation of large amounts of water in soil could lead to obtaining conclusive data for the parameters under consideration.

The landslide transducers are 8 m in length and 22 mm in diameter. They were mounted in drillings with diameter of 40 mm. The bottom side of the transducers

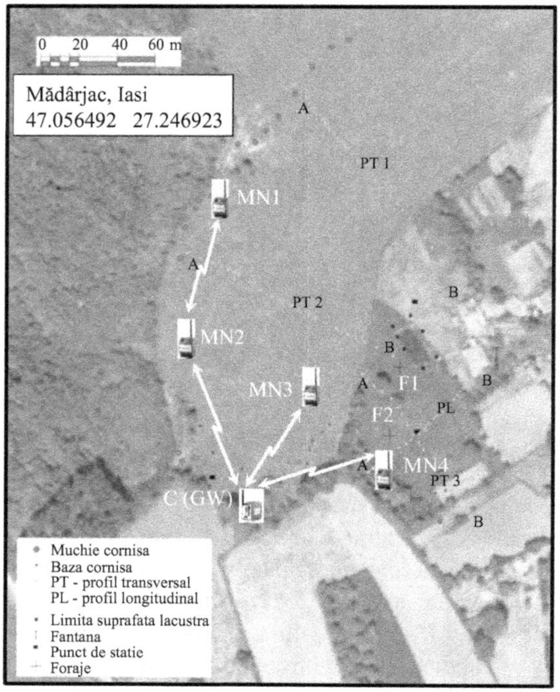

Figure 8.26 In-field deployment of WSN

was buried 0.75 m in the clay layer, so that it remains steady during the landslide. The space between the rod and the hole was filled with sand. Every EN was supplied with four AA alkaline batteries, whereas the RN was supplied with a 25 W solar panel buffered by a 7A/12 V lead battery. The gateway was supplied also by a solar panel similar to that of RN. The quantity of precipitation and the temperature are measured using an OneTemp digital rain and temperature gauge mounted in the neighbourhood of the gateway, to which it transmits serially the data. PWP is measured using the transducer described in Section 8.4.4, being mounted at the bottom of every node.

Some results representing the data recorded during the year 2017 are depicted in Figures 8.27–8.30.

Analysing the records, we may conclude that 2017 was roughly dry in the region. Between the end of April to mid-August, the highest quantity of precipitation was recorded, but only 2 days can be remarked with up to 50 mm/day. The temperature was quite high and the rainy period was superposed over high temperatures period, causing the water to rather evaporate. In January, when the temperatures are negative, no liquid precipitations happened and very low water accumulation has been recorded. Referring to PWP data, one may notice that some relatively low in amplitude increases matching the rainy days were recorded,. Some water accumulation in soil was also recorded during March, maybe due to

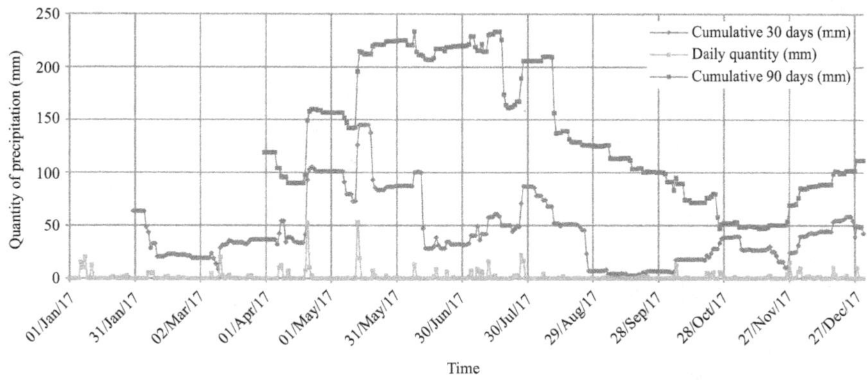

Figure 8.27 Quantity of precipitations recorded during the year 2017

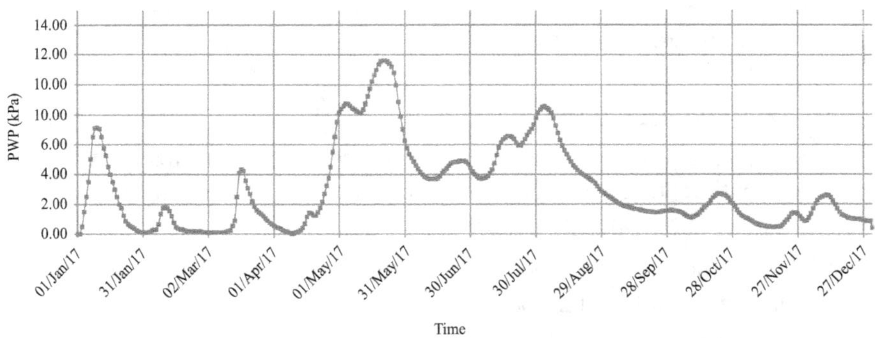

Figure 8.28 PWP recorded at sliding layer level

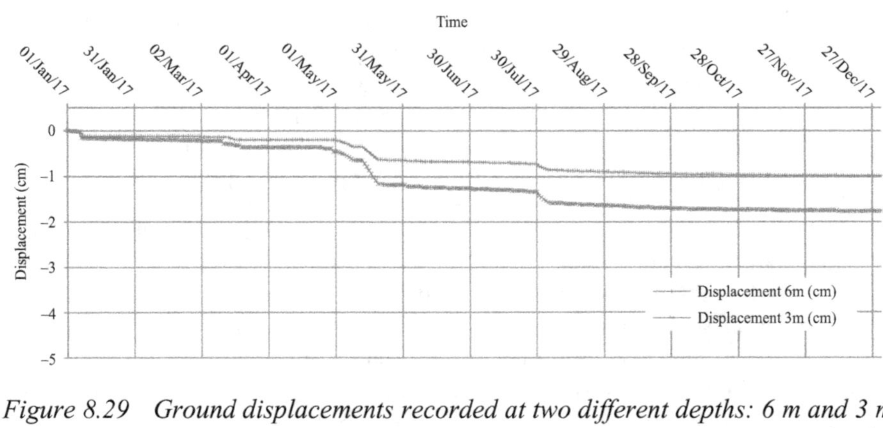

Figure 8.29 Ground displacements recorded at two different depths: 6 m and 3 m

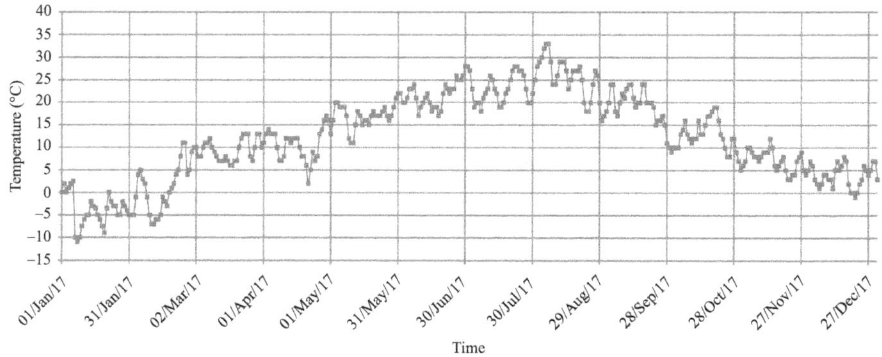

Figure 8.30 Temperature recorded during the year 2017

snow melting. These values of indirect parameters were not liable to trigger a warning and much less an alarm, since very small displacements, of about several millimetres all the year were displayed.

8.6.2 Graphic user interface

The coordinator sends regularly the recorded data at specified intervals of time (gateway scan time) to the cloud as spreadsheet text files in which the numbers are tab separated. A file is structured as follows:

- A header containing: the recording interval (date/time), scan time for gateway in hours and scan time for nodes in minutes;
- Quantity of precipitation integrated for 24 hours, in millimetres, recorded by the rainfall gauge mounted close to gateway;
- At every scan time:
  - Date, time, gateway power supply in volts and outside temperature in degrees centigrade;
  - Parameters for every measurement node: MN name, configuration (end node or router node), GPS coordinates, power supply in volts, pore water pressure value in kPa and status warning/alarm and
  - Parameters for every measurement point of the node: MP name, depth, displacement in millimetres, orientation in degrees versus North, temperature in degrees centigrade at the MP level and voltage supply in volts.

In Figure 8.31, a fragment of such a file is displayed. It is worth mentioning that no metadata has been added to the file, as it is converted into a MySQL file by the server, while the data are stored into a database.

The central server downloads the file from the cloud at regular intervals if time. At request, the monitoring centre, as well as any final user may open the graphic user interface (GUI) in a browser in order to inspect and analyse the recordings. Such an interface is presented in Figure 8.32.

```
Measurement data
14/12/2017 23.00 14/12/2017 23.00 24 15
2.4
14/12/2017 23.15 14.6 -2.7
MN1 END 47.056492 27.246923 11.8 1.5 0 0
MN1MP1 2 1.2 146 5.7
MN1MP2 4 0.9 146 5.7
MN1MP3 6 0.5 146 5.6
MN2 ROUT 47.056492 27.246923 12.3 0.6 0 0
MN2MP1 2 0.4 0 5.4
```

Figure 8.31 Fragment of a spreadsheet file sent by the gateway to the cloud

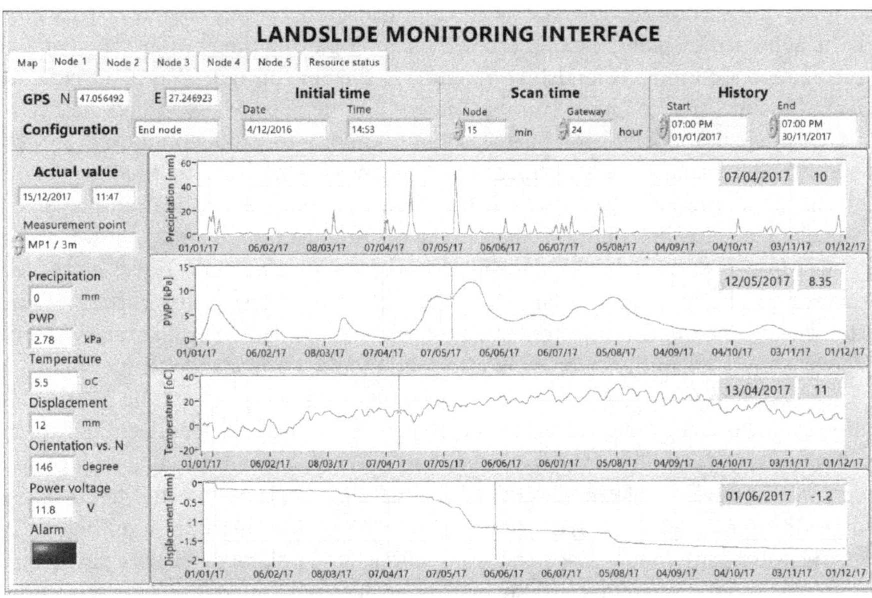

Figure 8.32 Graphic user interface for displaying the recorded data for landslide monitoring WSN

The GUI is organized as tabulated structures, one tab for one MN. A tab with the map containing the WSN deployment on terrain is accessible. Another tab showing the status of the coordinator of all MNs and MPs is also available. The interface displays both the actual values of the landslide parameters, and the history recordings for a certain period of time.

The values of displacement and orientation are indicated for every measurement point of the node. The power voltage is measured at the terminals of the buffer battery of the solar panel, but also the voltage values are given at every MP level in order to assess the resources status at every measurement session. As stated in Section 8.4.5, the PCM may be supplied also by alkaline batteries whose voltage is measured separately from the landslide transducer. The central server may receive by SMS a warning or an alarm at any time from the gateway, as soon as a parameter exceeds a limit or the alarm procedure is executed by WSN (see Section 8.4.3). In this case, the server downloads immediately the file from the cloud and the parameters are displayed on the interface, concomitantly with taking a decision by the user.

8.7 Conclusions

Landslides produce every year huge disasters and human losses all over the world. Developing monitoring actions for preventing landslide consequences are important goals for authorities charged with natural hazards in any country. This goal may be achieved by using the modern IoT technologies that may offer efficient and cost effective solutions. A landslide monitoring system configured as a WSN on an IoT architecture shall be based, in its structure, on devices of very good quality in terms of precision, operation robustness and immunity to harsh conditions, as well as on efficient management protocols and procedures in order to finally provide useful and reliable information about the phenomenon. The application presented in this chapter is a WSN built on a basic network architecture that may, however, be extended to a more complex monitoring system by adding as many nodes as needed for covering a large surface of land. The network may be easily customized for any particular terrain and can offer redundancy of collected data. Yet, the main benefit of the system is its ability to provide information about a landslide triggering from its very beginning due to the highly sensitive landslide transducers the network is equipped with. Using relatively low energy demand, it allows collecting data at fixed intervals of time for ordinary surveillance. When a small displacement of soil occurs, the system is able to adaptively change the scan time in order to correctly issue a warning or an alarm with a high level of trust. Another important benefit is that a measurement node may be endowed with a large palette of sensors devoted to gauging essential parameters that specialists decide might characterize a landslide. Consequently, one may assert that the main feature of this approach is its flexibility, because of its capability to be easily extended both as complexity and as recorded data relevance for correctly assessing the evolution of a possible landslide in a certain area of surveillance.

References

[1] Liu S. T. and Wang Z. W., 'Choice of surveying methods for landslides monitoring', *Landslides and Engineered Slopes* – Chen *et al.* (Eds.), Taylor & Francis Group, London, 2008, pp. 1211–1216

[2] Du J. C. and Teng H. C., '3D laser scanning and GPS technology for landslide earthwork volume estimation', *Automation in Construction*, 2007, 16(5), 657–663.

[3] Savvaidis P. D., 'Existing landslide monitoring systems and techniques', Proceedings of the Conference 'From Stars to Earth and Culture', Thessaloniki, Greece, 2003, 242–258.

[4] Zhang Y., Tang H., Li C., Lu G., Cai Y., Zhang J. and Tan F., 'Design and testing of a flexible inclinometer probe for model tests of landslide deep displacement measurement', *Sensors*, 2018, 18(1), 224

[5] Gian Q. A., Tran D. T., Nguyen D. C., Nhu V. H. and Bui D. T., 'Design and implementation of site-specific rainfall-induced landslide early warning and monitoring system: a case study at Nam Dan landslide (Vietnam)', *Geomatics, Natural Hazards and Risk*, 2017, 8(2), 1978–1996

[6] Higuchi K., Fujisawa K., Asai K., Pasuto A. and Marcato G., 'Application of new landslide monitoring technique using optical fiber sensor at Takisaka landslide', *Proceeding of 1st North American Landslide Conference*, 2007, Japan, pp. 1074–1083

[7] Nichol J., Shaker A. and Wong M.S., 'Application of high-resolution stereo satellite images to detailed landslide hazard assessment', *Geomorphology* 2006, 76, 68–75

[8] Massonnet D. and Feigl K.L., 'Radar interferometry and its application to changes in the earth's surface', *Reviews of Geophysics*, 1998, 36(4), 441–500

[9] Mishra S. B., Nagaraj C., Venkateswarlu V. and Adhikari G. R., 'Measurement and analysis of pore water pressure in large caverns and shafts in a hydro-electric project', *Geotechnical and Geological Engineering*, 2008, 26, 367–374

[10] Reid M.E., Baum R.L., LaHusen R.G. and Ellis W.L., 'Capturing landslide dynamics and hydrologic triggers using near-real-time monitoring', *Landslides and Engineered Slopes* – Chen *et al.* (Eds.), Taylor & Francis Group, London, 2008, pp. 179–191

[11] Wang G., 'GPS landslide monitoring: single base vs. network solutions – a case study based on the Puerto Rico and Virgin Islands permanent GPS network', *Journal of Geodetic Science*, 2011, 1(3), 191–203

[12] Maurizio B. and Fiani M., 'Monitoring of large landslides by terrestrial laser scanning techniques: field data collection and processing', 2013, *European Journal of Remote Sensing*, 46, 1, 126–151

[13] Pei H.-F., Yin J.-H. and Wei J., 'Development of novel optical fiber sensors for measuring tilts and displacements of geotechnical structures', Measurement Science and Technology, 2013, 24(9), 095202

[14] Vinoth S., Ajay Kumar L. and Mishra A. K., 'Status and developments of slope monitoring techniques in opencast mines', INDOROCK 2016: *6th Indian Rock Conference*, 17–18 June 2016, Mumbai, India, pp. 767–781

[15] Stuyfzand P. J. and van der Schans M. L., 'Effects of intake interruptions on dune infiltration systems in the Netherlands, their quantification and mitigation', *Science of The Total Environment*, 2018, 630, 757–773

[16] Ye X.W., Su Y.H. and Han J.P., 'Structural health monitoring of civil infrastructure using optical fiber sensing technology: a comprehensive review', *The Scientific World Journal*, Hindawi Publishing Corporation, 2014, Article ID 652329, http://dx.doi.org/10.1155/2014/652329

[17] http://www.ga.gov.au/scientific-topics/positioning-navigation/geodesy/geodetic-techniques/interferometric-synthetic-aperture-radar

[18] Stumpf A., Malet J.-P., Allemand P., Pierrot-Deseilligny M. and Skupinski G., 'Ground-based multi-view photogrammetry for the monitoring of landslide deformation and erosion', *Geomorphology*, 2015, 231, 130–145

[19] Qiao G., Lu P., Scaioni M., Xu S., *et al.*, 'Landslide Investigation with Remote Sensing and Sensor Network: From Susceptibility Mapping and Scaled-down Simulation towards in situ Sensor Network Design', *Remote Sensing*, 2013, 5(9), 4319–4346

[20] Lewis F. L. 'Wireless sensor networks', in Cook D.J. and Das S.K. (Eds.), *Smart Environments: Technology, Protocols, and Applications*, New York: John Wiley & Sons, 2004, pp. 11–46.

[21] Fosalau C., Zet C. and Petrisor D., 'Implementation of a landslide monitoring system as a wireless sensor network', *Proceedings of the 7th IEEE Annual Ubiquitous Computing, Electronics & Mobile Communication Conference*, 20–22 October 2016, New York, USA, pp. 695–700.

[22] Zet C., Fosalau C. and Petrisor D., 'Pore water pressure sensor for landslide prediction', *Proceedings of the IEEE Sensors 2015*, Busan Coreea, 1–4 November 2015, pp. 1–4.

[23] Fosalau C., Zet C. and Petrisor D., 'Multiaxis inclinometer for in depth measurement of landslide movements', *Sensor Review*, 2015, 35(3), 296–302

[24] Pallas-Areny R. and Webster J., *Sensors and Signal Conditionning*, 2nd Edn., New York: John Wiley & Sons, 2001

[25] Window A.L., *Strain Gauge Technology*, Dordrecht, Springer Netherlands, 1992

[26] Hÿtch M., Houdellier F., Hüel F. and Snoeck E., 'Nanoscale holographic interferometry for strain measurements in electronic devices', *Nature*, 2008, 453, 1086–1089

[27] Tanner J.L., Mousadakos, S, Broutas, P. *et al.*, 'Nanoparticle strain sensor', *Procedia Engineering*, 2011, 25, 635–638

[28] Maddipatla D. *et al.*, 'A novel flexographic printed strain gauge on paper platform', *Proceedings of IEEE Sensors 2015*, Busan, South Korea, 2015, pp. 1657–1660

[29] Yao S. *et al.*, 'Silver nanowire strain sensors for wearable body motion tracking', *Proceedings of IEEE Sensors 2015*, Busan, South Korea, 2015, pp. 1823–1826.

[30] Hogas I., C Fosalau., Zet C., 'A new strain sensor based on electrospinning and thin film technologies', *Proceedings of the 2016 International Conference and Exposition on Electrical and Power Engineering* (EPE 2016), Iasi, Romania, 20–22 October, 2016

[31] Hasegawa, R., 'Magnetic wire fabrication and applications', *Journal of Magnetism and Magnetic Materials*, 2002, 249, 346–350

[32] Ogasawara I. and Ueno S., 'Preparation and properties of amorphous wires', *IEEE Transactions on Magnetics*, 1995, 31(2), 1219–1222

[33] Phan M.-H. and Peng H.-X., 'Giant magnetoimpedance materials: Fundamentals and applications', *Progress in Materials Science*, 2008, 53, 323–420

[34] Morón C., Carracedo M.T., Zato J.G. and Garcia A., 'Stress and field dependence of the giant magnetoimpedance effect in co-rich amorphous wires', *Sensors and Actuators A: Physical*, 2003, 106, 217–220

[35] Bayri, N. and Atalay, S., 'Giant stress-impedance effect in Fe71Cr7Si9B13 amorphous wires', *Journal of Alloys and Compounds*, 2004, 381(S1–S2), 245–249

[36] Panina L. V., 'Magnetoimpedance (MI) in amorphous wires: new materials and applications', *Physics Status Solidi A* 2009, 206(4), 656–662

[37] Chiriac H. and Óvári T.-A., Giant magnetoimpedance effect in soft magnetic wire families, *IEEE Transactrions on Magnetics*, 2002, 38(5), 3057–3061

[38] Fosalau C., Damian C., Zet C., 'A high performance strain gage based on the stressimpedance effect in magnetic amorphous wires', *Sensors and Actuators A*, 2013, 191, 105–110

[39] Petrişor D., Foşalău C., Zet C. and Damian C., 'Measurement of landslide displacement and orientation using strain gauges based on amorphous magnetic microwires', *Proceedings of the 19th Symposium IMEKO TC 4 Symposium and 17th IWADC Workshop*, Barcelona, 18–19 July 2013, pp. 655–660

[40] Huanga A.-B., Leea J.-T., Hoa Y.-T., Chiub Y.-F. and Chengc S.-Y., 'Stability monitoring of rainfall-induced deep landslides through pore pressure profile measurements', *Soils and Foundations*, 2012, 52(4), 737–747

[41] Freescale datasheet, 'Differential pressure sensor MPX5999D'. Available from: http://cache.freescale.com/files/sensors/doc/data_sheet/MPX5999D.pdf

[42] Zet C., Fosalau C., Petrisor D. and Damian C., 'Landslide transducer based on amorphous wire strain gauges', *Proceedings of 19th IMEKO TC4 Symposium Measurements of Electrical Quantities*, Barcelona, Spain, 18–19 July 2013, pp. 286–289

[43] Zet C., Fosalau C., Petrisor D. and Hogas I., Study of the power consumption of a landslide sensor node, *Proceedings of the 7th IEEE Annual Ubiquitous Computing, Electronics & Mobile Communication Conference*, New York, USA, 20–22 October 2016, pp. 444–448

Chapter 9
Industrial Internet of the things

José Miguel Dias Pereira[1,3] *and Vítor Viegas*[2,3]

This chapter focuses on several topics related to the industrial Internet of the things (IIoT). Starting with current loop signalling in industrial plants and finishing with wireless networks, the main strengths, weaknesses, opportunities and threats of IIoT are underlined. The first part of the chapter includes a summary of the main protocols that are still used in instrumentation and control (I&C) networks and the second part of the chapter presents topics related to industrial Ethernet, real-time demands, collision avoidance, synchronization and time-sensitive requirements in industrial networks.

It is important to refer that smart sensing capabilities, including adaptability, self-testing, self-calibration, auto-ranging and errors correction capabilities, among others, contributed deeply to improve the accuracy and the performance of industrial measurement systems and to promote interoperability of sensing units and industrial devices. The implementation of these capabilities in industrial instruments, from different manufacturers, was very important since it enabled cost reductions in the maintenance and operation of I&C systems and promoted the quality of the manufactured products.

Finally, it is important to refer that IIoT is not a panacea that will solve all industrial problems in a short time without any drawbacks. New investments are required and the industrial staff must develop new skills to manage successfully IIoT and to extract the advantages that can be provided.

9.1 OSI model

Before starting the description of industrial protocols, it is important to present the seven layers of the open system interconnection (OSI) model defined by the International Organization for Standardization (ISO). As represented in Figure 9.1, the OSI model defines seven layers, namely, physical, data link, network, transport, session, presentation and application layers.

The physical layer defines the conditions that are required to transmit the signals in the physical medium. For example, devices like cables, connectors, hubs,

[1]ESTSetúbal/IPS, Setúbal, Portugal
[2]CINAV – Escola Naval, Base Naval de Lisboa, Alfeite, Almada, Portugal
[3]Instituto de Telecomunicações, Lisboa, Portugal

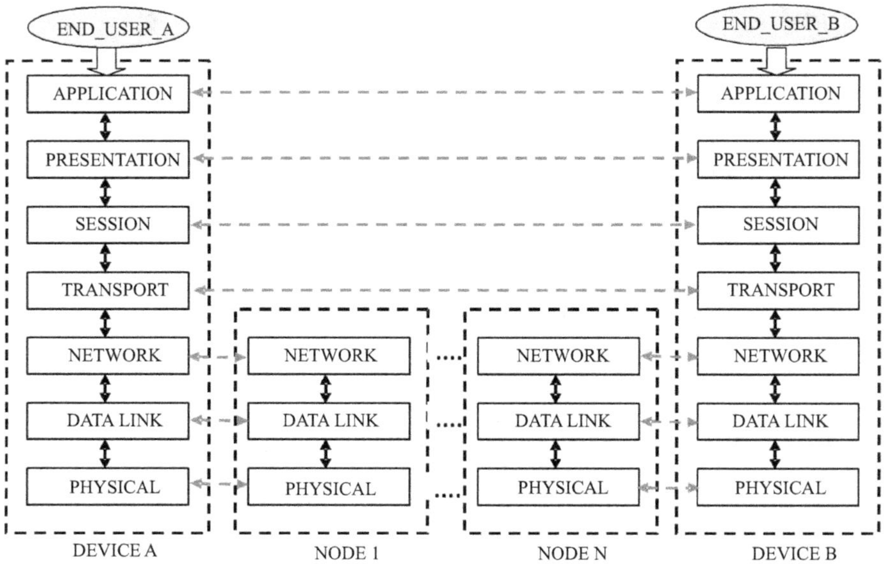

Figure 9.1 OSI model layers that are used to connect the network end devices A and B through network routers

repeaters and signals' characteristics, such as voltages, amplitudes, bitrates and coding schemes, are defined according to the physical layer specifications of the OSI model. The data link layer (DLL) provides node-to-node data transfer detecting, and eventually correcting, errors that may occur in the physical layer. In Ethernet networks, DLL is divided in two sublayers: media access control (MAC) and the logical link control (LLC) that checks for physical transmission errors using a frame check sequence (FCS). One of the main tasks of the DLL is related to the techniques that are used to share the physical medium between several network devices. Devices like bridges and switches operate in the DLL of the OSI model. The network layer uses the source and destination addresses of the DLL protocol data unit (PDU) to evaluate if the packet reaches its final destination. In affirmative case, it is sent to the transport layer, otherwise the frame destination address is updated and the packet is forwarded to a neighbour network node. In the Internet networks, routing is performed according to the logical Internet Protocol (IP) address. The transport layer performs tasks related to data flow control, re-transmission and error recovery. In Internet networks the transmission control protocol (TCP) and the user datagram protocol (UDP) are the most common protocols that are used in the transport layer. In this context, it is important to refer that in industrial applications the usage of TCP or UDP, as transport layer protocols, must be carefully evaluated considering the associated limitations in terms of real time requirements and data delivery reliability. UDP uses a connectionless transmission mode, with no end-to-end hand-shaking, or acknowledgement, and with no guarantees of packets' delivery sequence, so the usage of UDP is not acceptable in a large number of industrial applications that require packet delivery handshaking and preservation of data packet temporal sequence delivery. Thus, the advantages that could be associated with UDP, in terms of a lower ratio between the number of bytes used for the protocol control information (PCI) and for PDU, cannot be used in a large number of applications. The upper layers of the OSI model include the session, presentation and application layers. The session layer provides the mechanisms for opening, closing and managing applications that run between different end network users or network servers. Tasks that are related to authentication, authorization, session restoration and security are also implemented in this layer of the OSI model. The main tasks that are associated with the presentation layer are related to the data translation between the data format that is used in the application layer and the data format that is used in the network. The tasks performed in the presentation layer also include encryption and decryption. Finally, the application layer supports network services that are related to end-users' applications, such as remote file access, network management, directory services, network virtual terminals management and electronic messaging, among others.

Regarding the usage of the OSI model in industrial networks, it is important to refer that according to the applications' requirements, essentially in terms of real time and reliability, simplified versions of the OSI model are used. Some of the layers are not implemented, at all, and others could be compacted in a single layer. Basically, the objective for these simplifications is related to data protocol efficiency that is affected by the relation between the number of bytes associated with the service data unit (SDU) and the number of bytes associated with the PCI fields.

9.2 Industrial protocols: from current loops to fieldbuses

The first step towards standardization of industrial communications and protocols was performed using pneumatic and electrical signals with normalized amplitude levels. Concerning pneumatic signals, the amplitude values of the measuring variables were translated to a pneumatic signal with a pressure value that typically varied between 3 and 15 p.s.i., 0.2 and 1 kgf/cm^2 or between 0.2 and 1 bar. Regarding electrical signal amplitudes, there are also several voltage or current normalized amplitude levels that have been used but the most common, still used nowadays in a large number of instruments, is a current signal that varies between 4 and 20 mA. The minimum value of the normalized signal of the instrument, commonly denominated transmitter, is associated with the minimum value of the measuring quantity, also known as the lower range value (LRV), and the maximum value of the measuring quantity, also known as the upper range value (URV), is associated with the maximum value of the normalized output signal of the transmitter. The same type of signals was used to control actuators such as motors or valves. For example, if the actuator is a pneumatic control valve, a pressure amplitude of 3 p.s.i. can be associated to a null aperture of the valve (valve completely closed) and a pressure amplitude of 15 p.s.i. can be associated to a full aperture of the valve (valve completely open).

Signals' amplitude normalization was also the first step towards interoperability. If two manufacturers of the same type of instruments use identical signals amplitudes, it could be possible to exchange the instruments without modifying anything more in the process, namely in the I&C system, as long as both instruments can be calibrated for the same measuring range.

In the next paragraphs a summary of the main characteristics that are associated with the 4–20 mA current loop, Highway Addressable Remote Transducer (HART) and Foundation Fieldbus (FF) protocols are used to focus the main evolution phases that occurred, in the last years, in industrial process communications and protocols. It is important to underline that different fieldbus types, such as Profibus, could be used to present the main characteristics of these types of industrial communication protocols. The choice of FF was done mainly for two reasons. One of the reasons is FF provides a real-time clock enabling deterministic communications, timestamp and synchronized control execution, and the second reason comes from some practical experience acquired with FF by using a prototype that was implemented in our laboratory for testing purposes [1,2]. Figure 9.2 depicts the P&I diagram of the FF prototype that was implemented and that enables the establishment of different control strategies related to the following physical variables: water level, water temperature and output water flow.

9.2.1 Current loops

The current loop is a simple protocol based on analogue signalling [3]. It associates a given value, of a measuring quantity, to current value amplitude. This type of

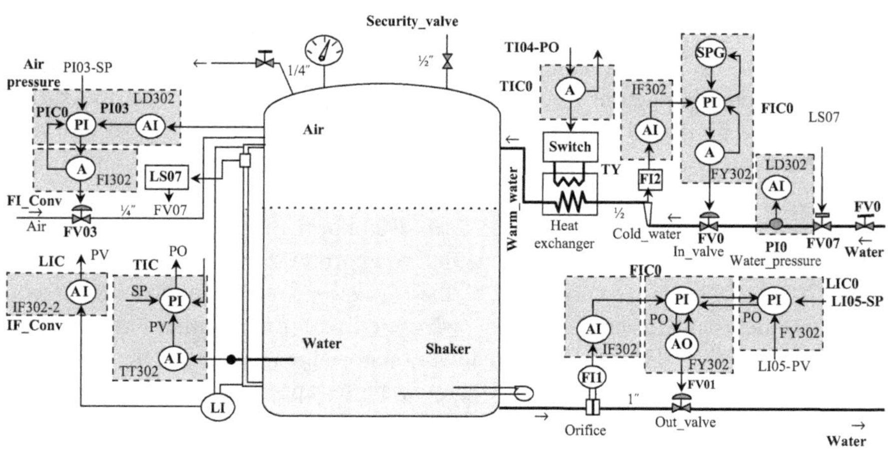

Figure 9.2 FF prototype used to perform laboratory tests of different control strategies (level, temperature and output water flow)

signalling uses a simplex communication mode, based on a point-to-point connection between the transmitter and the receiver, and the accuracy of the transmitted signal is not affected by the voltage drop in the interconnecting wiring, as long as, the maximum loop resistance value is not exceeded. The relationship between the output current of a 4–20 mA transmitter and the value of the quantity under measurement is given by

$$I_T = 4 + \frac{16}{\mathrm{URV}-\mathrm{LRV}} \cdot (\mathrm{X}-\mathrm{LRV})\,\mathrm{mA} \qquad (9.1)$$

where X represents the quantity under measurement and URV and LRV represent the URV and the LRV, respectively.

The current loop is established by a pair of wires that transport the measuring information from the transmitter to the receiver, closed by a simple resistor of 250 Ω that converts the 4–20 mA current range to a 1–5 V voltage range.

A single pair of wires can be used to power each instrumentation device and to carry, simultaneously, the measuring data. The distance between the instrumentation device and the power supply can be higher than some hundreds of metres depending, obviously, on the conductors' section and on the existence of current repeaters. Current, instead of voltage, signalling is much less sensitive to voltage drops and external noise interferences and the minimum value of current, typically equal to 4 mA, is associated with the minimum value of the quantity under measurement. This relationship between the minimum output current from the transmitter and the minimum value of the measuring quantity, known as live zero, is very important for maintenance and calibration purposes because it allows easy detection of transmission problems, such as broken wires or wrong connections in the junction boxes. Several manufacturers also use current amplitude values lower than 4 mA or higher than 20 mA to signalize fault or alarm conditions such as measurement values lower than the LRV or higher than the URV, respectively. Even if the working principle of this signalling mode is very simple, there are a large number of limitations which include, mainly, the need of a pair of wires for each instrument and measuring quantity; there is no maintenance task that can be performed remotely except the one that is related to the detection of broken wires in the connection between the transmitter and the receiver and it is not possible to access or modify remotely the LRV and URV or the transmitter and it is not possible to perform instrument asset management tasks, since it is not possible to remotely access the basic maintenance-related parameters, such as calibration date or data. Regarding the lack of capability to perform a remote identification of any instrumentation device, several maintenance and operational problems can occur and affect substantially industrial system performance. For example, if two current loops are exchanged, the information that is received is completely wrong because the control system associates the current value with a wrong quantity that is not the one which is under measurement. Additionally, if the current loop resistance exceeds its maximum value, the current decays and there is no chance to distinguish this decay from the one that is caused by a reduction in the quantity under measurement.

Regarding the interface that are frequently needed to convert current to pressure signals, it is important to refer that the 4–20 mA signalling enables an easy implementation of a current to pressure (I/P) conversion being possible to use I/P converters that translate the 4–20 mA signal into a pneumatic signal without the need of any electrical power source near the pneumatic device. The usage of these converters is common to interface pneumatic valves that are controlled by a 4–20 mA current signal. Another advantage of this signalling mode results from its simplicity and easy understanding by nonspecialized commissioning and maintenance personnel. It is important to underline that this elementary protocol does not have the capability to connect multiple transmitters to a given receiver, but one transmitter can send its measuring information to multiple receivers, as long as they are connected in series and the maximum loop resistance is not exceeded.

9.2.2 HART

With the advent of smart sensing it was possible to implement a large number of new capabilities in field instrumentation. Some of the new capabilities, depicted in Figure 9.3, include auto-configuration, Plug and Play, self-test, self-calibration, self-identification, error correction, auto-ranging, choice of engineering units, data logging, filtering, dithering and multivariable measurements, among others. The usage of these new capabilities implies a half-duplex communication mode between the instrumentation devices and the control unit (host device). The data

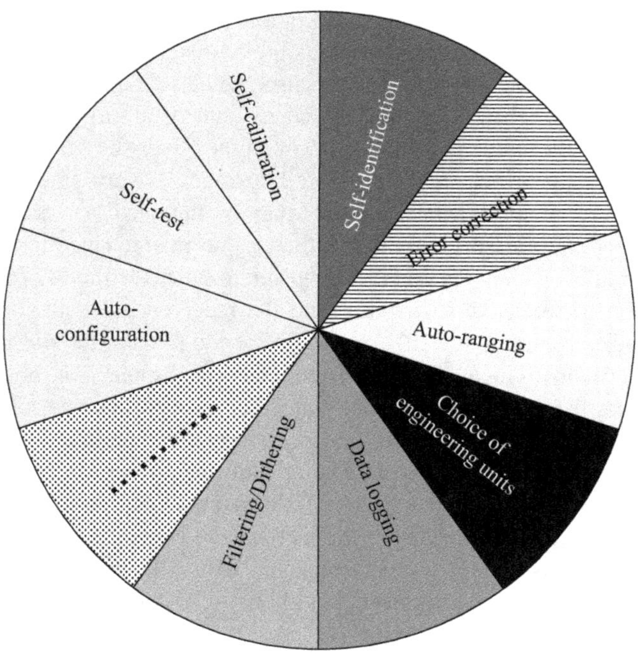

Figure 9.3 Enhanced capabilities of smart sensing devices

that come from the instruments are not a simple analogue current signal; instead, several digital messages are exchanged between the host, or HART communicator [4,5], and the HART instrumentation devices.

The HART signal [6–8] includes two superimposed current components. A direct current component, whose amplitude varies, typically between 4 and 20 mA, and a time variable component that uses a digital frequency shift keying (FSK) signal whose peak-to-peak amplitude varies between 0.8 and 1.2 mA. The first component corresponds to the traditional analogue current loop signalling signal, previously described, and the alternating current component is used to exchange digital data between the HART instrumentation device and the control unit. It is important to underline that the same wiring, used to transmit the 4–20 mA analogue signal, is used to carry the HART communication signal. This means that the remote access to measurement configurations, maintenance data or calibration parameters, of each instrument, can be performed remotely in a supervision centre using a HART communicator.

Each HART message sent by an instrument includes a device status information field consisting of two bytes that are used to report communication errors, the status of the received command and the operational state of the HART instrument. From the contents of status bytes it is possible to access several conditions, such as device faults, communication errors, maintenance status or calibration needs, among others. This type of information, not supported at all by the traditional current loop signalling, can be used to improve I&C systems performance, to trigger calibration tasks, if calibration is out of date, and to trigger predictive maintenance tasks, avoiding nonscheduled I&C system shutdowns.

Regarding the OSI model, as depicted in Figure 9.4, HART includes three protocol layers, namely, the physical layer, the DLL and the communication stack. As previously referred, the physical layer supports a half-duplex communication mode, using a master-slave protocol, and an FSK [7] data stream codification for the digital component of the signal. The data link layer defines the access control, communication services and the structure of the data stream. The HART user layer, not included in the OSI model, defines a set of commands that are common to almost all HART devices. These commands are known as the universal and command practice commands. Regarding device-specific commands, HART device descriptions (DD) can be written using a standardized device description language (DDL), being possible to take full advantage of all devices' capabilities and promoting interoperability of devices with different specific commands.

One drawback of the HART communication protocol is the low message-update rate that is about two messages per second. Even in the burst mode [9,10], supported by some HART devices, where the instrument is continuously broadcasting measuring and status data, the maximum update rate is only about three to four messages per second.

It is important to refer that the HART protocol also has the capability to connect multiple field devices to the same pair of wires. In multidrop applications, communication is limited to the master-slave protocol and up to 15 devices can be connected to a single HART line. In this case, if all HART devices are active, or if a

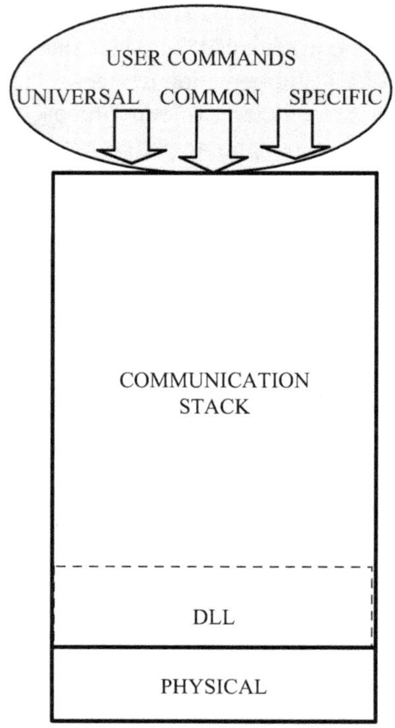

Figure 9.4 HART protocol layers

HART multiplexer is used to access multiple HART signal loops, the maximum message update rate, from a single device, can achieve values that are several tens per second.

A wireless version of the HART protocol, known as Wireless HART, brought to the wireless domain the multiple advantages of the HART communication protocol. This protocol uses a radio standard similar to IEEE 802.15.4 (ZigBee), with a 10 dBm adjustable transmission power, and the wireless transmission takes advantage of frequency hopping techniques, on a per packet basis, being this characteristic essential to improve transmission reliability which is a topic of major importance in industrial environments [11,12]. Similar to HART, Wireless HART uses a half-duplex transmission mode but additional features, such as the usage of superframe scheduling [13] and time division multiple access (TDMA), are used to reduce collisions and to improve communications' reliability. Regarding the OSI layers, Wireless HART implements all layers of the OSI model, except layers 5 and 6, session and presentation layers, respectively.

Concerning network topologies, Wireless HART supports mesh, star and combined mesh-star topologies, and alternative paths can be automatically activated in case of communication failures. Additionally, every wireless HART device can act as a router and wireless HART adaptors are available to connect traditional HART devices to a wireless HART network.

The main limitations of Wireless HART are associated with instrument devices powering, coverage range, reliability and message transmission rate. Concerning the transmission rate, a typical maximum value of 1 message per second can be achieved which could be a low value for some automation applications. Regarding battery-powering solutions of wireless HART devices, it must be referred that nowadays battery technology advances can assure more than 10 years of autonomy using adjustable transmission power levels techniques, sleep and power-down modes and, above all, avoiding the transmission of redundant data. A compromise between sampling data rate and feedback control errors, namely the integral of the absolute error, must be found in order to use proper sampling rate for each loop, avoiding the useless depletion of the battery of wireless devices. The drawback associated with transmission failures of Wireless HART devices can be minimized by using the model predictive control (MPC), which can predict lost analogue input values that occur during short fault periods of wireless transmission [14].

Finally, it is important to refer that, according to the FieldComm Group, almost 85% of the installed HART devices are not integrated into an industrial network and their capabilities are partially explored, being a great part of useful information lost. In this context, field device tools (FDT) and device type management (DTM) solutions [15,16], which promote a centralized configuration, operation and maintenance, through a standardized user interface, compatible with different communication protocols, are not being useful in a large number of factories.

9.2.3 Foundation fieldbus

Foundation Fieldbus (FF) includes two types of implementations that use different physical media and communication speeds. Type one implementation is associated with H1 network segments, while type two implementation is associated with H2 network segments, also known as High-Speed Ethernet (HSE). The low-speed segments (H1) are used to connect field devices to the FF interfaces, while the high-speed segments (H2) are used to connect different H1 segments and to establish the connection with other networks and supervision centres. Data rates used in H1 and H2 segments are equal to 31.25 kbps and 100 Mbps, respectively. As an example, Figure 9.5 represents a FF network that includes three FF/H1 segments, one H2 segment, a set of personal computers that are used for centralized configuration, operation and maintenance, and a gateway that is used to connect the FF network to other networks that use different communication protocols [17–20].

Regarding the new capabilities that are offered by FF, it must be underlined that this protocol can share deterministic and random traffic in the same transmission frame. Deterministic traffic, usually associated with data acquisition from sensors and data transmission to actuators, is assigned to reserved time slot intervals of the H1 data frame. Synchronization and time distribution functions are coordinated by the link access scheduler (LAS), which can be replicated in several devices of the FF network. The bus access congestion is avoided by using a token-based protocol. Compel data (Cp_D) tokens are used for deterministic or cyclic

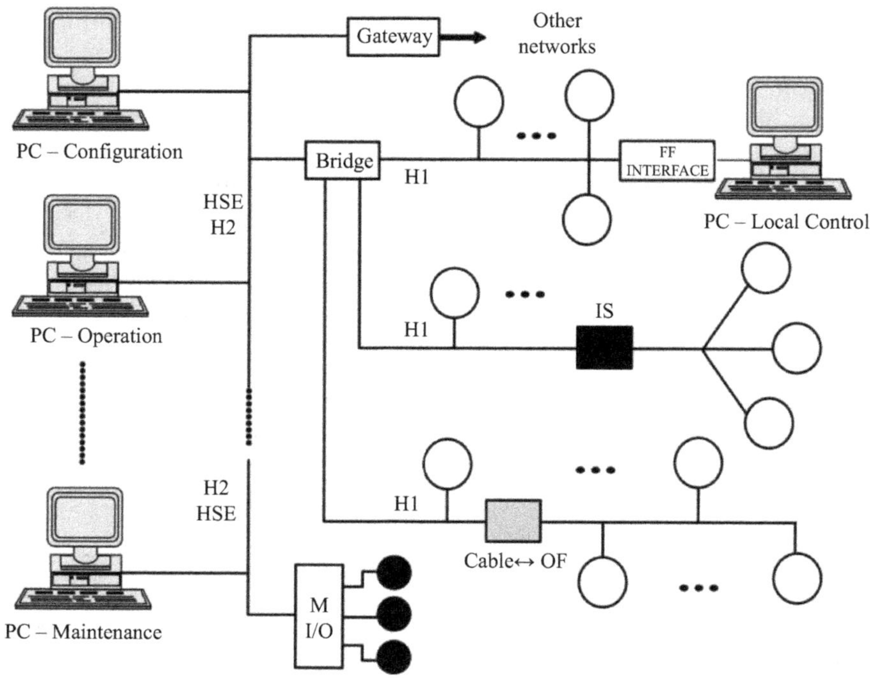

Figure 9.5 Example of a FF network that includes three H1 segments and a single H2 segment. HSE: high-speed Ethernet; M I/O: input-output multiplexer; IS: intrinsically safe barrier; OF: optical fibre)

data, and pass tokens (PT) are used for nondeterministic, or acyclic, data. Token distribution and network synchronization is centrally performed by the LAS.

As represented in Figure 9.6, FF H1 uses a simplified version of the seven layer OSI model. The physical layer of FF uses Manchester encoding being each bit associated with a positive or a negative signal amplitude transition in the middle of the bit period. Thus, the FF H1 signal is a self-clocking signal with no DC component, being galvanic isolation easily implemented by inductive coupling. Clock recovery techniques are used to extract a timing reference value from the physical signal that has a nominal period of 32 μs with a 10% jitter uncertainty.

The protocol layers of FF H1 do not include OSI layers 3 to 6 and the application layer is standardized in two sublayers, namely, the Fieldbus Access Sublayer (FAS) and the Fieldbus Message Specification (FMS) sublayer. By its turn, the protocol layers of FF H2 contain a more complete version of the OSI model, since specific DLL, IP and TCP/UDP layers are included in the model to provide Ethernet medium access control (MAC), IP networking and data transport capabilities, respectively. It is also important to refer that the FF H1 protocol model includes a version that includes safety instrumented functions that support a safety integrity level (SIL) up to class 3 [21], associated with a dangerous failure per hour probability between 10^{-7} and 10^{-8}. The field device agent (FDA), included in the

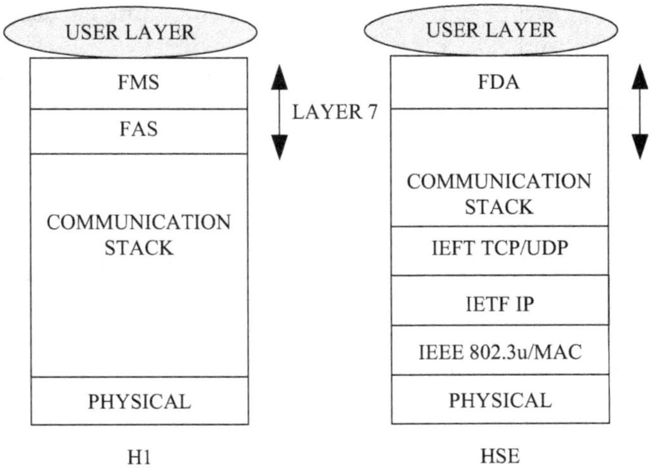

Figure 9.6 FF H1 and FF H2 protocol layers. FMS: Fieldbus message specification; FAS: Fieldbus access sublayer; IETF: Internet engineering task force [22]; IP: Internet protocol; MAC: medium access control; HSE: high-speed Ethernet; FDA: field device agent

top of the FF HSE protocol layers, provides FMS and system management over TCP/UDP to support link devices that can be simultaneously connected to H1 and HSE devices.

Concerning the user layer, FF standardizes a large number of function blocks (FB) that are recognized by all certified FF manufacturers. Using defined FB, the FF applications are easily developed using user-friendly graphical programming languages. This standardization has a huge impact in which concerns hardware (HW) and software (SW) interoperability since the user programs can run in different instrumentation devices, from different manufacturers, using open standards, and without need to perform any SW changes.

Another advantage that is associated with the FF is its capability to implement Plug and Play (P&P) functionalities. The LAS uses a probe node PDU to search for new devices that connect to the network, thus supporting hot swap capability. FF can also implement stand-alone control in the field (CIF) [23,24] solutions, being several control tasks, such as PID control, performed in instrumentation devices without any need of I/O data centralization such as the one provided by programmable logic controllers (PLC) based solutions. In this way, with the usage of CIF capabilities, it is possible to assure field operation without intervention of the distributed control system (DCS). FF also assures maintenance advances in instrumentation asset management system (IAMS) tasks, being possible to access complete information about the identification (tag numbers) of all the devices that are connected to the FF network, their state, maintenance and calibration data, among other data. Reliability improvements are also achieved by using redundant

I&C tasks that run in different FF devices connected to different segments of the communication network.

Using predictive maintenance capabilities, it is possible to anticipate potential risk failures, detecting, in advance, equipment's conditions that require the attention of maintenance personnel.

Regarding protocol capabilities, it is important to underline that although the FF protocol enables the sharing of scheduled and unscheduled communication in the same data stream, the macrocycle period (T_{MC}) is in the order of a few hundreds of ms, being only possible to assure a real-time (RT) behaviour when process time constants are much higher, at least five to ten times higher, than T_{MC}.

In order to estimate the T_{MC} values that can be achieved with FF, suppose that the PDU has an average length equal to PDU_{av}, the number of loops under control, in a given H1 segment, is equal to NL, the number of scheduled time slots per loop is equal to NTS_L and the time ratio between unscheduled and scheduled traffic, in the MC, is equal to US_R. Based on these parameters, the period of T_{MC} is given by

$$T_{MC} = \frac{NL \cdot NTS_L \cdot PDU_{av}}{1 - US_R} \cdot \frac{8{,}000}{31{,}250} \text{ ms} \tag{9.2}$$

Figure 9.7 represents T_{MC} as a function of the number of loops defined in a given H1 segment, considering that each loop has a number of scheduled TS equal to three, the PDU_{av} value is equal to 50 bytes and the time ratio between

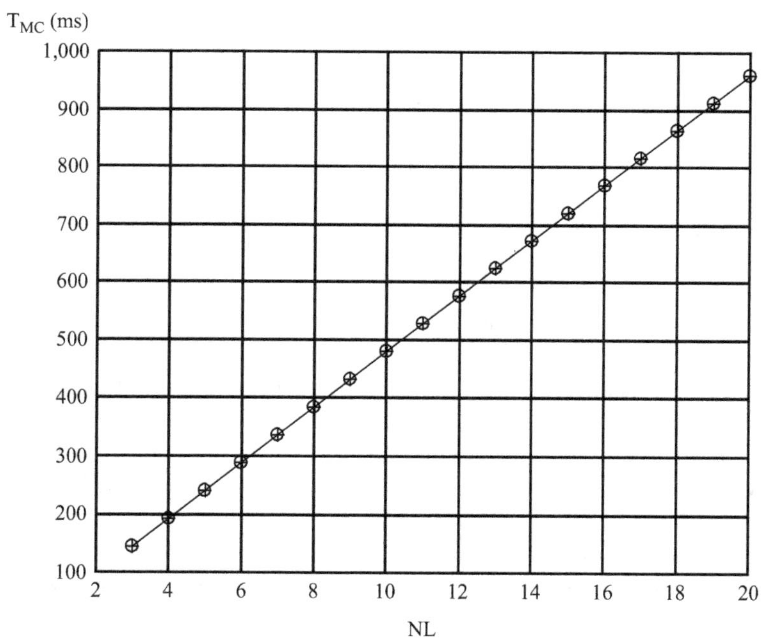

Figure 9.7 Macrocycle period as a function of the number of loops defined in a given FF H1 segment (NTS = 3; PDUav = 50 bytes; USR = 20%)

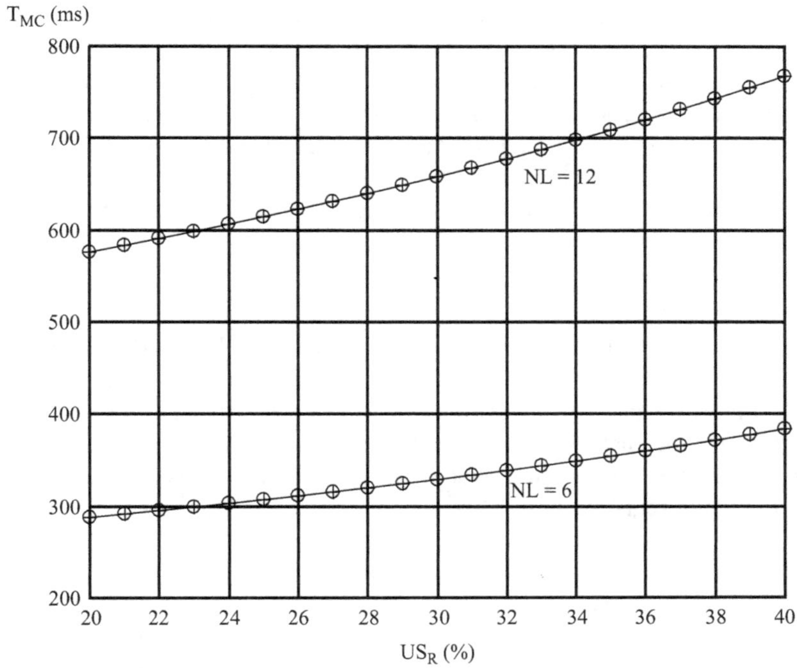

Figure 9.8 Macrocycle period as a function of unscheduled versus scheduled traffic ratio (NTS = 3; PDUav = 50 bytes; NL = 6 and NL = 12)

unscheduled and scheduled traffic is equal to 0.2 (20%). As expected, the dependence of T_{MC} in relation to NL is linear and when NL is about 20 the T_{MC} is almost equal to 1 s. This is typical of HART message update rate.

By its turn, the lower and upper curves of Figure 9.8 represent the variation of T_{MC} as a function of US_R for a FF segment with 6 and 12 loops, respectively. As expected, in this case, the value of T_{MC} increases with US_R and for a given value of US_R the value of T_{MC} duplicates when the NL is duplicated.

The values that were estimated for TMC must be carefully analysed in every industrial application since they have a direct impact on control loop performance and on real-time limitations of FF network.

9.3 Industrial Ethernet

Ethernet, as defined by IEEE standard 802.3, is a nondeterministic protocol and, thus, not suitable for applications with RT demands. The MAC of Ethernet is based on the well-known Carrier Sense Multiple Access with Collision Detection (CSMA/CD). The random occurrence of collisions is not compatible with the achievement of deterministic latency times to exchange data in the network. However, Industrial Ethernet should ensure RT demands, in terms of latency and jitter, and must enable data transmission of non-RT data that share the same transmission support [25–27]. The first approach to reach these objectives was materialized by using proprietary solutions, such as Profinet IRT, Ethercat, SERCOS III and Ethernet Powerlink [28–30], among others. However, these solutions are not open and require special interfaces and gateways to connect different network types from different manufacturers. In this section several topics related to the implementation of time-sensitive networking (TSN) [31] are addressed, starting with a brief introduction of real-time definitions.

9.3.1 Real time

Definitions about RT are sometimes not consistent in the literature. As a matter of fact, RT is not necessarily associated with fast execution times but is more related to determinism and synchronization of tasks' execution in a given system. Considering a job that must be processed in a computer processing unit (CPU), Figure 9.9 represents the timings that are associated with its execution. The release time (T_{RL}) represents the time when the job is available to the system, the ready time (T_{RD}) represents the time when the job can start execution, the end of execution time (T_{EE}) represents the time when the job finishes execution and the deadline (T_{E_max}) represents the time when the job execution must be finished. According to these time marks, the execution time is the time interval between T_{EE} and T_{RD} and the response time is the time interval between T_{EE} and T_{RL}.

The deadline limit that is associated with a given job is directly related to its RT requirements. According to the RT requirement, it is common to consider four RT categories, namely, Hard Real Time (HRT), Soft Real Time (SRT), Firm Real Time (FRT) and Weakly Hard Real Time (WHRT). HRT requirements are associated with mission and safety-critical systems where failures can result in catastrophic events, such as loss of life or property. Systems that have HRT requirements use typically hardware redundancy techniques and, if considered for industrial applications, must be redundant in terms of hardware, software and networking, using, if necessary, RT operating systems (RTOS) to assure the right outputs within stringent deadlines. Examples of HRT requirements in the industrial field are automatic control and robotics applications that must preserve the security of equipment and safety of the people that work nearby those systems.

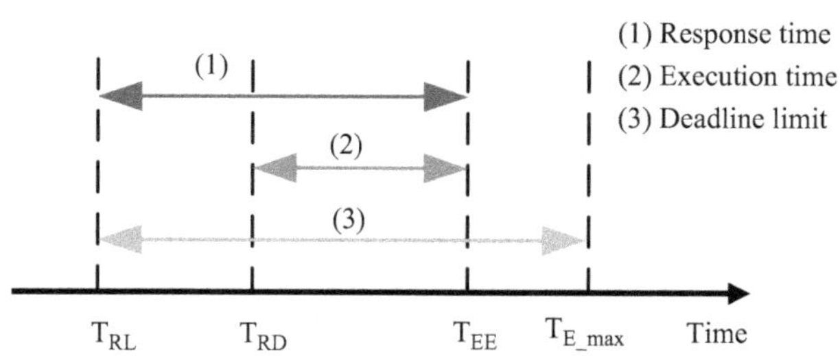

Figure 9.9 Time marks of a job execution. TRL: release time; TRD: ready time; TEE: end of execution time; TE_max: deadline time

Another example, in process industries, can be found in the ethylene production by naphtha cracking technology where pressure values of the process can achieve values over 1,000 bar and where timing requirements are very critical. By its turn, in SRT systems deadline overruns, within certain limits, are not catastrophic and are not associated with critical failures. Real-time performance of SRT systems is usually measured by a quality of service (QoS) index that can be used to evaluate system performance. Concerning industrial applications, SRT requirements could, for example, be associated with the temperature control of a fluid that is contained in a tank, since the temperature adjustment is usually a very slow process and deadline overruns, within certain limits, are not critical and don't cause any reduction of system performance. In FRS systems or applications, with high tolerance for deadline overruns, there are no critical problems that arise when deadlines are not fulfilled but the usefulness of tasks' completion decreases and could be of no value if the task is not completed within a maximum period of time. In the industrial field, considering an example about the measurement of the water temperature contained in a boiler, and supposing that a set of four temperature sensors are used to take an average value of the water temperature, if the temperature acquisition from one of the sensors is not obtained before the predefined deadline, the deadline overrun has no impact in the system performance, since the average value of the temperature acquired from the other sensors is still meaningful to evaluate the water temperature. Regarding WHRT systems or applications, they can tolerate a maximum number of deadline overruns within a certain time period. For example, in the industrial field, a WHRT task could be associated with a hypothetical wireless measurement system that tolerates a maximum number of transmission errors. If MPC is used, the real measurement data that are not received can be replaced by simulated data during a predefined maximum time period but, after that period, system critical faults can appear. Obviously, in this case the choice of a wireless transmission system could not be the best choice or, at least, other redundant mechanisms should be implemented.

Concerning RT requirements in industrial systems, it is also important to underline that as the hardware price of the instruments decrease, and their computing power increase, industrial systems are enlarging their distributed computing characteristics, increasing the implementation of control functions tasks in the field devices, which implies an increase of the performance of communication networks and protocols, namely in which concerns their RT performance. Thus, RT demands of the industrial applications must fulfil latency and jitter requirements not only in local computing tasks but, above all, those requirements and tasks' scheduling (synchronization) must be fulfilled in the industrial data networks that support the communication between network nodes.

Nowadays, industrial Ethernet solutions can assure cycle times of some tens of microseconds [28–30], much lower than the cycle times of FieldBus solutions whose typical values are above several tens of ms. Thus, Fieldbus solutions, such as FF, can only be used for low-speed sensing and drive control systems, and are not suitable for applications that require cycle's times lower than ms, namely, applications related to motion control, high-speed devices interfacing and transient fault detection, among others. For these type of applications, industrial Ethernet and other specific automation protocols, not covered in this chapter, are commonly used.

9.3.2 Collision avoidance and synchronization

Collision avoidance in Ethernet networks can be achieved by using fully Ethernet switches instead of Ethernet hubs or bridges. The usage of fully Ethernet switched networks enables full-duplex communication, between all network nodes, and as long as the switch provides the required number of ports according to the maximum network traffic, there are no collisions, nor data loss and, at the same time, the communication bandwidth is doubled, since for a given link, between two nodes, the communications in both senses can occur simultaneously. Moreover, besides the segmentation of the collision domains provided by full-duplex switches, these network devices can handle prioritization according to the different QoS levels for the different types of frames that are transmitted in the network. Ethernet frames also include a set of bytes to support virtual local area networks (VLAN) that minimize broadcast traffic in the LAN. In this way, as long as the switching time over the network is predictable, it is possible to avoid missing hard deadlines improving RT performance of the data network.

Besides collision avoidance and prioritization, industrial Ethernet solutions must synchronize network nodes using clock synchronization protocols. One of these protocols is the Precision Time Protocol (PTP) [32,33] that can work on the top of different protocols, not only Ethernet, and enables that different clocks in a heterogeneous network, including, obviously, network nodes, with different accuracies, resolutions and stability, to be synchronized to a grandmaster clock. With a network that supports PTP it is possible to achieve a sub μs timing precision that fulfils the requirements of most industrial measurement, control and automation applications.

The proprietary Ethernet solutions that are used nowadays can support the PTP but, as already referred, the different solutions don't assure interoperability of equipment and the integration of the different network levels, from manufacturing unit to office networks, can only be achieved with the usage of network interfaces and gateways which introduce additional costs, limits network flexibility and, above all, compromises network performance in terms of latency and jitter.

9.4 Time-sensitive networking

To promote network standardization, real-time capabilities and interoperability in an integrated Ethernet network, a novel version of the IEEE 802 standards is being developed [34–40]. To fulfil real-time demands in factory networks, several mechanisms are being considered for TSN. One of these mechanisms, very similar to the one that is used with FF, but running at a much higher bit rate, is the usage of TDMA within each data stream macrocycle (MC). The Time-Aware Scheduler (TAS), together with network timing synchronization mechanisms, assures a deterministic end-to-end latency and a maximum jitter uncertainty for time-sensitive data. Without the implementation of the TAS, low-priority traffic, without any real-time requirements, could prevent the on-time delivery of time-sensitive data. The TAS assigns specific time slots, in each MC, to carry time-sensitive data throughout the data network without the need of any token-based mechanism. As represented in Figure 9.10, TDMA enables data transmission of scheduled data in a deterministic and periodic way. Each MC contains several timeslots, some of them are assigned to time-sensitive data and others to best-effort data that has not end-to-end hard deadline requirements, in terms of latency and jitter. In this way, as long as there is no timing overlap between time slots, it is possible to assure a timely transmission of RT data without collisions.

Another mechanism that must be considered to avoid that the best-effort data overlaps the scheduled time-critical data is based on the usage of guard bands. As represented in Figure 9.11, these guard bands, whose time duration must be higher than the maximum unscheduled packet duration ($\Delta Tmax_{USF}$), avoid the occurrence of timing overlaps associated with noncritical data with TS reserved for critical data. However, used in this way, this mechanism has an inherent waste of transmission bandwidth because there is always a nonefficient usage of the channel data transmission capability whenever the unscheduled data packet has a time duration lower than $\Delta Tmax_{USF}$.

In order to avoid the loss of useful bandwidth caused by the usage of guard bands, IEEE proposes another mechanism that is based on the usage of framelets. This mechanism enables that best-effort data frames are segmented into framelets that can have a minimum size of 64 bytes, being possible to interrupt, without data loss, best-effort frames to transmit scheduled data with real-time requirements. This mechanism is denominated frame pre-emption. Figure 9.12 depicts the

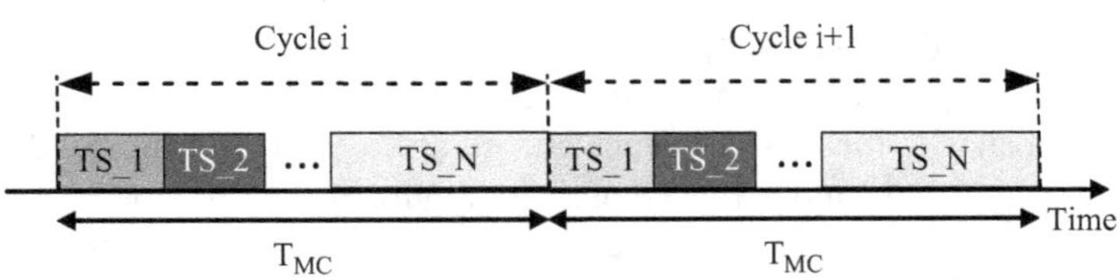

Figure 9.10 Assignment of time slots using a periodic TDMA mechanism within the TSN macrocycle. T_{MC}: macrocycle period

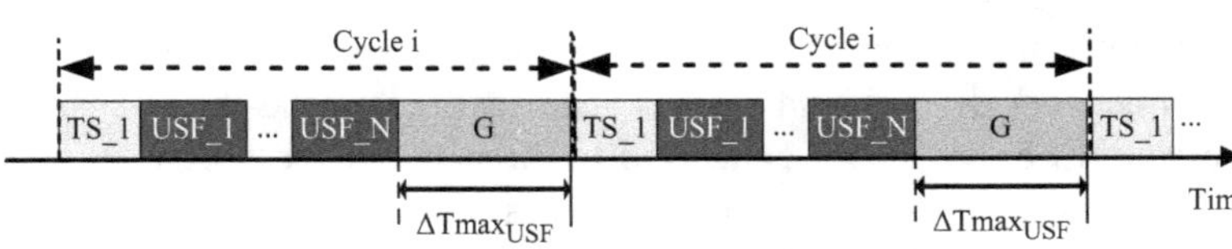

Figure 9.11 Usage of guard bands to prevent unscheduled data packets to extend into a TS used for critical data. $\Delta Tmax_{USF}$: maximum unscheduled packet time duration

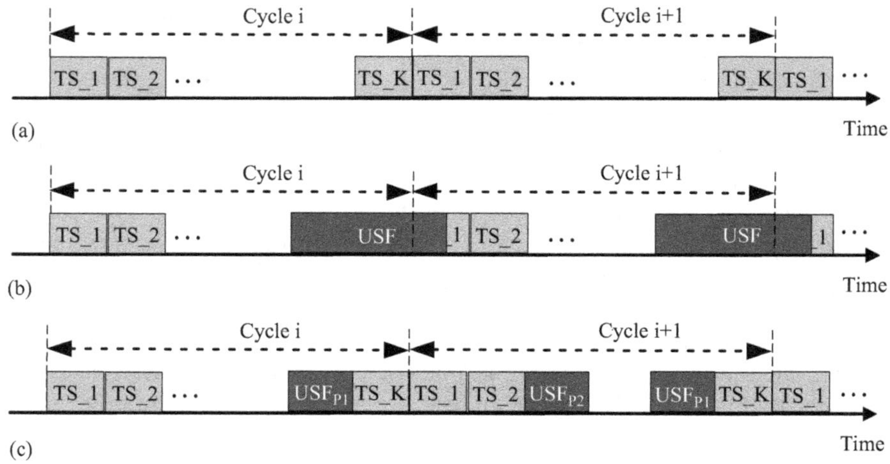

Figure 9.12 Frame pre-emption mechanism: (a) time slots associated with time-sensitive data; (b) overlap caused by unscheduled packet and (c) usage of the framelet segmentation mechanism to prevent timing overlaps and to optimize network performance

segmentation of an unscheduled frame (USF) in two framelets (USF_{P1} and USF_{P2}) in order to assure the correct timing requirements associated with the scheduled data (TS1, TS2 and TSK), being possible to reduce the guard band bandwidth waste to a maximum size of 64 bytes.

In the network nodes TSN also implements Time-Aware Gate (TAG) mechanism that uses the priority code point (PCP) of the IEEE 802.1Q frame header to prioritize the traffic queue that is transmitted in a specific time slot within the MC. Each output port of any switch (Out_P) has an associated gate schedule control that manages the traffic queues according to their priorities and assigned TS [41–43]. In the example represented in Figure 9.13, all input gates of the transmission multiplexer are closed, expect gate 1 that is open to transmit time-sensitive data buffered in traffic queue 1. Regarding data transmission priorities, different centralized or hybrid approaches [31] can be used to assign the correct data Class of Service (CoS) priority in order to ensure that data frames are transmitted according to their latency and jitter requirements.

Another issue that is of major importance in TSN is synchronization. Clock distribution mechanisms must be implemented in order to achieve a common time base for all network devices, end devices and network switches that transmit time-sensitive data. In every device connected to the network the time marks that are used to define the start and the end of each time slot must be synchronized to assure that the time-sensitive data are transmitted according to its scheduled TS. Different types of time synchronization protocols, such as the PTP [32,33], previously referred, can be implemented in order to assure clock accuracy values better than 1 μs, between all the network devices. The White Rabbit (WR) project proposes

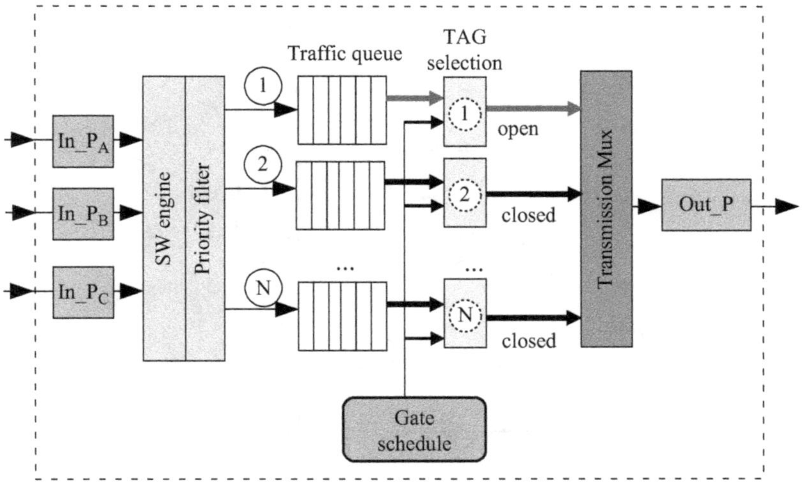

Figure 9.13 Block diagram of a TSN switch with gate schedule capabilities

enhancements of the IEEE 1588 PTP that supports clocks' synchronization in the sub-nanosecond level [44] that is more than enough for TSN requirements. It is important to refer that the time accuracy values are measured against the network reference clock, usually referred as Grand Master Clock (GMC) that synchronizes all the network devices clocks after measuring transmission message delays between networks nodes. As represented in Figure 9.14, there are four different messages that are used for network synchronization purposes, namely, synchronization, follow-up, delay request and delay response messages. With this set of messages, it is possible to compensate clock offsets caused by propagation delays, random encapsulation times and internal queuing delays of the synchronization messages. Considering the offset timing values that are associated with each pair of synchronization messages, the synchronization of the slave clock is based on the average time offset of the timing stamps of the messages that are exchanged between the master and slave clocks:

$$\Delta T = \frac{\Delta T_{MS} + \Delta T_{SM}}{2} \tag{9.3}$$

where ΔT_{MS} is the time offset that is associated with the master to slave synchronization messages and ΔT_{SM} is the time offset that is associated with the slave to master synchronization messages.

Standard IEEE 1588-2008 [32] enables that the GMC performs up to ten synchronization cycles per second in order to assure a tight synchronization of the network clocks.

Concerning network reliability, that is another issue of major importance for almost all TSN applications, IEEE is also developing standards to improve redundant routing mechanisms [45]. Taking in mind that replication only makes sense for

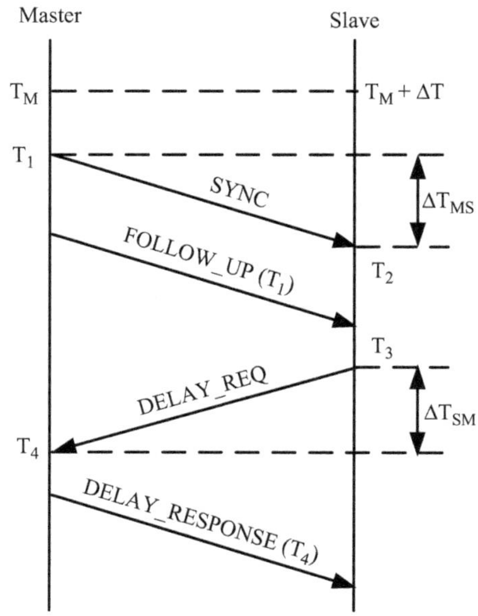

Figure 9.14 Messages used to perform clock synchronization with PTP. SYNC: synchronization message; FOLLOW_UP: follow-up message; DELAY_REQ: delay request message; DELAY_RESPONSE: delay response message

critical data and not for best-effort data, data replication mechanisms, through redundant path, can be implemented in the end devices, or if they don't support redundant network connections they can be implemented in the network switches. As an example, Figure 9.15 represents a message replication that is implemented in the first network node (NN1) of the TSN. The time-critical data frame is replicated and transmitted over two redundant paths that are connected to NN1. The first frame received in the destination switch (NN3) is forward to the destination end node (EN3) and the second one is discarded, being this process transparent to the higher OSI layers. Even if it is possible to improve reliability by expanding this mechanism to multiple redundant paths, it is important that each redundant path fulfil latency and jitter requirements of the running applications and an obvious compromise must always be achieved since network performance is reduced if redundancy is not used judiciously. To minimise network congestion IEEE P802.1CB [37] enables a conditional replication of packets according to destination address, traffic class and data routing information.

Several other standards for TSN are still under development. In this section we focused only on the main mechanisms that were being considered to fulfil the RT requirements. Table 9.1 summarizes the main IEEE TSN standards related with the latency and deterministic response required for RT applications and also with the achievement of the network synchronization, security and reliability.

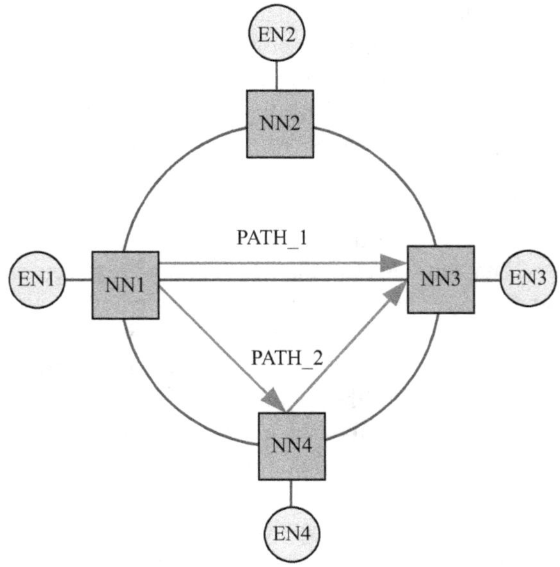

Figure 9.15 Message replication of critical data in TSN. EN: end node; NN: network node; PATH_1: direct path; PATH_2: redundant path

Table 9.1 IEEE TSN standards

IEEE standard/ draft	Title	Main purpose
802.1AS	Network Time Synchronization	Common clock for all network nodes
802.1Qbv	Scheduled Traffic	No collision of scheduled Ethernet frames
802.1Qci	Filtering & Policing	Network security
P802.1CB	Seamless Redundancy	Active switch-over with no data loss
P802.1Qcc	Stream Reservation	Centralized network configuration and management
802.1Qbu/802.3br	Frame Pre-emption	Bandwidth optimization using framelets

9.5 Industrial Internet of the things

The main techniques and technologies that are nowadays applied in machine-to-machine (M2M) network applications are basically the same that will be applied in IIoT. In both cases, devices are used to capture data, such as physical measurements, that is transmitted over a wired, wireless or hybrid network, and processed by an application that runs remotely. However, even if it seems similar, the main and substantial difference between M2M technologies and IIoT is related to the scale of integration. In IIoT the network integration enables that a huge quantity of data can be analysed online, in distributed cloud storage systems, and that a large number of services are integrated to optimize process efficiency. Large amounts of data, from different equipment, running in the same or different manufacturing units (MU), can be processed together improving the overall performance and efficiency of a global production system. This capability can justify the implementation of IIoT due to the cost reductions and products' quality improvements that can be achieved. It is important to refer that advanced algorithms that are used to process the Big Data can detect hidden correlations between data that can't be found by processing, locally, captured data for a single device, or from a group of devices, eventually working in different MU. In this scenario, diagnostic data from sensors, actuators and PLC, as well as historical data trends of the different industrial variables, can provide the required data to establish a successful predictive maintenance, reducing maintenance costs, avoiding superfluous preventive maintenance tasks and costly unscheduled production shutdowns. Predictive analytics tools can be used to identify the precursors of unplanned adverse events, determine the most profitable way to handle ageing and degrading of assets and to optimize the usage of maintenance resources [46–48]. In summary, it can be referred that IIoT promotes the integration of devices, systems and resources using, as much as possible, standard communications and protocols to integrate companies' distributed systems and to increase the entire value chain of their business.

Regarding IIoT protocols, even if the purpose is always standardization, it is obvious that the different real time and security requirements in the MU floor and in the cloud are completely different. Thus, different protocols, more or less compatible with the OSI model, are implemented to provide the desired global interconnectivity between MU and services. Concerning transmission modes, an increased usage of Wireless and 5G technologies, to provide remote and centralized data visualization and to support monitoring and maintenance activities, is obviously expected. IPv6 and network security are also two important topics that must be considered for the success of IIoT.

9.5.1 Smart factory

Smart factory (SF) is a new concept that aims to integrate manufacturing Operational Technologies (OT) with Internet Technologies (IT). OT are focused on field instrumentation, alarms, events and fault detection issues, among others, and, traditionally, uses proprietary data format and protocols. By its turn, IT are focused on cost, productivity, compliance, efficiencies and logistics, among others, using data analytics based on historical data, internet protocols and cloud computing services. As depicted in Figure 9.16, smart factories include not only the MU but also other manufacturing and customer services such as M2M communication, data storage and analysis of large amount of data, usually referred as Big Data Analytics (BDA), remote and centralized data visualization, using wireless devices, logistics and maintenance. Regarding maintenance, the implementation of IAMS provides a

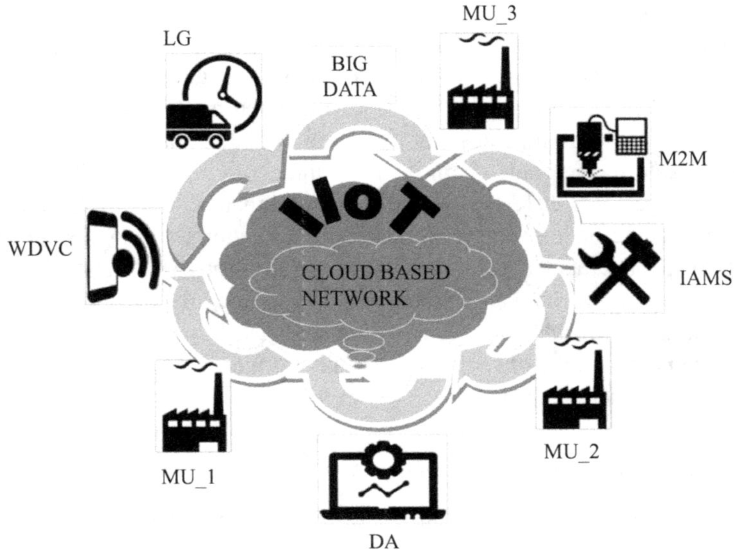

Figure 9.16 Smart factory integration. MU: manufacturing unit; DA: data analytics; IAMS: instrument asset management system; LG: logistics; M2M: machine-to-machine communication; WDVC: wireless data visualization and control

reduction of maintenance and operational costs, avoiding potential unscheduled shutdown risks of the MU and enabling a predictive online surveillance of all factory equipment.

Smart factories enable the implementation of flexible and adaptive production systems that will be crucial in the near future of manufacturing units. As an example, and considering only the connection between manufacturing and logistics, if MU A and MU B are competing for a similar product, with identical production costs and quality, the choice of the right logistic solution to deliver the products to the consumers, according to the required quantities and expected delivery time, can dictate which company will be successful in the market. Another interesting example, related to the implementation of IIoT, is the integration of different MU that are working in different geographical places and manufacturing identical and/or different components that are assembled to build a given final product. If one of the factories can make a component cheaper, in terms of production costs, because that MU acquired raw material at a lower cost, or has as excessive storage stock that must be reduced, the production level of each MU can be adjusted to maximize the overall productivity of the set of MU. Data sharing between different MUs can also be useful to implement a flexible and dynamic production planning that optimizes the global production efficiency. For example, if one of the MUs has a production problem related to an equipment fault or loss of raw materials that could cause an unscheduled production shutdown, another MU can assume the production during the shutdown period, avoiding potential losses of productivity or product shortages in the market.

Concerning industrial models, IIoT replaces the traditional automation pyramid model, depicted in Figure 9.17, by a distributed model based on an Internet hyper-connectivity that includes not only the MU but also other services and resources related to the value chain of the manufactured products. These services and resources perform different MU related tasks that have no real-time requirements, such as enterprise resource planning (ERP), manufacturing execution (ME), information management (IM), supervision and control (S&C) and virtual PLC, with no RT requirements, among others. The usage of dynamic and flexible industrial network models with cloud virtualized resources can enable an optimization of the performance of multiple MU that are working in a collaborative way to build a given final product.

9.5.2 Wireless protocols

Wireless capabilities are also a significant driving force for the adoption of IIoT. The huge technical evolution around the mobile phones almost make forget their initial goal, to provide wireless voice communications [49]. Wireless communications provide a ubiquitous capability to interact with Internet and thus with the IIoT. Concerning wireless communications, several wireless protocols were proposed with different capabilities in terms of bandwidth, coverage area, power consumption and cost. For industrial applications, on the factory floor, besides Wireless HART, previously introduced ZigBee seems to be a promising solution. ZigBee is based on the IEEE wireless standard (IEEE 802.15.4) [50] but provides a standardization that extends to higher layers of the OSI model. Layers 3 and 4,

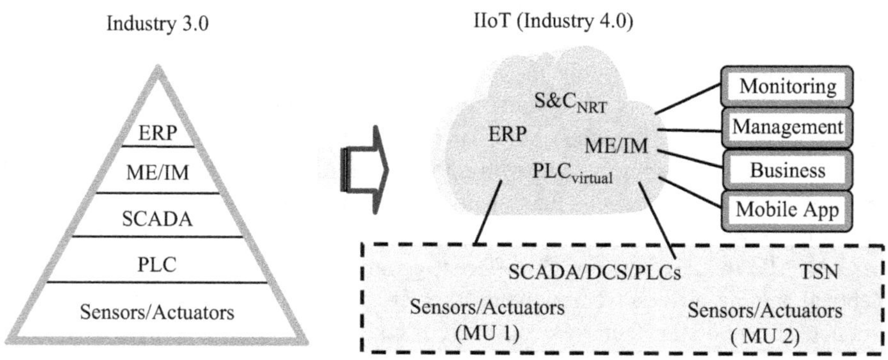

Figure 9.17 Industrial models: Industry 3.0 and IIoT. ERP: enterprise resource planning; ME: manufacturing execution systems; IM: information management; SCADA: supervision control and data acquisition; S&C: supervision and control with no RT requirements; PLC: programmable logic controller; MU: manufacturing unit; TSN: time-sensitive networking

associated with the network and transport layers, respectively, are standardized in order to provide additional features related to network authentication and routing capabilities that support different network topologies, namely mesh networking that is essential to assure reliability in wireless industrial applications. Moreover, ZigBee's advantage relative to other wireless solutions is the standardization it provides for higher OSI layers, promoting interoperability of devices from different manufacturers. This is a key issue in which concerns industrial applications since it makes possible to upgrade industrial wireless instrumentation devices without obligations to use devices from specific manufacturers. Other advantage of ZigBee is related to the low power consumption of the transceiver that enables extended battery-powering autonomy, particularly if power harvesting solution is implemented [51]. Other advantages of ZigBee are related to its immunity to external noise sources, always present in any industrial environment, and also with the protocol capability to change routing paths when failures or low QoS levels are detected in certain network link(s). However, due to the low power consumption of ZigBee devices, the coverage range that is provided by ZigBee, in point-to-point connections, is low, typically, in the order of a few tens of metres. However, ZigBee devices can work as routers and repeaters extending substantially the overall coverage range of the wireless network and improving, at the same time, its reliability. Network configuration can be modified dynamically, without the need of any human intervention, according to the QoS that is provided by the different wireless links of the data network. Regarding transmission throughput, in the 2.5 GHz band, ZigBee has a maximum data transmission rate of 250 Kbps, which is suitable for transmission of intermittent or monitoring data, but it is not acceptable for data transmission in industrial applications with real-time requirements, such as, control applications that require synchronized timings for data acquisition, data processing and data delivery.

Regarding the addressing capabilities required for the implementation of IIoT, ZigBee protocol has a limitation related to the number of bits used for addressing purpose. If short addressing mode is used, there are only 16 address bits that are not enough for Internet addressing purposes. Thus, an improved version of ZigBee, Known as ZigBee IP was defined to surpass this limitation but this protocol is not compatible with Ethernet and gateways are required to interconnect ZigBee IP networks with Ethernet networks.

9.5.3 5G networks

The next generation of cellular technology is focused not only on smartphone Internet applications but also on device-to-device communications, M2M communications and RT applications related to IIoT [52–54]. To fulfil the IIoT requirements, 5G networks are being designed to provide a data rate in the range of 10 Gbps with an end-to-end latency time lower than 1 ms. Actual 4G networks, depending on network types, provide a typical download speed of about 20 Mbps, although its maximum theoretical throughput is 150 Mbps, but these values are peak data rate values, and the network does not provide latency, and jitter, guarantees.

The 5G network protocol stack includes an open wireless architecture, an upper and a lower network sublayers, an open transport protocol layer and an applications of service layer (Figure 9.18).

Another feature of 5G networks is related to the usage of a flat IP architecture [55] that provides a way to identify devices by symbolic names, instead of identification based on common IP addresses. The main advantages of this addressing mode are related to a lower implementation cost of network services, reduced network latency and jitter, promoting a decoupling between radio access and core network and expanding network reconfiguration capabilities [56].

5G networks will enable the connection of IoT nodes directly to 5G base stations avoiding the need for gateways and other network interfaces that affect negatively network planning and reduce its performance in terms of latency and availability. Other advantage of 5G networks is their capability to interface a large number of machine networks, sensors and actuators, wired or wirelessly connected, by using the denominated capillary networks. These networks can use different protocols being the connection to the 5G network performed through gateways, such as in wireless applications, Bluetooth or ZigBee to 5G network gateways. However, as almost always happens, there is a compromise between advantages and drawbacks, and the usage of gateways, as protocol converters, increases latencies and makes almost impossible, with nowadays technologies, to achieve a maximum 1 ms latency when the end devices are connected to capillary networks. It is important to take in mind that these timing values could only fulfil the requirements of applications that use low-speed sensors and control systems where real time issues and synchronization are not important in terms of applications' performance, such as monitoring and centralized management tasks of geographically distributed systems.

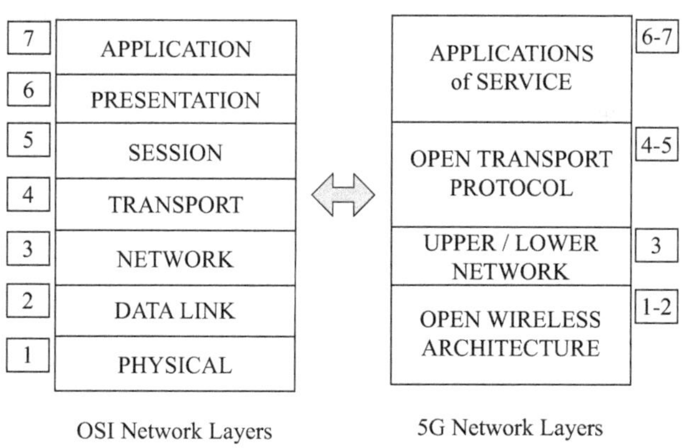

Figure 9.18 Relationship between the 5G network layers and the seven layer OSI model

9.5.4 IPv6 and security

The adoption of IPv6 is being delayed, year by year, because the migration from IPv4 to IPv6 requires investments, the payback period of the investments can be large, there are potential risks during the transition phase and the advantages of IPv6, for some Internet users, are not obvious [57,58]. However, IPv6 will be essential for the success of IIoT. The main reasons for the adoption of IPv6 are related to the extension of the addressing capabilities, network security, connectability and scalability. Due to the large number of devices that will be connected to the Internet network, the actual 32 bits IP addressing field used by IPv4 is not enough for IIoT, and will be extended to 128 bits with IPv6. With this huge addressing capability single network devices, for example, a vibration sensor, connected to a mechanical rotating machine, used for machine surveillance purposes, can have an exclusive IP address without need to use Network Address Translation (NAT) routers and firewalls to convert private intranet addresses to a common and public IP address. With the extended address capability that will be provided by IPv6, an expected number of 25 billion of 'things' can be connected to the Internet by the year 2025 using their own IP address without any foreseen restrictions in the medium–long term. It is still important to underline that the address extension capability actually provided by the NAT is not sufficient for IIoT users and applications. Moreover, small and low-cost end-user devices, like sensors or actuators that require Internet access, don't have the capabilities to implement the protocol mechanisms that are required to exchange routing information with NAT firewalls or routers.

Regarding network security, it is important to underline that network security issues are essential for the success of IIoT. There are two main reasons that justify this importance, one is related to the huge enlargement of the network that will include a much larger number of network devices, users and applications, and the other reason is related to the critical consequences for MU, people and economy that can result from hackers' attacks. Thus, IPv6 together with the Internet Key Exchange (IKE) protocol improve the IP network security by using more efficient authentication and encryption algorithms. The usage of the IP security (IPSec) [59,60] with the IPv6 site-to-site tunnel mode capability reinforces and extends the security mechanisms that are actually provided by VPNs, being the encryption and integrity checking of VPNs always implemented in all IPv6 connections.

Another advantage of IPv6 is its address structure. An IPv6 address contains different sub-address fields that are associated with link-local, site-local and global scope addresses. The first 48 bits are associated with the public site prefix and the next 16 bits are associated with the private subnet identification (ID). The last 64 bits are associated with the interface ID, also known as token. The first group of 64 bits includes the 48 bits unique MAC of the network interface and this first group defines the global access, which is equivalent to an IPv4 public address that enables communications of different hosts over a shared link. A link-local address based on the interface ID enables multiple addresses for each network node being this capability a great advantage in which concerns IIoT applications. This 64-bit interface identifier can be automatically generated from the interface's MAC address or assigned manually.

9.6 Conclusions

Several topics related to IIoT were underlined in this chapter. It is important to refer that IIoT can be considered as a hub for mechanical, electrical and maintenance industries, among others, enabling the development of new synergies and business models that result from data integration. In this context, it is important to refer that advanced algorithms must be developed for each application and the quality of the results that can be achieved implies multidisciplinary work teams that must include, at least, process engineers, programmers and data scientists, to implement data analytics based on large amount of data that must be transformed in useful information aligned with the industrial efficiency goals, in terms of products' profit and quality. Nowadays, schools are still graduating students to be programmers or traditional engineers, focused in specific knowledge fields, rather than data scientists with engineering knowledge in a broad and complementary set of areas. Teaching must be focused on multidisciplinary topics, creative thinking, critical reasoning and trainability. Change in teaching methods and promoting the development of students' creative initiatives is not an easy task and is a process that needs time to achieve the desired results. Basically, teacher-centred approaches must be replaced by student-centred approaches [61].

Regarding data security, a particular attention must be dedicated to IIoT systems. All security gaps must be closed and this is not an easy task when different interfaces and protocols are used, and particularly, when data are transmitted over the Internet. A major requirement is related to the compatibility of the software used by all the partners that will be involved in the IIoT and, whenever possible, requiring that these partners assure the interoperability of their devices, or systems, based on open standards without any limitations in terms of performance or reliability. Without winning these challenges, IIoT can't be successful and the value-add that is envisaged with its implementation could be a mirage or, at least, be delayed in time.

Abbreviations and symbols

BDA	Big Data Analytics
Cp_D	Compel Data
CD	Collision Detection
CIF	Control in the Field
CoS	Class of Service
DD	Device Description
DDL	Device Description Language

DLL	Data Link Layer
DTM	Device Type Manager
ERP	Enterprise Resource Planning
FAZ	Field Access Sublayer
FB	Function Block
FCS	Frame Check Sequence
FDA	Field Device Agent
FDT	Field Device Tools
FF	Foundation Fieldbus
FMS	Field Message Specification
FRT	Firm Real Time
FSK	Frequency Shift Keying
GMC	Grand Master Clock
HART	Highway Addressable Remote Transducer
HRT	Hard Real Time
HSE	High-Speed Ethernet
HW	Hardware
I&C	Instrumentation and Control
I/P	Current to Pressure
IAE	Integral of Absolute Error
IAMS	Instrumentation Asset Management System
ID	Identifier
IoT	Internet of the Things
IIoT	Industrial Internet of the Things
IKE	Internet Key Exchange
IM	Information Management
IP	Internet Protocol
IPSec	Internet Protocol Security
IPv4	Internet Protocol Version 4
IPv6	Internet Protocol Version 6
IRT	Ishocronous Real Time
IS	Intrinsically Safe
ISO	International Organization for Standardization
IT	Internet Technologies
LAN	Local Area Network
LAS	Link Access Scheduler
LLC	Logical Link Control
LRL	Lower Range Limit

LRV	Lower Range Value
M2M	Machine to Machine
MAC	Medium Access Control
MC	Macrocycle
ME	Manufacturing Execution
MPC	Model Predictive Control
MU	Manufacturing Unit
NAT	Network Address Translation
NRT	Non Real Time
OSI	Open Systems Interconnection
OT	Operational Technologies
P&I	Process and Instrumentation
P&P	Plug and Play
p.s.i.	Pound per Square Inch
P2P	Peer to Peer
PCI	Protocol Control Information
PDU	Protocol Data Unit
PLC	Programmable Logic Controller
PT	Pass Token
PTP	Precision Time Protocol
QoS	Quality of Service
RT	Real Time
RTOS	Real-Time Operating System
S&C	Supervision and Control
SCADA	Supervision Control and Data Acquisition
SDU	Service Data Unit
SF	Smart Factory
SIL	Security Integrity Level
SRT	Soft Real Time
SW	Software
TAG	Time-Aware Gates
TAS	Time-Aware Scheduler
TCP	Transmission Control Protocol
TDMA	Time Division Multiple Access
T_{MC}	Macrocycle Period
TSN	Time-Sensitive Networking
UDP	User Datagram Protocol
URL	Upper Range Limit

URV	Upper Range Value
USF	Unscheduled Frame
WHRT	Weakly Hard Real Time

References

[1] Pereira J.M. 'A fieldbus prototype for education purposes'. *IEEE Instrumentation & Measurement Magazine*. 2004;7(1):24–31

[2] Viegas V., Pereira J.M. 'Quimera: the easy way to simulate foundation fieldbus applications'. *Computer Applications in Engineering Education*. 2016;24(6):914–925

[3] ISA, Compatibility of Analog Signals for Electronic Industrial Process Instruments – ANSI/ISA-S50.1, 1982.

[4] Pereira J.M., Postolache O., Girão P.S., Ramos H.G. 'Replacing Hart Hand-Held Terminals by PC Based Virtual Instruments'. Proceeding of the XVII IMEKO World Congress, Metrology in the 3rd Millennium. Dubrovnik, Croatia, June 2003, pp. 620–624

[5] Pereira J.M., Postolache O., Girão P.S. 'HART protocol analyzer based in LabVIEW'. *International Scientific Journal of Computing*. 2004;3(2):39–42

[6] HART Communication Foundation, HART Line Journal, No. 1, HART Enables New Class of Multivariable Instruments, FieldComm Group, Austin, USA, 2000.

[7] Bell System, Technical Reference: Data Communications Using Voiceband Private Line Channels, AT&T, USA, 1973

[8] Bell Telephone Company, Bell System Technical Reference: PUB 41212, Data Sets 202S and 202T Interface Specification, AT&T, USA, 1976

[9] HART Communication Foundation, Application Guide – HCF LIT 34, HART Comm. Foundation, Austin, USA, 1999

[10] Fisher-Rosemount, HART Field Communications Protocol – A Technical Overview, HART Comm. Foundation, Austin, USA, 2002

[11] Pereira J.M., Postolache O., Girão P.S. 'Spread spectrum techniques in wireless communication'. *Instrumentation & Measurement Magazine*. 2009;12(6):21–24

[12] Pereira J.M. 'Spread spectrum techniques in wireless communication'. *Instrumentation & Measurement Magazine*. 2009;13(1):8–14

[13] Saputra O., Shin S. 'Deadline monotonic scheduling to reduce overhead of superframe in ISA100.11a'. Proceeding of the Summer Domestic Conference the Korean Institute of Communication and Information Sciences. Jeju Island, Korea, 2014

[14] Blevins T., Chen D., Nixon M., Wojsznis W. Wireless Control Foundation: Continuous and Discrete Control for the Process Industry. ISA, Research Triangle Park, USA, 2015. p. 256

[15] Sisson A. 'FDT Technology: What Are DTMs?'. Magazine Control – Promoting Excellence in Process Automation [online]. 2009. Available from

https://www.controlglobal.com/articles/2009/fdtdtm0905/ [Accessed April 2018]

[16] Richard Zurawski (Ed.). *Industrial Communication Technology Handbook.* Taylor & Francis Group; 2017. p. 1756

[17] Fieldbus Foundation, Tecnhical Overview FD-043, Fieldbus Foundation, Austin, USA, 1998

[18] Fieldbus Foundation, Fieldbus Installation and Planning Guide AG-165, Fieldbus Foundation, Austin, USA, 1998

[19] ISA, Introduction to Fieldbuses for Process Control, ISA, Research Triangle Park, USA, 2002

[20] Fieldbus Foundation, Network Management Specification FF-801, Fieldbus Foundation, Austin, USA, 1998

[21] IEC. Std 61508-1:2010 – Standard Functional safety of electrical/electronic/programmable electronic safety-related systems – Part 1: General requirements [online]. Available from http://www.iec.ch/functionalsafety/standards/page2.html [Accessed April 2018]

[22] Network Working Group. A Standard for the Transmission of IP Datagrams over Ethernet Networks [online]. Available from https://tools.ietf.org/html/rfc894 [Accessed April 2018]

[23] Pang Y, Yang SH, Nishitani H. 'Analysis of control interval for foundation fieldbus-based control systems'. *ISA Transactions.* 2006;45(3):447–458

[24] Hong, S. H. 'Scheduling algorithm of data sampling times in the integrated communication and control systems'. *IEEE transactions Control Systems Technology*. 1995;3(2):225–230

[25] Doyle P. 'Introduction to real-time Ethernet I'. *The Extension – A technical Supplement to Control Network.* 2004;5(3):1–4

[26] Doyle P. 'Introduction toreal-time Ethernet I'. *The Extension – A technical Supplement to Control Network*. 2004;5(4):1–4

[27] Crockett N. 'Introduction industrial Ethernet is ready to revolutionize the factory – connecting the factory floor'. *Manufacturing Engineer*. 2003;82(3):41–44

[28] PROFIBUS and PROFINET International. Specifications and Standards. Available from https://www.profibus.com/download/specifications-standards/ [Accessed April 2018]

[29] Beckhoff EtherCAT. Application Notes. Available from http://www.beckhoff.com/english/default.htm?ethercat/default.htm [Accessed April 2018]

[30] ETHERNET Powerlink. Powerlink Technology. Available from https://www.ethernet-powerlink.org/powerlink/technology [Accessed April 2018]

[31] Belden Inc., TSN – Time Sensitive Networking, Netherlands, 2016

[32] IEEE. Clock Synchronization in Telecommunications via PTP [online]. Available from http://ieeexplore.ieee.org/document/4319093/ [Accessed April 2018]

[33] IEEE. Std 802.1Q-2014 (Revision of IEEE Std 802.1Q-2011) – IEEE Standard for Local and metropolitan area networks–Bridges and Bridged Networks. Available from https://standards.ieee.org/findstds/standard/802.1Q-2014.html [Accessed April 2018]

[34] IEEE. Std 802.1AS-2011 – IEEE Standard for Local and Metropolitan Area Networks – Timing and Synchronization for Time-Sensitive Applications in Bridged Local Area Networks [online]. Available from https://standards.ieee.org/findstds/standard/802.1AS-2011.html [Accessed April 2018]

[35] IEEE. Std 802.1Qbv-2015 (Amendment to IEEE Std 802.1Q-2014 as amended by IEEE Std 802.1Qca-2015, IEEE Std 802.1Qcd-2015, and IEEE Std 802.1Q-2014/Cor 1-2015) – IEEE Standard for Local and metropolitan area networks – Bridges and Bridged Networks – Amendment 25: Enhancements for Scheduled Traffic [online]. Available from https://standards.ieee.org/findstds/standard/802.1Qbv-2015.html [Accessed April 2018]

[36] IEEE. Std. for Local and metropolitan area networks–Bridges and Bridged Networks–Amendment 28: Per-Stream Filtering and Policing [online]. Available from http://ieeexplore.ieee.org/document/8064221/ [Accessed April 2018]

[37] IEEE. Std 802.1CB-2017 – IEEE Standard for Local and metropolitan area networks–Frame Replication and Elimination for Reliability [online]. Available from https://standards.ieee.org/findstds/standard/802.1CB-2017.html [Accessed April 2018]

[38] IEEE. Std 802.1Qcc – IEEE Draft Standard for Local and metropolitan area networks – Media Access Control (MAC) Bridges and Virtual Bridged Local Area Networks Amendment: Stream Reservation Protocol (SRP) Enhancements and Performance Improvements [online]. Available from https://standards.ieee.org/develop/project/802.1Qcc.html [Accessed April 2018]

[39] IEEE. Std 802.1Qbu-2016 (Amendment to IEEE Std 802.1Q-2014) – IEEE Standard for Local and metropolitan area networks – Bridges and Bridged Networks – Amendment 26: Frame Preemption [online]. Available from https://standards.ieee.org/findstds/standard/802.1Qbu-2016.html [Accessed April 2018]

[40] IEEE. Std 802.3br-2016 (Amendment to IEEE Std 802.3-2015 as amended by IEEE Std 802.3bw-2015, IEEE Std 802.3by-2016, IEEE Std 802.3bq-2016, and IEEE Std 802.3bp-2016) – IEEE Standard for Ethernet Amendment 5: Specification and Management Parameters for Interspersing Express Traffic. Available from https://standards.ieee.org/findstds/standard/802.3br-2016.html [Accessed April 2018]

[41] Craciunas S., Olivier R. 'Combined task- and network-level scheduling for distributed time-triggered systems'. *Real-Time Systems*. 2016;52(2):161–200

[42] Abdelzaher T., Shin K. 'Combined task and message scheduling in distributed real-time systems'. *IEEE Transactions on Parallel and Distributed Systems*. 1999;10(11):1179–1191

[43] Baruah S., Rosier L., Howell R. 'Algorithms and complexity concerning the preemptive scheduling of periodic, real-time tasks on one processor'. *Real-Time Systems*. 1990;2(4):301–324

[44] Ronen O., Lipinski M. 'Enhanced synchronization accuracy in IEEE1588'. Proceeding of the IEEE International Symposium on Precision Clock

Synchronization for Measurement, Control, and Communication. Beijing, China, October 2015, pp. 76–81

[45] Tongji S., Luo F., Xu J. 'An Analysis of Frame Replication and Elimination for Time-Sensitive Networking'. Proceeding of the 2017 VI International Conference on Network, Communication and Computing. Kunming, China, Dec. 2017, pp. 166–170

[46] Gartner Inc. Using Advanced Analytics to Predict Equipment Failure. Available from https://www.gartner.com/doc/3065317/using-advanced-analytics-predict-equipment [Accessed April 2018]

[47] SAS Institute Inc. Channeling Streaming Data for Competitive Advantage. Available from https://www.sas.com/en_us/whitepapers/channeling-data-streams-107736.html [Accessed April 2018]

[48] Forrester Research Inc. The Forrester Wave: Big Data Streaming Analytics. Available from https://www.sap.com/documents/2016/06/92d1323c-797c-0010-82c7-eda71af511fa.html [Accessed April 2018]

[49] Sharma, P. 'Evolution of mobile wireless communication networks – 1g to 5g as well as future prospective of next generation communication network'. *International Journal of Computer Science and Mobile Computing*. 2013;2 (8):47–53

[50] IEEE. Std 802.15.4-2015 (Revision of IEEE Std 802.15.4-2011) – IEEE Standard for Low-Rate Wireless Networks. Available from https://standards.ieee.org/findstds/standard/802.15.4-2015.html [Accessed April 2018]

[51] Kuorilehto M., Kohvakka M., Suhonen J., Hamalainen P., Hannikainen M., Hamalainen T. *Ultra-Low Energy Wireless Sensor Networks in Practice: Theory, Realization and Deployment*. John Wiley & Sons Ltd, West Sussex, England, 2007. p. 372

[52] Bangerter B., Talwar S., Arefi, R., Stewart K. 'Networks and devices for the 5g era'. *IEEE Communications Magazine*. 2014;52(2):90–96

[53] Tehrani M., Uysal M., Yanikomeroglu H. 'Device-to-device communication in 5g cellular networks: Challenges, solutions and future directions'. *IEEE Communications Magazine*. 2014;52(5):86–92

[54] Institute for Communication Systems 5G Innovation Centre – University of Surrey. The Flat Distributed Cloud (FDC) 5G Architecture Revolution. Available from https://www.surrey.ac.uk/sites/default/files/5G-Network-Architecture-Whitepaper-(Jan-2016).pdf [Accessed April 2018]

[55] Agyapong P., Iwamura M., Staehle D., Kiess W., Benjebbour A. 'Design considerations for a 5g network architecture'. *IEEE Communications Magazine*. 2014;52(11):65–75

[56] Sexton C., Kaminski N., Marquez-Barja J., Marchetti N., DaSilva L. '5G: adaptable networks enabled by versatile radio access technologies'. *IEEE Communications Surveys & Tutorials*. 2017;19(2):688–720

[57] Zhou X., Jacobsson M., Uijterwaa H., Mieghem P. 'IPv6 delay and loss performance evolution'. *International Journal on Communication Systems*. 2008;21:643–663

[58] Cho K., Luckie M., Huffaker B. 'Identifying IPv6 network problems in the dual-stack world'. Proceeding of the ACM SIGCOMM Workshop on Network Troubleshooting; Portland, Oregon, U.S.A., September 2004, pp. 283–288

[59] Lobyal D., Mansotra V., Sing U. (Eds.). 'Next generation networks'. Proceedings of CSI-2015. New York: Springer; 2018. pp. 256–274

[60] Ahmad N., Yaacob A. 'IPSec over heterogeneous IPv4 and IPv6 networks: issues and implementation'. *International Journal of Computer Networks & Communications*. 2012;4(5):57–71

[61] Pysarevskyi M. Education in Finland: schools without walls, learning with no marks. Available from https://innovationhouse.org.ua/en/statti/obrazovanye-v-fynlyandyy-shkoly-bez-sten-ucheba-bez-otsenok/ [online]. 2017. [Accessed April 2018]

Chapter 10
Internet of things for cargo ports

Yongsheng Yang[1], Meisu Zhong[1], Haiqing Yao[1], Fang Yu[1], Xiuwen Fu[1], Chao Mi[1], and Octavian Postolache[2]

Internet of Things (IoT) is a very large domain, with a great variety of technologies and applications involved, spanning across a variety of business sectors. The goal of the present work is to model an IoT architecture for cargo ports. In this chapter the main objects/things and their roles in a cargo port, different sensors and instrumentation that was used or might be used for sensing and control in a cargo port and some examples of IoT technologies that might contribute for improving the safety of processes and the security of the ports are presented. The analysis of the data on information and communication technologies that were implemented or having potential application to cargo port logistic suggests that IoTs might contribute to increase the efficiency and the quality of services in cargo ports, the operational safety and the security of the ports, by increasing the accessibility and accuracy of relevant information.

10.1 Introduction

Internet of Things (IoT) technologies by connecting sensors, devices and intelligent operations (i.e., intelligent cargo movement) provide clear business benefit in terms of asset tracking and supply-chain management. This is achieved by means of continuous process synchronization of supply-chain information, and seamless real-time tracking and tracing of objects [1].

Described as a twenty-first-century silk road, made up of a 'belt' of overland corridors and a maritime 'road' of shipping lanes, the multibillion dollar Belt and Road Initiative (BRI) or *yi dai yi lu* promises to better connect Asia, Africa and Europe. BRI includes 71 countries that account for half the world's population and a quarter of global GDP [2]. In this context the importance of various cargo ports is increasing as the trade is expected to increase. Million per year of containers pass through the port in many cargo ports (more than 40 million TEUs in Shanghai port, ~34 million TEUs in Singapore port, ~14 million TEUs in Rotterdam port and more than 9 million TEUs in Los Angeles port) [3].

[1]Institute of Logistics Science and Engineering, Shanghai Maritime University, Shangai, China
[2]Instituto de Telecomunicações, Lisboa and ISCTE-Instituto Universitário de Lisboa, Lisbon, Portugal

Container ports become more and more important and have very fast yearly growth rate. China, United States, South Korea, Japan, Indonesia, Thailand, Germany, India, Vietnam and Brazil were the top 10 exporters of containerized cargo in 2014 [4].

Goods management and security in the cargo ports are increasingly challenging. The expected increase of trade in the following years, the limited possibility to increase the space and traffic in ports and the need for improved security in the ports (that is, high resources and time-consuming) require solutions for better management of port asset and processes. IoT technology already is a solution adopted by some ports to increase the efficiency of management. Most of the information systems for cargo ports were done for optimization and acceleration of container stacking. In this chapter, we present the work on modelling the IoT framework for cargo ports. What measures and what IoT architecture that might improve the asset and process management in cargo ports are mainly investigated. In the following sections of the paper we present what are the main objects/things and their roles in a cargo port (Section 10.2), what kind of sensors and instrumentation was used or might be used for sensing and control in a cargo port (Section 10.3), and some examples of IoT technologies that might contribute for improving the safety of processes and the security of the ports (Section 10.4).

10.2 Port objects/things and their roles

Modelling of an IoT architecture for a domain require a thorough understanding of that domain and conceptualization/generalization of the domain processes. Identification of reference model can lead to a better understanding of the targeted system and is a tool for building architecture where design choices are based on best practices, in terms of functionality and information usage, to deal with conflicting requirements regarding functionality, performance, deployment and security. Following we describe a cargo port, listing several objects/things that generally can be found in a cargo port and their main role.

Worldwide, different types of cargo ports are built. Cargo ports are specialized in handling different types of cargo, which should be loaded and unloaded by very different mechanical means. The port may handle grains, liquid fuels, wood, automobile, chemicals, animals, etc., or only one type of cargo as, for example, petroleum dock port or fishing port. The ports that handle containerized cargo are known as container ports. A port of call is an intermediate stop for a ship on its itinerary, where the ships may take on supplies or fuel, as well as unloading and loading cargo. The cargo port also can be classified into seaport – that handle ocean-going vessels; river port – used for river traffic; inland port – a port on a navigable lake, river or canal with access to a sea or ocean; dry port – a port situated in the hinterland servicing an industrial/commercial region connected with one or several ports with rail or road transport and is offering specialized services between the dry port and the overseas destination [5].

10.2.1 Hinterland

Hinterland is defined as port's area that serves both import and export of goods, the inland region lying behind the port or the area from which products are delivered to a port for shipping elsewhere. The hinterland includes the port authority (e.g., custom authority), logistic centre, the hinterland transportation infrastructure and storage yard. Different services are provided in the hinterland – custom clearance, administration of shipping documents, special treatment of goods, special inspection of containers, quarantine, shipping and receiving containers, maintenance of the containers, loading and unloading of trucks, trains, feeders to/from the hinterland distribution, empty container handling and storage of goods. The facilities in the hinterland are generally equipped to deal with everything from import of product to inspection, fumigation, customs clearance, cargo sorting, cargo inspection and labelling of imported variety goods and food items. Packing that suits the needs of particular types of cargo can be conducted at packing sites, where equipment such as overhead travelling cranes are installed.

10.2.2 Logistic centre

Logistic centre is a place in a defined area within which all activities relating to transport, logistics and the distribution of goods (for both national and international transit) are carried out by various operators on a commercial basis. The logistic centre must be open to allow access to all companies involved in the activities of the port [5]. It includes offices of shipping companies, freight forwarders, stevedoring firm, waterway authorities, transportation and storage companies. It should comply with international and regional standards and quality performance to provide the framework for commercial and sustainable transport solutions. At the terminal controls tracking of things from ship unloading progress to positioning of containers for shipping, launching and scheduling for taking import containers should be integrated. At the terminal operators the cargo is managed to be sent to distribution centres. Many customers depend on schedule and organization of the players in the chain, with all the security buffer stocks. However, some players organize the chain themselves by integrating and optimizing many of the logistic steps.

10.2.3 Carousel and carriers

In a cargo port many types of carousel and carriers can be found: automated guided vehicle (AVG), belt system, chain-connected chassis system, overhead grid rail system, semitrailer truck, trucks, trains and straddle carriers.

10.2.4 Cranes

Many types of cranes (see ELS classification of crane [6]) can be found in a port. These may be classified taking into account the location where it is placed (i.e., quay crane and storage yard crane) or the main functionality of the crane (i.e., goliath gantry crane, transfer crane, block handling crane, floating dock crane, mobile harbour crane, lift trucks – reach stackers, forklift trucks and container lift truck). Figure 10.1 presents the images of quay and storage yard from two cargo ports: (a) Sines, Portugal, and (b) Shanghai, China, where different type of cranes can be observed. In several ports the quay crane is guided and controlled from central control office with cameras (nobody is on board of the crane, so the crane-crab can be accelerated faster and the cycle time reduced). Many quay cranes pick only one container in a stroke. In a few ports (e.g., Shanghai port) there are quay cranes that can pick up 2 containers in a stroke and thereby double the capacity. With the continuous growth of container throughput and the increasing size of ships, many automated container terminals have been built and are under construction in the world. The Chinese company ZPMC has a demonstration system in Shanghai with a fully automated quay crane system and an automated container yard stacking-system in place. The IV phase of Yangshan Shanghai port, as the latest built 'fully automated' container port in China, is operating in Shanghai from 2018. It has a large area of container port, the line length of the port is 2,350 m, constructing 50,000–70,000 ton berths and the annual capacity can afford 6.3 million TEU. The fully automated process mode is 'double trolley quay crane + automated guided vehicle (AGV) + automated rail crane (ARMG)'.

Currently, several cargo ports have fully automated container terminals, including six in Europe, four in Asia, two in North America and one in Australia. The details are listed in Table 10.1.

10.2.5 Containers

There is a great variety of containers. This variety consists of the most common ones – the dry containers, and those being designed for the special kinds of goods.

(a) (b)

Figure 10.1 (a) Port quay and (b) storage yard

Table 10.1 Automated container terminals

Region	Port name	Location	Completion (year)
The European	ECT port	Rotterdam, the Netherlands	1993
	CTA port	Hamburg, Germany	2002
	Euromax port	Rotterdam, the Netherlands	2010
	London port	London, England	2013
	APMT MVII port	Rotterdam, the Netherlands	2014
	RWG port	Rotterdam, the Netherlands	2011
The Asian	Wanhai port	Nagoya, Japan	2008
	Xiamen Ocean gate port	Xiamen, China	2014
	Qingdao port	Qingdao, China	2017
	Shanghai Yangshan port	Shanghai, China	2018
North America	Long Beach China Ferry port	Long Beach, United States	2016
	TraPac port	Los Angeles, United States	2016
Oceania	Patrick port	Brisbane, Australia	2013

The containers can be classified as dry containers, open type containers, open side platform containers, refrigerated containers, platform containers, live animal shipping containers and tank containers. In Figure 10.2 a simple model of containers data process flow is presented. The complexity of the containers' data flow greatly increases when more detailed data are included, for example: (i) container history information – vectors, logistic companies that have transported the container until the present port; nodes, destination points before entering the present port; vendor, regions and previous country passed before entering the present port; (ii) container configuration information – container type, good type, good characteristics transported inside the container, manifest of nonconformity noticed on container and security NC and (iii) alert information – security level inside the port, intelligence police relative reports about security issues, port location and ship entering the port [7].

10.2.6 Buffer zones

Various buffer zones can be found in a cargo port: loading buffer in front of the crane; buffer truck docking station; buffer storage.

10.2.7 Storage area

The containers are stored at the appropriate temperature required by different goods (i.e., at ambient temperature, fixed temperature and cold storage/refrigerated). Facilities for different goods storage exists in many cargo ports (e.g., for liquid fuel, grains, etc.).

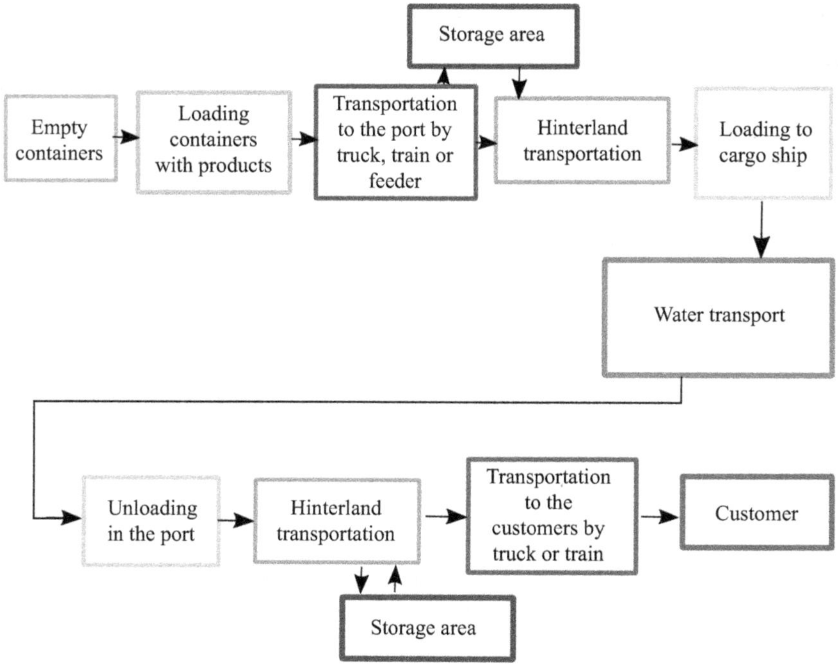

Figure 10.2 Model of containers process data flow

10.3 Measurements

10.3.1 The sea gauges and weather conditions

The weather conditions, particularly wind, ice and high sea waves, have great importance during operations of ship berthing and loading. The instrument and equipment used for weather conditions can also be used to estimate the environmental impact of ports (e.g., pollution by noxious gases) and alert on natural disaster. The meteorological station, either on land or on the sea, should have instruments for accurate weather data (i.e, barometer, anemometer, thermometer, rain gauge, SODAR – radar or sound waves that determine the wind speed and direction at various elevations above the ground; weather radar that locates precipitation, calculates its trajectory and estimates its types, direction and speed of wind).

In Las Palmas de Gran Canaria's Port, Spain, an information system stores and process data acquired from the measurement channels that includes sea tide measurement gauge and sensors for whether conditions (e.g. wind speeed, temperature, humidity, solar radiation, atmospheric pressure) [8]. The goal of the system was to display the current state of all the sensors available on the port's surroundings as well as their historical evolution. The system might also show the geographical location of all the resources available to the port authority and the rest of users of the port and the routes of the nearby ships. The system might support control tasks, through an alert management that monitors the value of many sensors.

The information on weather condition might be used in IoT system to predict the function of other connected sensors. For example, a radar with a good weather range of 1,500 m (i.e., a 1 m^2 target will have a 95% probability of detection at this range) might have a usable range in 20 mm/hour rain of only 1,000 m for a similar target [9]. Moreover, the importance of meteorological station in the port is depicted in Shibata *et al.* [10], where disaster information networks that combine the wired, wireless and satellite networks to realize the never-die-network environment on both normal and urgent cases are proposed.

10.3.2 Position and displacement detection

The object's coordinates (linear and angular) with respect to a selected reference is detected for position measurements. A position sensor is often a linear device whose output signal represents a distance of the object from a certain reference point. A proximity sensor is a kind of position sensor that generates the output signal when a certain distance to the object is detected and used for control (e.g., as end switch). Position change for a specific distance or angle gives information on displacement. This is measured when an object is referenced to its own prior position rather than to an external reference. The measurement of position and displacement of physical objects is essential in many port scenarios (e.g., container management, crane controls, intrusion detection system and transportation traffic control).

Many cargo ports are characterized by a wide area for container stacking (see Figure 10.1). Intermediate stacking in the port and multiple handling of containers (an average of three or four times) until the container leaves the port, and storage of containers for many days, can be observed in many cargo ports. Position of the containers that could be managed using a RFID systems [11,12] might serve as an intelligent decision for storage (in which multiple handling might be avoided), for monitoring, for real-time tracking of the containers, for detection of critical events and also to stop goods' smuggling and theft. Inductive and magnetic sensors (e.g., linear variable differential transformer, rotary variable differential transformer, eddy current sensors, Hall effect sensors, magnetoresistive sensors and magnetostrictive detector) and acoustic or microwave-based sensors have applications for sea tide gauge measurement, liquid level in a tank and elevator displacement. Optical sensors that are insensitive to magnetic fields and electrostatic interferences have also many applications in a port. A vision-based container horizontal positioning method during automated process of container stacking was presented by [13]. Cameras are installed at the bottom of the crane crossbeam so that one corner of the container on the spreader could be taken when it is near every handling location. A laser-based system was proposed by [14] for positioning of vessels during berthing. Laser docking systems are usually used in specialized cargo terminals such as LNG or crude oil, where it is obligatory according to different technical societies. Compared to the other laser-based berthing systems, the smart system proposed by the authors is a smart sensor with integrated devices capable of working as a part of a larger setup or as a standalone device. The lasers and processing units should be located at the ship to shore (STS) crane, just above the wheels or at about 2 m height. The system includes a ranging sensor Riegl LD90 Laser Distance Meter (diode based), a docking system Jenoptik LDM301 Laser Distance Meter (diode based), the Digi XBee-PRO XBP24 ZigBee to SIO module for remote tidal sensor data transmission, MaxBotix XL-MaxSonar WR1 ultrasonic distance meter for tide measurement and for wind velocity measurement the ultrasonic 2D (Gill WindObserver 65) and 3D (Gill WindMaster PRO) anemometers. A Simultaneous Localization and Mapping (SLAM) algorithm based on laser measurements system and encoders, and mathematical kinematic modelling of the AVG was described by [15] for positioning and obstacle avoidance of automatic guided vehicle (AVG) in partially known environment. The encoders were used to estimate the motion state of the AGV. The D* Lite algorithm was used to generate path from the start point to the goal point for AGV and to avoid unknown obstacles detected by laser scanner. A system based on optical character recognition for automatic number plate recognition (ANPR) and automatic freight-code recognition (AFCR) was proposed by [16]. The system uses this information to tag the photographic evidence of the corresponding vehicles. The system also includes colour-based indexing scheme, which automatically identifies the representative colour of each vehicle captures and better vehicle boundary detection. The proposed system uses a seven-point camera network for monitoring a single lane of traffic: ANPR cameras – WVGA infrared spectrum cameras with built-in infrared illuminators, dedicated to capturing images of front-mounted licence plates; AFCR cameras – WVGA visible-spectrum cameras with built-in 'white-light' illuminators, dedicated to capturing rear-mounted freight-code information and general purpose visible-spectrum cameras (termed 'overview' cameras) with built-in infrared for night time operation targeted at capturing multipurpose visible-spectrum images. As is expected the authors reported that the system might fail when 'tailgating' is present (i.e., where a vehicle closely follows another vehicle to the extent that the licence plate is obscured). Poor performance of classification using optical sensors for trucks detections and monitoring can also be attributed to (i) the varying illumination conditions associated with outdoor capture; (ii) the complications associated with reflective vehicle surfaces; (iii) the proliferation of glass constituting the objects (windscreen and side windows) and (iv) the segmentation quality provided by the preprocessing step [16].

Detection of position and displacement of things and people in a port has high importance for safe operations in port and security of port. Issues related to security of a cargo port and safety of processes regard several aspects such as perimeter security, internal security and operative controls, maritime security, decision support systems security, prevention and emergency management, etc. For example, detection of illegal ships and oil slicks as well as protection of the ports from effects of explosives or other weapons of mass destruction are the key issues for reducing economic, environmental and social impact. Synthetic Aperture Radar (SAR) system was proposed for detection of ships and/or oil slicks for maritime surveillance [17]. The system might function well day and night, in stormy weather or rainy days. CetusII AUV and MIRIS high-resolution sonar (i.e., equipped with side-scanning sonar) were proposed for inspection of the ship hull and berthing [18]. A robot equipped with DIDSON for short-range imaging sonar and M900-2250 for long-range imaging sonar was developed and tested by [19]. The robot searches and recognizes objects by the dual imaging sonar. They suggest that effective underwater exploration might be carried out by using short-range imaging sonar with frequencies of 1 MHz or higher (for good quality sonar image) together with long-range imaging sonar with frequencies bellow 1 MHz. DoD's Architecture Framework that employs multisensory approach (i.e., ground surveillance radars and cameras) was proposed to detect, asses and track perimeter incursions [9]. Using these devices, the proposed system can perform the following processes: (i) threat analysis – what threats are relevant to the facility, ordering these by their impact, criticality, etc.; (ii) operational security – providing real-time situational awareness and incident resolution (i.e., intrusion detection and tracking, intrusion assessment and intrusion response).

In Table 10.2, differences between some features of the laser, radar and ultrasound devices for position detection are presented.

The sensors in Table 10.3 can be used to measure distance and position in smart port. Laser and Lidar sensors have the shortest response time and the highest response accuracy. It is a good choice if the price is not a consideration. They are used in anticollision systems of spreaders, AGV and RMG in smart port. Clear and dry environment is necessary when they are used. Ultrasonic sensors have high precision; they can be used in the dry environment regardless of the response time and they are also used in the anticollision system of quayside crane and RMG in track. Infrared radiation sensors are used in the anticollision system of spreader, portal trolley in quayside crane without considering of accuracy level. Electromagnetic induction sensors are cheap and have short response time; regardless of accuracy it can also be used in the position system in smart port.

The sensors in Table 10.4 can be used for navigation in smart port. RFID system, Differential GPS and Laser-based navigation system sensors can gain absolute position. Differential GPS and Laser-based navigation system have higher accuracy level but are more expensive. If we want absolute position and do not

Table 10.2 Characteristics of three types of sensing devices for objects/things position detection

	Ranging method		
	Laser diode	**Radar**	**Ultrasound**
Range	To 300	To 2 Km	To 10 m
Emission power	10 mW (laser)	10 mW (96 GhZ)	1 W (sound)
Beam	Very narrow (<2°)	Narrow (<5°)	Wide (>5°)
Weather dependability	Low	Very low	Significant
Price	Average	Very high	Low
Method of measurement	FMCW	Pulse	Continuous-interrupted
Refresh rate	0.1 KHz	1 KHz	0.001 KHz
Lifetime	Average	Average	High

Table 10.3 The comparisons between the distance measurement sensors used in smart port

Type \ Parameter	Relative cost	Effective range	Environmental adaptability	Applications
Ultrasonic sensor	$$	High	Sensitive to water, humidity and wind interference; long response time	Anticollision of quayside crane and RMG in track
Laser and Lidar	$$	Very high	Sensitive to dust, water and oil interference; very short response time	Anticollision of spreader, main trolley, portal trolley, quayside crane, AGV and RMG
Electromagnetic induction sensor	$	Very low	Resistant to dust, water, humidity, wind and oil interference; short response time	Anticollision of quayside crane and RMG in track
Infrared radiation sensor	$	Low	Sensitive to sun and reflector interference; short response time	Anticollision of spreader, main trolley and portal trolley in quayside crane

Table 10.4 The comparisons between the navigation sensors used in smart port

Type \ Parameter	Relative cost	Accuracy	Measurement of the position	Environmental adaptability
HF and UHF RFID solutions	$$	High	Absolute position	Sensitive to foundation settlement
Differential GPS	$$$	Very high	Absolute position	Sensitive to metal and other shelter
Laser-based navigation system	$$$	Very high	Absolute position	Sensitive to dust, water, humidity, oil, sun and reflector interferences
Inertial navigation system	$$	High	Relative position	Sensitive to cumulative error, vibration and slip
Encoders	$	Low	Relative position	Sensitive to cumulative error, vibration and slip

need too high precision, RFID is a better choice. Inertial navigation system and encoders can gain the relative position: they are all sensitive to cumulative error, vibration and slip and inertial navigation system is a better choice without considering of prize.

These features are important for planning the devices' network for tracking different objects or things in a cargo port.

Wireless sensor networks have been recognized as a feasible means for object tracking and its research can be traced back to 1980 when a program called Distributed Sensor Networks (DSN) was started [20]. Wireless sensor networks based on multisensor device Waspmote, developed by Liblium Company, was proposed for optimal management of containers [21]. The authors suggest that by using these devices various container information could be obtained: (i) position and displacement of container; (ii) container doors opening unexpectedly; (iii) temperature; (iv) air pressure and (v) illegal container content. Currently, different technologies for wireless sensor network are commercialized (e.g., Tiny-mesh and Wirepas mesh). However, over the years the major concern related to wireless sensor networks is the power consumption. Efforts have been made by various researchers to derive ways to extend the lifetime of power storage or to minimize the use of energy in the tracking process. A method for reduction of the power consumption in wireless sensor network, which was inspired on Immune Network Theory, was proposed by [22]. The algorithm determines the best sleeping policy in order to optimize the power consumption of the entire sensor network. By using artificial immune system, when the number of sensors deployed in a sensor field and thus the corresponding density of sensor nodes increases, the average power consumption per sensor decreases. The authors proposed a control algorithm to enable sensor nodes in a wireless sensor network to individually determine their working status. By reducing the number of actives nodes, communication and hence collision are reduced. The average power consumption was reduced when the number of sensors was increased in the proposed framework. Increasing the density of sensor node and using the artificial immune method put the sensors in the low-power states more frequently. Compared to the always-on network, the average power consumed for each sensor node was significantly lower [22].

10.3.3 Velocity and acceleration detection

Sensors for velocity and acceleration detection might have many applications in port (e.g., for traffic control). The advantage of MEMS technology contribute to the development of new solution in the field of Inertial Measurement Units that provide information about acceleration, velocity or direction of port operation equipment (e.g. automated guided vehicle). For example, the Bosh CISS (Connected Industrial Sensor solution) is a multisensor device based on MEMS technology that can detect acceleration, vibration, rotation, as well as environmental conditions. This device allows to activate or deactivate sensors and to configure the sampling rates according to specific needs (i.e., data streaming mode, event detection or time aggregation).

Detection of the speed and location of illegal ships by using wireless sensor networks equipped with Global Positioning System (GPS) was proposed by [23].

10.3.4 Chemical, biological and radiation sensors

Multisensor technology might be used for real-time and accurate information on cargo (i.e., type of goods, good characteristics transported inside the container). The digital data on different aspects related to logistic of a cargo port in conjunction with data from various sensors from port can support the risk evaluation of a cargo that might be used to plane the container inspection (i.e., chemical screening, biological screening or radiation screening). A system for detection of hazardous chemical and biological contents of containers as well as contraband was proposed by [24]. Two screening methods, one external to the container while it is being moved from conveyance to ship or vice versa, and an alternative *in situ* method to sample, detect and report dangerous cargo from inside a sealed container while in transit was described. The external screening method that is noninvasive acquires air sampling from shipping containers while the container is being loaded onto or unloaded from a ship. The container headspace air was used as a carrier gas for presentation of diffused agents to chemical analysis and biological particle detectors. Standard shipping containers have vents which provide a mechanism for extraction of the headspace air. Air from the shipping container was evacuated by a high-volume vacuum pump, and a low-volume vacuum pump performed a subsampling of the evacuated air. The sensors were protected by an environmental enclosure that is also characterized by shock and vibration isolators. The sensors might detect in air samples the chemicals vapours and particulate matter from the sealed container. They suggest the necessity for a smart collector – one that can reject innocuous sample backgrounds, yet collect and concentrate substances of high interest, and prepare samples with minimal clutter and possibility of saturation. Detection of illegal nuclear materials' transportation using a structural fingerprints of containers was proposed by [25]. The microstructure of the containers material as well as of the dedicated weld seam applied to the lid and the container body provides a unique fingerprint of the full container, which can be reproducibly scanned using an appropriate technique. A Russian universal industrial ultrasonic flaw detector USD60 that generates an ultrasonic pulse by piezo-electric conversion was used. The amplitudes of A-scan (the received electric time signal, the received pulse) indicate the intensity of reflection that in some way but not in a defined manner correlates with reflector dimension. The arrival time of reflector amplitudes was used for their localization assuming straight propagation along the centre line of the wave field intensity profile.

10.3.5 Port infrastructure and machinery health monitoring

With increasing trade, the usage of port machinery is increasingly continuous, which might increase the risk of equipment's damage. To ensure the safety of the goods in port the status of port machinery should be measured by placing sensors on key location of port machinery for nondestructive detection of materials. Nondestructive evaluation techniques allow detection of the presence of cracks, discontinuities, mechanical fatigue and other imperfections without material damage. For example, electrical conductivity can be used as an index of mechanical fatigue. The planar meander and mesh-type sensors respond very well to conducting and magnetic materials but no so well to dielectric materials. However, the interdigital sensor responds very well to dielectric materials [26]. A real-time strain monitoring system based on distributed Fibber Bragg Grating sensors was proposed by [27] for monitoring the health of port machinery. Fibber grating strain sensors are ideal choices for large structure and with severe electromagnetic interference (EMI) [27]. A container crane movement analysis and assessment based on structural health monitoring is described in [28]. The accelerometers are used in this system to monitor the dynamic crane behaviour. The 3D finite element model (FEM) and multi-input single output nonlinear autoregressive neural network with external output were used to identify static and dynamic displacement of the crane under different load cases [28].

Sensing using IoT technology allows to know what needs repairing, what needs replacing and where stock is held, and may convert the current maintenance model from repair and replace, to predict and fix, which may reduce the likelihood of any failure. The IoT technologies may continually analyse the changes in objects/things of cargo port. Any changes can trigger a predetermined process of actions or activities. Alerts may also be sent directly to the operator device, keeping all relevant parties informed and aware. IoT technologies for health maintenance of port infrastructure and port machinery might (i) reduce equipment downtime (i.e., repair service can be performed before the equipment failure); (ii) increase productive uptime (i.e., by predictively identifying problems, so repairs can be scheduled during production downtime, rather than peak periods); (iii) reduce maintenance costs; (iv) increase asset lifespan (i.e., by performing predictive maintenance before costly failures occur) and (v) improve worker safety. These inbuilt predictive features might improve the capacity of port businesses to work proactively rather than reactively to maintain the port infrastructure and port machinery health.

10.4 IoT for cargo ports

'Full Automation' is the brightest label of Shanghai Yangshan Port. The efficiency of Yangshan Port (IV phase) will reach 40 cases/hour, far exceeding the efficiency of manual terminals, and the labour force will be reduced by 70%. The deployment of IoT technologies has a great importance for 'full automation' and for improving efficiency of the processes in cargo ports. IoT community has developed many architecture, platforms, standards, frameworks and solutions for IoT-based information systems.

IoT is a powerful tool for things management in a cargo port by having the capacity to gather insights from the field using sensor technologies as well as by supporting data analytics.

Information from sensors might be used in a cargo port (i) for sea gauges and weather condition monitoring and prediction; (ii) detection of position of the ships and containers; (iii) intrusion detection system; (iv) monitoring displacement of the cranes, trucks or AVG; (v) monitoring the level of fuel oil in storage tanks; (vi) detection of illegal or hazardous chemical, biological and radioactive things;

(vii) evaluation of structure and functioning of port machinery; (viii) detection of damaged containers; (ix) detection, prediction or prevention of failure of port machinery and communication network systems, among many other things. The recent development of sensors having low power consumption, low cost, small size, limited radio range and capacity for long-term working in harsh environments increase the likelihood of the development of IoT systems' development for cargo ports, adapted to the various conditions and features of that port. Multimodel sensing that refers to sensing with different modalities and then exploiting the inter-dependence of these modalities to extract more useful data should be considered in IoT system for cargo port. Information from sensors distributed in the port environment might be improved by data fusion and data meaning algorithms.

The communication of data in IoT framework might be supported by different types of wired or wireless technologies such as CAN bus, ZigBee, Bluetooth, Wi-Fi and long-range communication LoRa. Cellular connectivity like LTE is expected to increase in the future, especially with 5G and NarowBand-IoT. Several characteristics of different types of communication technologies are presented in Table 10.5. There are several hardware platforms available for sensor network deployment such as TelosB, Mica, IRIS, Waspmote, XBee ZB module, LoRa, BTLE, Tiny-mesh, Wirepasmesh and Xelba xOne Hub. The power of communication locations should consider buildings, height restrictions and other topographic constraints.

Cloud computing, Hybrid Cloud or Microservices might be adopted for IoT architecture for cargo port for computing, storage, analytics and data visualization. Cloud Computing provides functional and also nonfunctional support (i.e., Service Level Agreements, low-latency fault-tolerance, cost-optimization and security). These capabilities are provided as services (Platform-as-Service, Infrastructure as-a Service, Software-as-a-Service, etc.).

Hybrid Cloud blends internal information system with cloud-powered services. It might integrate private, public and managed cloud with the existing on-premises information system (e.g., Fujitsu MetaArc and Cloud Service K5). This may improve the agility, accessibility and security of the IoT systems for cargo ports.

Both ETSI [29] team and Microsoft Azure [30] team recommend the use of Microservices in IoT architecture. Microservices is an architectural approach for developing applications as a set of small services, where each service is running as a separate process, communicating through simple mechanisms [29]. The main advantage of Microservice-based architecture is that one can develop, deploy, upgrade and scale every Microservice independent of others. Moreover, developing

Table 10.5 The comparison of different communication technologies

	ZigBee	**Wi-Fi**	**LoRa**	**NB-IoT**	**4G/5G**
Speed	250 kbps	300 Mbps	9.6 kbps	200 kbps	100 Mbps/1 Gbps
Relative cost	$	$$	$$	$$	$$
Frequency	784 MHz	2.4 GHz/5 GHz	250 kHz	200 kHz	2.4 GHz
Range (outdoor)	100 m	100 m	20 km	10 km	–
Compatibility	802.15.4	802.11ac/n	802.11ac		802.16m

Microservices separately enable the use of different technologies for each Microservice and to have distinctive Microservices for devices that communicate using different protocols. Also, this may support the safety and security of the system. For example, if a Microservice which communicates with a certain group of sensors crashes, the application provided by Microservices which communicate with other sensors will still be up and running. Microservices seems to be the technology that might integrate various technologies and services without having to worry about compatibility issues. However, the communication mechanisms used by Microservices may not be supported in the legacy systems.

Several commercial technologies might be used for developing IoT in cargo ports (e.g., Telit Communications, HiReply). For example, HiReply incorporates services, devices and middleware for IoT. It uses sectional, configurable modules to provide all the elements necessary for the creation of vertical solutions.

Reference Architecture for IoT was described by the team of IoT-A European project [1], ETSI [29], Microsoft Azure [30] and CREATE-IoT European project [31]. ETSI team identified several functional and nonfunctional requirements for IoT (i.e., multitenancy, massive data processing, high-throughput, high-availability, low latency and security). A single instance of a software application should serve multiple customers (multitenancy). For example, part or all containers data from storage yard should be available to port authorities involved in goods management. When all of the customer data are handled using the same software resources, the architecture design should be extremely rigorous and should prevent data leakage [29]. High-volume and high-velocity data are produced, analysed and used to trigger action in IoT systems (massive data processing). The processing of data can be done in real time (streaming) or is collected over time and then fed into an analytics system (batch processing model). By building data streams, one can feed data into analytics tools as soon as it is generated and get near-instant analytics results. For streaming data processing, complex event processing (CEP) might be used with the goal to identify significant events and respond fast. Batch processing is adequate in situations where nonreal-time analytics results are needed, and when it is more important to process large volume of information than to get fast analytics results. Most often it is used when dealing with extremely large amounts of data, and/or when data sources are legacy systems that are not capable of delivering data in streams. Service provided by IoT should be resilient to failures and able to restart on another machine. A failure can be when the machine where the service is running fails. It can be produced by a service-internal problem or during an application upgrade. The running service should determine whether it can continue to move forward to the newer version or to roll back to a previous stable version to maintain a consistent state. The service should be designed so that the process can be restarted at any time with no data loss. Low latency is also an important requirement for IoT. The ETSI team suggests that this can be achieved by brokers capable of routing IoT application messages in (near) real time: MapReduce – which may make use of massive in-memory databases capable of meeting low-latency requirements and Edge Computing that allows reduction of network latency by moving computing resources closer to the field domain where an action takes place [29]. For example, in computer vision applications, such as for monitoring traffic in port it would be desirable to extract meaningful information from the video right at the image sensor rather than in the cloud to reduce the communication cost. For the automated guided vehicle from the port, local processing is desired since the latency and security risks of relying on the cloud are too high. Advanced technologies can facilitate moving computation closer to the source by embedding computation near or within the sensor and the memories. The use of memristors for analog computation and 3D stacked memory might also help on complex signal processing in Edge Computing. Making the sensor more intelligent so that only useful and relevant data are communicated is also important for IoT architecture. For example, in most of the signals the information rate is actually much smaller than that suggested by the bandwidth (i.e., the signals are sparse in a particular domain). By compressive sensing the useful information content from the signal is extracted in a condensed and energy efficient manner.

IoT for cargo port should have an adaptive framework that offers the possibility to easily introduce new applications and to ensure that they are properly connected to the existing information systems and to process enormous quantity of data. In Figure 10.3, a model of IoT technology for cargo port is presented. A function

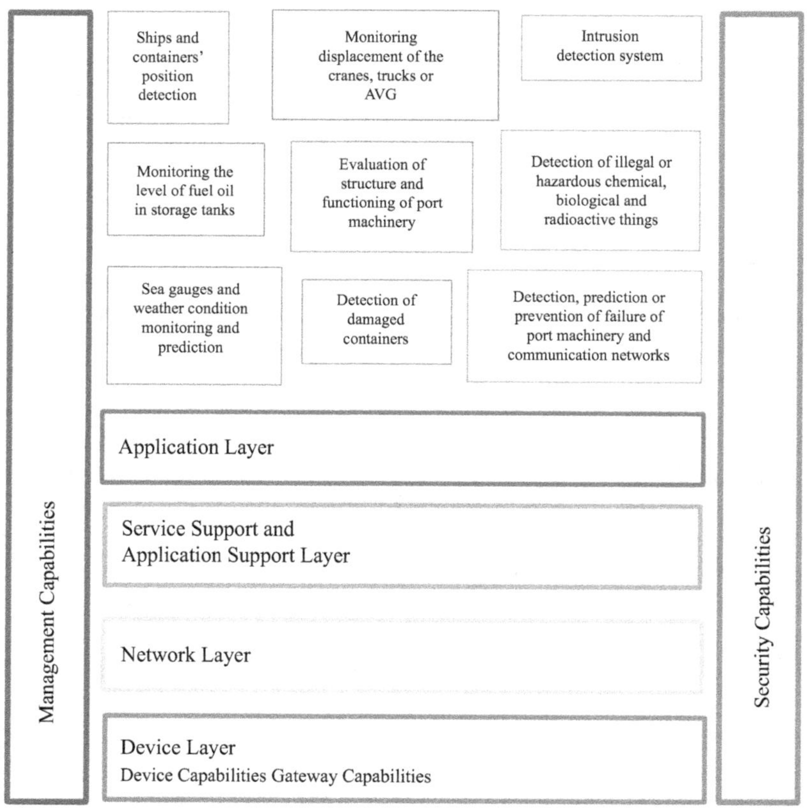

Figure 10.3 Model of IoT for cargo port

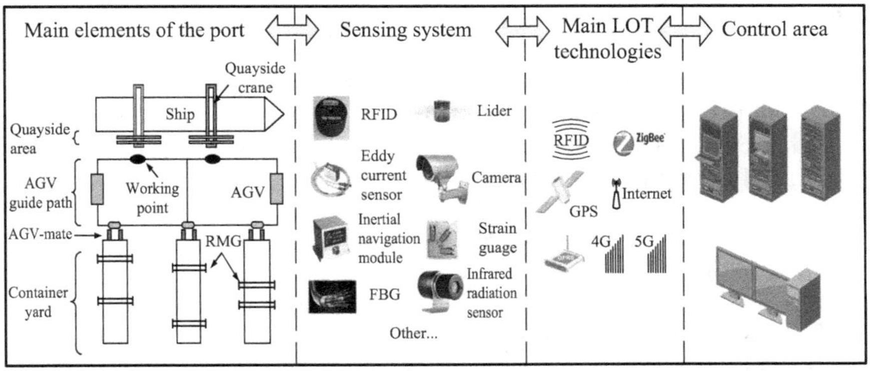

Figure 10.4 Layers of IoT for cargo ports with representation of several included elements

architecture that refers to ITU-T Y.2060 [31] and several required measurements in the cargo port were considered (Figure 10.4).

The amount of data that can be acquired by IoT technologies and the information provided by analytics tool in IoT frameworks are enormous. Nowadays, IoT data and information are susceptible to malicious usage not only by hackers but also by unauthorized competitors. Therefore, security of IoT system (i.e., protection of the confidentiality, integrity and availability of information asset) should be assured by design at every layer of the system architecture.

10.5 Conclusions

There are few studies on implemented information and communication technologies for cargo ports. IoT technology may bring many benefits for the cargo port information systems. It may contribute to increase the efficiency and quality of services in cargo ports, the operational safety and security of the ports, by increasing the accessibility and accuracy of relevant information.

The amount of data that can be acquired using advances in sensor technologies and IoT technologies and the information provided by analytics tool in IoT frameworks are enormous. More researches are needed for better understanding what are the optimal model and architecture for IoT platforms in cargo port, what sensors and instrumentation and what communication and analytics technologies may contribute to better logistic (including security) of cargo port. IoT for cargo ports should have an adaptive framework, which should allow multimodalities sensing, modularity (i.e., flexibility to introduce new applications), ensure adequate and safe connectivity to the existing information systems and process enormous quantity of data. As IoT data and information are nowadays susceptible to malicious usage not only by hackers but also by unauthorized competitors, design for cargo port information systems that support security of information at every layer of the information system architecture is required.

References

[1] IoT-A. Internet of things architecture. www.iot-a.eu/arm

[2] What Is China's Belt and Road Initiative? www.theguardian.com/cities/ng-interactive/2018/jul/30/ what-china-belt-road-initiative-silk-road-explaine

[3] List of Busiest Container Ports. https://en.wikipedia.org/wiki/List_of_busiest_container_ports

[4] Trade Statistics. www.worldshipping.org/about-the-industry/global-trade/trade-statistics

[5] Feasibility Study on the Network Operation of Hinterland Hubs (Dry Port Concept) to Improve and Moderate Ports' Connections to the Hinterland and to Improve Networking. https://web.archive.org/web/20080413200609/http://inloc.info/internal/wp1/activity_15_feasibility_study_dry_ports.pdf

[6] Engineered Lifting Systems and Equipment. https://www.engliftsystem.com

[7] Bocca E., Viazzo S., Longo F., Mirabelli G. 'Developing data fusion systems devoted to security control in port facilities.' *Proceedings of the Winter Simulation Conference*, M.E. Kuhl, N.M. Steiger, F.B. Armstrong, and J.A. Joines (Eds.), Orlando, FL, USA, 2005, pp. 445–449

[8] Fernandez P., Santana J.M., Ortega S., *et al.* 'SmartPort: a platform for sensor data monitoring in a seaport based on FIWARE.' *Sensors*, 2016; 16(417):1–24

[9] Hennin S., Germana G., Garcia L. 'Integrated perimeter security system'. *Proceedings of the IEEE Conference on Technologies for Homeland Security*, Woburn, MA, USA, 16–17 May 2007, pp. 70–75

[10] Shibata Y., Uchida N., Shiratori N. 'A resilient network and information management system for large scale disaster.' *Proceedings of the 30th International Conference on Advanced Information Networking and Applications Workshops*, Crans-Montana, Switzerland, 23–25 March 2016, pp. 656–660

[11] Kadir E.A., Rosa SL, Gunawan H. 'Application of RFID technology and e-seal in container terminal process.' *Proceedings of the 4th International Conference on Information Communication Technology (ICoICT)*, Bandung, Indonesia, 25–27 May 2016

[12] Bauk S., Gonzales D.G., Schmeink A. 'Examining some ZigBee/RFID safety system performances at seaport.' *Proceedings of the International Symposium ELMAR*, Zadar, Croatia, 12–14 September 2016

[13] Fu Y., Wang X., Mi C, Bian Z. A container horizontal positioning method with image sensors for cranes in automated container terminals. *Sensors & Transducers*, 2014; 166(3)190–196

[14] Perkovic M., Gucma M., Luin B., Gucma L., Brcko T. Accommodating larger container vessels using an integrated laser system for approach and berthing. *Microprocessors and Microsystems*, 2017; 52:106–116

[15] Pratama P.S., Nguyen T.H., Kim H.K., Kim D.H., Kim S.B. 'Positioning on obstacle avoidance of automatic guided vehicle in partially known environment.' *International Journal of Control, Automation and Systems*, 2016; 14(6): 1572–1581

[16] Sadlier D.A., Fergunson P., O'Conaire C.O., O'Connor N.E., Doyle K. 'Image-based vehicle indexing for a seaport transportation surveillance system.' *Proceedings of the 8th IEEE International Conference on Advanced Video and Signal-Based Surveillance*, Klagenfurt, Austria, 30 August–2 September 2011, pp. 367–372

[17] Jarabo-Amores P., Gonzalez-Bonilla M.J., Mata-Moya D., Martin-de-Nicolás J., Palma-Vazquez A. 'Demonstrator of maritime SAR applications: Automatic ship detection results.' *Proceedings of the IEEE International Geoscience and Remote Sensing Symposium*, Munich, Germany, 22–27 July 2012

[18] Trimble G.M., Belcher E.O. 'Ship berthing and hull inspection using the Cetus II AUV and MIRIS high-resolution sonar.' *Proceedings of the OCEANS'02 MTS/IEEE*, Biloxi, MI, USA, 29–31 October 2002

[19] Lee Y., Choi J., Jung J., Kim T., Choi H-T. 'Underwater robot exploration and identification using dual imaging sonar: Basin test.' *Proceedings of the IEEE Conference on Underwater Technologies*, Busan, 21–24 February, 2017

[20] Chong C.Y., Kumar S.P. 'Sensor networks: evolution, opportunities, and challenges.' *Proceedings of IEEE*, 2003; 91 (8):1247–1256

[21] Rezapour T.Y., Atani R.E., Abolghasemi M.S. 'Secure positioning for shipping containers in ports and terminals using WSN.' *Proceedings of the 11th International ISC Conference on Information Security and Cryptology*, Tehran, 3–4 September 2014, pp. 10–14

[22] Lai W.C., Lau H.Y.K. 'A bio-inspired object tracking algorithm for minimizing power consumption.' *Proceedings of the IEEE/ACM International Conference on Green Computing and Communication,* Hangzhou, China, 30 October–1 November 2010, pp. 355–360

[23] Arifin A.S., Firdaus T.S. 'Ship location detection using wireless sensor networks with cooperative nodes.' *Proceedings of the Ninth International Conference on Ubiquitous and Future Networks (ICUFN)*, Milan, Italy, 4–7 July 2017

[24] Rudzinski C., Masters D., Buck A., Wall M., Tremblay D., Wack E. 'Screening maritime shipping containers for weapons of mass destruction.' *Proceedings of the IEEE International Conference on Technologies for Homeland Security*, Waltham, MA, USA, 8–10 November 2010, pp 460–466

[25] Kroening M., Demyanuk D., Lider A., Sedney D. 'Structural Fingerprints of containers with nuclear materials for material control and accounting purposes.' *Proceedings of the 7th International Forum for Strategic Technology (IFOST)*, Tomsk, Russia, 18–21 September 2012

[26] Mukhopadhyay S.C., Gooneratne C.P., Gupta G.S., Yamada S. 'Characterization and comparative evaluation of novel planar electromagnetic sensors.' *IEEE Transactions on Magnetics*, 2005; 41(10): 3658–3660

[27] Wu W., Liang L. 'Long-term real-time monitoring system for port machinery.' *Proceedings of the 1st International Conference on Intelligent Networks and Intelligent Systems*, Wuhan, 1–3 November, 2008

[28] Kaloop M.R., Sayed M.A., Kim D. Kim E. 'Movement identification model of port container crane based on structural health monitoring system. *Structural Engineering and Mechanics*, 2014; 50(1):105–119

[29] ETSI Technical Report TR103527. SmartM2M: Virtualized IoT Architectures with Cloud Back-ends. https://etsi.org/news-events

[30] Microsoft Azure IoT Reference Architecture. https://azure.microsoft.com/en-us/blog/azure-iot-reference-architecture-update

[31] CREATE-IoT. Cross Fertilisation Through Alignment, Syncronisation and Exchanges for IoT: Recommendations for Commonalities and Interoperability Profiles of IoT Platforms. https://european-iot-pilots.eu/project/create-iot/

Index

www.ingramcontent.com/pod-product-compliance
Lightning Source LLC
LaVergne TN
LVHW011914200926
848484LV00002B/619
* 9 7 8 1 7 8 5 6 1 6 3 4 1 *